How to Integrate It

A Practical Guide to Finding Elementary Integrals

While differentiating elementary functions is merely a skill, finding their integrals is an *art*. This practical introduction to the art of integration gives readers the tools and confidence to tackle common and uncommon integrals.

After a review of the basic properties of the Riemann integral, each chapter is devoted to a particular technique of elementary integration. Thorough explanations and plentiful worked examples prepare the reader for the extensive exercises at the end of each chapter. These exercises increase in difficulty from warm-up problems, through drill examples, to challenging extensions that illustrate such advanced topics as the irrationality of π and e, the solution of the Basel problem, Leibniz's series, and Wallis's product.

The author's accessible and engaging manner will appeal to a wide audience, including students, teachers, and self-learners. It can serve as a complete introduction to finding elementary integrals, or as a supplementary text for any beginning course in calculus.

How to Integrate It
A Practical Guide to Finding Elementary Integrals

SEÁN M. STEWART
Omegadot Tuition, Sydney

CAMBRIDGE
UNIVERSITY PRESS

University Printing House, Cambridge CB2 8BS, United Kingdom

One Liberty Plaza, 20th Floor, New York, NY 10006, USA

477 Williamstown Road, Port Melbourne, VIC 3207, Australia

314-321, 3rd Floor, Plot 3, Splendor Forum, Jasola District Centre, New Delhi - 110025, India

79 Anson Road, #06-04/06, Singapore 079906

Cambridge University Press is part of the University of Cambridge.

It furthers the University's mission by disseminating knowledge in the pursuit of education, learning and research at the highest international levels of excellence.

www.cambridge.org
Information on this title: www.cambridge.org/9781108418812
DOI: 10.1017/9781108291507

First published 2018

A catalogue record for this publication is available from the British Library

ISBN 978-1-108-41881-2 Hardback
ISBN 978-1-108-40819-6 Paperback

Contents

Preface

Calculus occupies an important place in modern mathematics. At its heart it is the study of continuous change. It forms the foundation of mathematical analysis while the immense wealth of its ideas and usefulness of the tools to have emerged from its development make it capable of handling a wide variety of problems both within and outside of mathematics. Indeed, the sheer number of applications that the calculus finds means it continues to remain a central component for any serious study of mathematics for future mathematicians, scientists, and engineers alike.

The material presented in this volume deals with one of the major branches of calculus known as the *integral* calculus – the other being the *differential*, with the two being intimately bound. The integral calculus deals with the notion of an integral, its properties, and method of calculation. Our word for 'integrate' is derived from the Latin *integratus* meaning 'to make whole'. As calculus deals with continuous change, integration, then, is a general method for finding the whole change when you know all the intermediate (infinitesimal) changes.

A precursor to the concept of an integral dates back to the ancient Greeks, to Eudoxus in the fourth century BCE and Archimedes in the third century BCE, and their work related to the method of exhaustion. The method of exhaustion was used to calculate areas of plane figures and volumes of solids based on approximating the object under consideration by exhaustively partitioning it into ever smaller pieces using the simplest possible planar figures or bodies, such as rectangles or cylinders. Summing its constituent parts together then gave the area or volume of the whole. Integration thus renders something whole by bringing together all its parts. Its modern development came much later. Starting in the late seventeenth century with the seminal work of Newton and Leibniz, it was carried forward in the eighteenth century by Euler and the Bernoulli brothers, Jacob and Johann, and in the nineteenth century most notably by Cauchy before the first rigorous treatment of the integral was given by Riemann during the middle part of that century. Since this time many other notions for the integral have emerged. In this text we focus exclusively on the first and perhaps simplest of these notions to emerge, that of the *Riemann* integral.

Unlike differentiation, which once learnt is merely a skill where a set of rules are applied, as there is no systematic procedure for finding an integral, even for functions that behave 'nicely', many look upon integration as an 'art'. Finding an integral tends to be a complicated affair involving a search for patterns and is hard to do. The unavailability of a mechanical approach to integration means many different techniques for finding integrals of well-behaved functions in terms of familiar functions have been developed. This makes integration hard and is exactly how it is perceived by most beginning students. When encountering integration for the first time one is often bewildered by the number of different methods that need to be known before the problem of integration can be successfully tackled.

This book is an attempt at taking some of the mystery out of the art of integration. The text provides a self-contained presentation of the properties of the definite integral together with many of the familiar, and some not so familiar, techniques that are available for finding elementary integrals. Prerequisites needed for the proper study of the material presented in the text are minimal. The reader is expected to be familiar with the differential calculus including the concept of a limit, continuity, and differentiability, together with a working knowledge of the rules of differentiation.

The book takes the reader through the various elementary methods that can be used to find (Riemann) integrals together with introducing and developing the various properties associated with the definite integral. The focus is primarily on ideas and techniques used for integrating functions and on the properties associated with the definite integral rather than on applications. By doing so the aim is to develop in the student the skills and confidence needed to approach the general problem of how to find an integral in terms of familiar functions. Once these have been developed and thoroughly mastered, the student should be in a far better position to move onto the multitude of applications the integral finds for itself. Of course this is not to say applications for the integral are not to be found in the text. They are. While the text makes no attempt to use integration to calculate areas or volumes in any schematic way, applications developed through the process of integration that lead to important results in other areas of mathematics are given. As examples, the proof of the irrationality of the numbers π and e is presented, as are the solutions to the Basel problem, Leibniz's series, and Wallis's product.

Most chapters of the book are quite short and succinct. Each chapter is self-contained and is structured such that after the necessary theory is introduced and developed, a range of examples of increasing level of difficulty are presented showing how the technique is used and works in practice. Along the way, various strategies and sound advice are given. At the conclusion of each chapter, an extensive set of exercises appear. The text can serve as a complete introduction and guide to finding elementary integrals. Alternatively, it can serve as a resource or supplemental text for any beginning course in calculus by allowing students to focus on particular problem areas they might be having by working through one or more relevant chapters at a time.

In contemplating the material presented it cannot be overstated how important it is for one to attempt the exercises located at the end of each chapter. For those hoping to become fluent in the art of integration, this proficiency is best gained through perseverance and hard work, and in the practice of answering as many different and varied questions as possible. Indeed, a large portion of the text is devoted to such exercises and problems; they are a very important component of the book. To help aid students in their endeavours an attempt has been made to divide the exercises that appear at the end of the chapters into three types: (i) warm-ups, (ii) practice questions, and (iii) extension questions and challenge problems. The warm-ups are relatively simple questions designed to gently ease the student into the material just considered. The practice questions consolidate knowledge of the material just presented, allowing the student to gain familiarity and confidence in the workings of the technique under consideration. Finally, the extension questions and challenge problems contain a mix of questions that are either simply challenging in nature or that extend, in often quite unexpected ways, the material just considered. It is hoped many of the questions found in this last group will not only challenge the reader but pique their interest as more advanced results are gradually revealed. Of course, judging the perceived level of difficulty is often in the eye of the beholder so one may expect some overlap between the various categories. For problems considered more difficult, hints are provided along the way in the form of interrelated parts that it is hoped will help guide the student towards the final solution. In all, well over 1,000 problems relating to finding or evaluating integrals or problems associated with properties for the definite integral can be found dispersed throughout the end-of-chapter exercise sets.

Chapter 1 introduces formally what we mean by an integral in the Riemann sense. The approach taken is one via Darboux sums. The fundamental theorem of calculus, which we divide in two parts, is also given. Properties for the definite integral are given in three chapters (Chapters 2, 4, and 16). Sixteen chapters (Chapters 3 and 5–19) are then devoted to either a particular method that can be used to find a given integral or a particular class of integrals. Here methods including standard forms, integration by substitution, integration by parts, trigonometric and hyperbolic substitutions, a tangent half-angle substitution, trigonometric and hyperbolic integrals, integrating rational functions using partial fractions, integrating inverse functions, and reduction formulae can be found. The penultimate chapter, Chapter 20, introduces the improper integral. The field of improper integrals is immense and all we can do here is touch upon this important area. The final chapter, Chapter 21, is devoted to considering two very important improper integrals that arise in applications known as the Gaussian integral and the Dirichlet integral.

While all the familiar techniques of integration one would normally expect to find in any standard introductory calculus text are to be found here, the approach we take is somewhat different. Other treatments tend to be brief and hurried while the questions asked are often repetitive and uninteresting. In the present volume, as each chapter is devoted to a particular technique our focus is more concentrated and allows one to methodically work through each of these techniques. At the same time an abundance of detailed worked examples are given, and different and varied question types are asked. We also offer other useful methods not typically found elsewhere. These include integrating rational functions using the Heaviside cover-up method and Ostrogradsky's method (Chapter 11), tabular integration by parts (Chapter 7), and the rules of Bioche (Chapter 15). Finally, we provide two appendices. The first, Appendix A, on partial fractions is given for anyone who has either not encountered this topic before or is in need of a brief review. The second, Appendix B, contains answers to selected questions asked.

The genesis of this book grew from the large number of requests the author received over the years from students he taught introductory calculus to. Many students wanted additional material and questions to consolidate and test their growing skills in finding integrals and asked if a short text could be suggested to meet such a need. These many requests drove the author to seek out and create ever more varied and interesting problems, the result of which you now hold before you.

Inspiration for many of the exercises found in the text has been drawn from a wide variety of sources. Articles and problems relating to integration found in the journals *The American Mathematical Monthly*, *Mathematics Magazine*, *The College Mathematics Journal*, and *The Mathematical Gazette* have proved useful, as have online question-and-answer sites devoted to mathematics such as *Mathematics Stack Exchange* and *The Art of Problem Solving*. Joseph Edwards's *A Treatise on the Integral Calculus* (Volume 1), G. H. Hardy's *A Course of Pure Mathematics*, Michael Spivak's *Calculus*, and the Soviet text *Problems in Mathematical Analysis* edited by Boris Demidovich have also proved useful sources for questions. Answers to almost all exercises appearing in the text are given in Appendix B. While every care has been taken to ensure their accuracy, errors are regrettably unavoidable and the author would be most grateful if any errors found, could be brought to his attention.

In closing perhaps something needs to be said about why one should bother to learn any of the techniques of integration at all. After all, powerful computer algebra systems now exist that can find almost all of the integrals appearing in this text. Such a question is of course a bit like asking why bother to learn to add when all of arithmetic can be handled by a calculator. Understanding why things are the way they are is important. If nothing else, integration is incredibly useful. It is a standard topic in any introductory course on calculus and an important gateway to many areas of more advanced applied mathematics. Many of the techniques of integration are important theorems in themselves about integrable functions, providing a foundation for higher mathematics. While having the ability and the insight to see into an integral, and turn it from the inside out, may not be a very convincing reason for many, intellectually it is the most compelling reason of them all.

1

The Riemann Integral

There are subjects in which only what is trivial is easily and generally comprehensible. Pure mathematics, I am afraid, is one of them.

— G. H. Hardy

There are many ways of formally defining an integral, not all of which are equivalent. The differences exist mostly to deal with special cases for functions that may not be integrable under other definitions. The most commonly used definition for the integral is the so-called *Riemann integral*.[1] It is the simplest integral to define and is the type of integral we intend to consider here. The central idea behind the concept of a definite integral is its connection to an area defined by a region Ω bounded between the graph of a function $y = f(x)$, the x-axis, and the vertical lines $x = a$ and $x = b$. As this idea is central to the concept of the definite integral, we begin by carefully considering it in some detail.

The integrability of a function depends on a property of functions known as *boundedness*. What we exactly mean by this is given in the following definition.

Definition 1.1 (Boundedness) A function f is said to be *bounded* if there exists a number $M > 0$ such that $-M \leqslant f(x) \leqslant M$ for all $x \in [a, b]$.

Geometrically, the graph of a bounded function on the interval $[a, b]$ will lie between the graphs of the two constant functions having values $-M$ and M, respectively (see Figure 1.1). A function that is bounded then has all of its values lying between two finite limits.

Consider a non-negative ($f(x) \geqslant 0$) bounded function defined on the interval $[a, b]$. If we assume the area of a rectangle is given by its length multiplied by its height, the problem of finding the area of the region Ω enclosed by the graph of the function and the x-axis between $x = a$ and $x = b$ can be approached by subdividing the interval $[a, b]$ into a finite number of

1

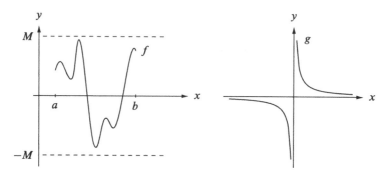

Figure 1.1. An example of a bounded function f to the left and an unbounded function g to the right.

sub-intervals. On summing the areas of each of the rectangles together, an approximation to the area of the region can be found. Choosing rectangles that are *inscribed* within the region and that *circumscribe* the region, as in Figure 1.2, and then summing their areas, one should expect the areas of the inscribed rectangles to be less than the expected area of the region Ω while the sum of the circumscribing rectangles ought to be larger than it. As the size of the subdivisions is made ever smaller, these two sums will come ever closer to the expected value for the area of the region Ω.

Now consider that the interval $[a, b]$ $(b > a)$ is subdivided or *partitioned* into a finite number of sub-intervals $I_k = [x_{k-1}, x_k]$ such that the end-points of each sub-interval is given by

$$a = x_0 < x_1 < x_2 < \cdots < x_{n-1} < x_n = b.$$

The size of each sub-interval need not be the same. The partition $\mathcal{P}$ can be denoted by either its sub-intervals

$$\mathcal{P} = \{I_1, I_2, \ldots, I_n\},$$

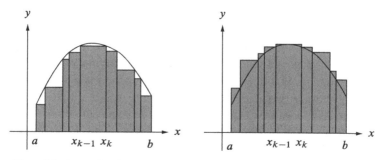

Figure 1.2. Approximating the area under a curve by means of inscribed (left) and circumscribing (right) rectangles.

or by the set of end-points that define the sub-intervals of the partition, that is by

$$\mathcal{P} = \{x_0, x_1, x_2, \ldots, x_{n-1}, x_n\}.$$

The length of the interval $I = [a, b]$ is given by $b - a$. You should also notice that the sum of the lengths of each sub-interval in a partition $\{I_1, I_2, \ldots, I_n\}$ of an interval I is equal to the length of the total interval as

$$\sum_{k=1}^{n}(x_k - x_{k-1}) = (x_1 - x_0) + (x_2 - x_1) + \cdots$$
$$\cdots + (x_{n-1} - x_{n-2}) + (x_n - x_{n-1})$$
$$= x_n - x_0 = b - a.$$

We now try to put in place an appropriate mathematical structure that will help guide us towards a definition for the area under the graph of a function. We do this by introducing the notion of *Darboux* sums.[2] Let $f : [a, b] \to \mathbb{R}$ be a bounded function and let $M_k(f)$ denote the maximum value of the function f on the kth sub-interval $[x_{k-1}, x_k]$ and $m_k(f)$ denote the minimum value of the function f on the kth sub-interval $[x_{k-1}, x_k]$.[3] We define the *upper Darboux sum* $U(f; \mathcal{P})$ for the function f with respect to the partition $\mathcal{P}$ by

$$U(f; \mathcal{P}) = \sum_{k=1}^{n} M_k(f)(x_k - x_{k-1}).$$

Likewise, we define the *lower Darboux sum* $L(f; \mathcal{P})$ for the function f with respect to the partition $\mathcal{P}$ by

$$L(f; \mathcal{P}) = \sum_{k=1}^{n} m_k(f)(x_k - x_{k-1}).$$

To understand these sums, geometrically if f is non-negative we see that the lower Darboux sum corresponds to the sum of the areas of the inscribed rectangles while the upper Darboux sum corresponds to the sum of the areas of the circumscribing rectangles depicted in Figure 1.2.

We are now in a position to give the definition for the Riemann integral. To help motivate this definition, if a bounded function $f : [a, b] \to \mathbb{R}$ is positive, we can see that if we wish to define the area of the region enclosed between the curve of the graph $y = f(x)$, the x-axis, and the lines $x = a$ and $x = b$ using inscribed and circumscribing rectangles, then this area must be at least $L(f; \mathcal{P})$ and at most $U(f; \mathcal{P})$. Thus one would expect the lower Darboux sum to correspond to the greatest lower bound for the area of the required region we seek, while the upper Darboux sum should correspond to the least upper bound for the area of this region. If the two are equal, that is, if $L(f; \mathcal{P}) = U(f; \mathcal{P})$, then it would seem reasonable to define the area of the required region to be this common value and to define Riemann integrability of a bounded function by this equality.

Definition 1.2 (Definition of the Riemann integral) Suppose that a function f is bounded on the interval $[a, b]$. If there exists a unique real number I such that

$$L(f; \mathcal{P}) \leqslant I \leqslant U(f; \mathcal{P}),$$

for every partition $\mathcal{P}$ on $[a, b]$ then f is said to be *Riemann integrable* on the interval $[a, b]$.

The unique real number I is called a *definite integral* and is denoted by

$$I = \int_a^b f(x)\, dx.$$

The function f is called the *integrand* of the definite integral while a and b correspond to the lower and upper limits of integration, respectively.

If f is non-negative the area of the region enclosed by the curve given by $y = f(x)$ and the x-axis from $x = a$ to $x = b$ is defined to be the Riemann integral of f. From this definition, a number of basic properties for the Riemann integral can be found. These we, however, defer until Chapters 2 and 4.

We now apply the definition of the Riemann integral to a number of simple functions.

Example 1.1 Show that the constant function $f(x) = 1$ on the interval $[a, b]$ is Riemann integrable, and $\int_a^b f(x)\, dx = b - a$.

Solution To show this let $\mathcal{P}$ be any partition of $[a, b]$ with end-points

$$\{a = x_0, x_1, x_2, \ldots, x_{n-1}, x_n = b\}.$$

Since f is constant we have $m_k(f) = M_k(f) = 1$ for every k (that is, for every sub-interval) between 1 and n. Thus

$$
\begin{aligned}
U(f; \mathcal{P}) &= \sum_{k=1}^{n} M_k(f)(x_k - x_{k-1}) \\
&= \sum_{k=1}^{n} (x_k - x_{k-1}) \\
&= (x_1 - x_0) + (x_2 - x_1) + (x_3 - x_2) + \cdots \\
&\quad \cdots + (x_{n-1} - x_{n-2}) + (x_n - x_{n-1}) \\
&= x_n - x_0 = b - a.
\end{aligned}
$$

A similar calculation for $L(f; \mathscr{P})$ also leads to a value of $b - a$. Since $\mathscr{P}$ is an arbitrary partition of $[a, b]$, we have

$$U(f; \mathscr{P}) = b - a = L(f; \mathscr{P}),$$

for every partition of $\mathscr{P}$ of $[a, b]$. Thus f is integrable and its Riemann integral is equal to $b - a$, as is required to show. ▶

In this example we see the answer agrees with what one intuitively expects to find if the definite integral is interpreted as corresponding to the area of the region enclosed by the curve and the x-axis between $x = a$ and $x = b$. Here the region is a rectangle, of unit height and of length $(b - a)$, which has an area of $(b - a)$ units squared.

Example 1.2 Consider the identity function $f(x) = x$ on the interval $[a, b]$. We will try to show this function is Riemann integrable from the definition of the Riemann integral.

Solution Let $\mathscr{P}$ be any partition of $[a, b]$ with end-points

$$\{a = x_0, x_1, x_2, \ldots, x_{n-1}, x_n = b\}.$$

Since f is the identity function, we have $m_k(f) = f(x_{k-1}) = x_{k-1}$ and $M_k(f) = f(x_k) = x_k$ for every k (that is, for every sub-interval) between 1 and n. Thus

$$U(f; \mathscr{P}) = \sum_{k=1}^{n} M_k(f)(x_k - x_{k-1}) = \sum_{k=1}^{n} x_k(x_k - x_{k-1}),$$

and

$$L(f; \mathscr{P}) = \sum_{k=1}^{n} M_k(f)(x_k - x_{k-1}) = \sum_{k=1}^{n} x_{k-1}(x_k - x_{k-1}).$$

Finding these sums, as we are about to see, is no easy task.

Subtracting the upper and lower Darboux sums, one has

$$U(f; \mathscr{P}) - L(f; \mathscr{P}) = \sum_{k=1}^{n} [x_k(x_k - x_{k-1}) - x_{k-1}(x_k - x_{k-1})]$$

$$= \sum_{k=1}^{n} (x_k - x_{k-1})(x_k - x_{k-1})$$

$$= \sum_{k=1}^{n} (x_k - x_{k-1})^2,$$

while adding

$$U(f;\mathcal{P}) + L(f;\mathcal{P}) = \sum_{k=1}^{n} [x_k(x_k - x_{k-1}) + x_{k-1}(x_k - x_{k-1})]$$

$$= \sum_{k=1}^{n} (x_k + x_{k-1})(x_k - x_{k-1})$$

$$= \sum_{k=1}^{n} (x_k^2 - x_{k-1}^2).$$

As the latter series telescopes, a simple expression for this sum can be found. On writing the terms out in the sum we have

$$U(f;\mathcal{P}) + L(f;\mathcal{P}) = (x_1^2 - x_0^2) + (x_2^2 - x_1^2) + (x_3^2 - x_2^2) + \cdots$$

$$\cdots + (x_{n-1}^2 - x_{n-2}^2) + (x_n^2 - x_{n-1}^2)$$

$$= x_n^2 - x_0^2 = b^2 - a^2.$$

It therefore follows that

$$U(f;\mathcal{P}) = \frac{b^2 - a^2}{2} + \frac{1}{2} \sum_{k=1}^{n} (x_k - x_{k-1})^2$$

and

$$L(f;\mathcal{P}) = \frac{b^2 - a^2}{2} - \frac{1}{2} \sum_{k=1}^{n} (x_k - x_{k-1})^2.$$

As no simple closed form is known for the sum that appears in the above expressions for the lower and upper Darboux sums, the best one can do is to try and bound each sum using inequalities. While it will not be formally shown here, it can be seen that as the size of the sub-intervals approaches zero the sum can be made as small as we wish while remaining positive. In this case it follows that

$$U(f;\mathcal{P}) \leqslant \frac{b^2 - a^2}{2} \quad \text{and} \quad L(f;\mathcal{P}) \geqslant \frac{b^2 - a^2}{2}.$$

But from the definition of the Riemann integral we must have $L(f;\mathcal{P}) \leqslant U(f;\mathcal{P})$. This shows that

$$U(f;\mathcal{P}) = \frac{b^2 - a^2}{2} = L(f;\mathcal{P}),$$

and thus f is integrable and $\displaystyle\int_a^b f(x)\,dx = \frac{b^2 - a^2}{2}.$ ▶

The answer found here agrees with what one intuitively expects to find if the definite integral is interpreted as corresponding to the area of the region enclosed by the curve $y = x$ and the x-axis between $x = a$ and $x = b$. Here the region is a trapezium, with parallel sides of lengths a and b and separated by a distance $(b - a)$, the area of which, according to the well-known formula for the area of a trapezium, is

$$\text{Area} = \frac{1}{2}(b - a)(a + b) = \frac{b^2 - a^2}{2},$$

agreeing with the value found for the definite integral.

What the foregoing example shows is that determining whether a bounded function on the interval $[a, b]$ is integrable from the definition of the Riemann integral is no easy task. In a moment we will show how considerable progress in determining the integrability of a bounded function, and in finding the value for its definite integral, can be made.

It turns out there are a wide class of functions that are Riemann integrable. Throughout most of this text we restrict ourselves to functions from one of the largest and most useful classes – the class of functions that are bounded and *piecewise continuous*. Recall that geometrically a function is said to be *continuous* if in the whole of its domain for which the function is defined the graph of the function consists of a single, unbroken curve with no 'holes' or 'gaps' in it. More formally, a function f is defined to be continuous at a point as follows.

Definition 1.3 (Continuity at a point) Suppose f is defined on some open interval containing the point a. Then f is continuous at the point a in the open interval if

$$\boxed{\lim_{x \to a} f(x) = f(a)}$$

The formal notion of continuity at a point can be extended to continuity on an interval if the function is continuous at every point in the interval. Piecewise continuity in a function can then be defined as follows.

Definition 1.4 (Piecewise continuity) A function $f : [a, b] \to \mathbb{R}$ is said to be piecewise continuous if it is continuous on the interval $[a, b]$ at all except perhaps a finite number of points.

A piecewise continuous function is therefore a function that is continuous at all but a finite number of points within its domain. Geometrically a piecewise continuous function will be made up of a finite number of curves or 'pieces', with each piece being continuous. In Figure 1.3 we illustrate piecewise

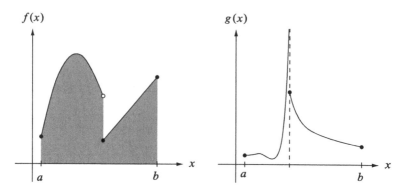

Figure 1.3. Example of two piecewise continuous functions on the interval $[a, b]$. The function f to the left is bounded while the function g to the right is unbounded.

continuity in two functions f and g on the interval $[a, b]$. To the left the function f is both bounded and piecewise continuous, while to the right, the function g is piecewise continuous but is not bounded.

We are now in a position to give a very important theorem that tells us exactly under what circumstances we have integrability in a function. By integrability we mean a function is integrable, that is, it can be integrated in the Riemann sense.

Theorem 1.1. *Let $f : [a, b] \to \mathbb{R}$ be a bounded and piecewise continuous function. Then f is integrable on the interval $[a, b]$.*

As the proof of this theorem is rather technical and involved, it will not be given here. The bounded and piecewise continuous functions cover a large and very important class of useful functions, examples of which are those functions you have already encountered in your previous mathematics studies. What Theorem 1.1 shows us is integrability is a far less restrictive condition on a function than differentiability. Recall that differentiability in a function at a point requires the function to not only be continuous at the point but also to be 'smooth' there, that is, the graph of the curve of the function should not contain any sharp edges or points where it is to be differentiated.

What Example 1.2 showed is that proving a function is Riemann integrable from the definition using Darboux sums is often a very difficult task, while the calculation of its definite integral is usually all but impossible. However, from Theorem 1.1, knowing in advance when a function f is Riemann integrable on $[a, b]$ often means the calculation of its definite integral becomes a far easier proposition. This is the case due to the following theorem.

Theorem 1.2. *Suppose $f : [a, b] \to \mathbb{R}$ is a bounded function that is Riemann integrable. Then there is a sequence of partitions $\{\mathcal{P}_n\}$ of $[a, b]$ such that*

$$\lim_{n \to \infty} [U(f; \mathcal{P}_n) - L(f; \mathcal{P}_n)] = 0.$$

In this case

$$\lim_{n \to \infty} L(f; \mathcal{P}_n) = \int_a^b f(x) \, dx = \lim_{n \to \infty} U(f; \mathcal{P}_n).$$

The proof of this theorem will not be given here. What this result shows us is as the maximum size s_n of the sub-intervals generated by $\mathcal{P}_n$ goes to zero, that is, as $\lim_{n \to \infty} s_n = 0$, the definite integral becomes 'squeezed' between the lower and upper Darboux sums.

Example 1.3 Show that the function $f(x) = x$ on the interval $[0, 1]$ is Riemann integrable and $\int_0^1 f(x) \, dx = \frac{1}{2}$.

Solution Note that f is integrable on $[0, 1]$ since it is bounded and continuous on $[0, 1]$.

To find the value for the definite integral we are free to partition the interval $[0, 1]$ in any manner we wish. For convenience the interval will be partitioned into n equally sized sub-intervals each of length $1/n$.

$$\mathcal{P}_n = \left\{ 0, \frac{1}{n}, \frac{2}{n}, \frac{3}{n}, \ldots, \frac{n-2}{n}, \frac{n-1}{n}, 1 \right\}.$$

Since $f(x) = x$ is increasing on $[0, 1]$, the minimum value of f on any particular sub-interval is given by its starting point, while the maximum value of f is given by its end-point. For the kth sub-interval in $\mathcal{P}_n$, since the starting point is $x_{k-1} = (k-1)/n$ and the point at the end is $x_k = k/n$, we have

$$m_k(f) = f(x_{k-1}) = \frac{k-1}{n} \quad \text{and} \quad M_k(f) = f(x_k) = \frac{k}{n}.$$

Thus the lower Darboux sum on the partition $\mathcal{P}_n$ is given by

$$
\begin{aligned}
L(f; \mathcal{P}_n) &= \sum_{k=1}^{n} m_k(f) \cdot (x_k - x_{k-1}) \\
&= \sum_{k=1}^{n} \frac{k-1}{n} \cdot \left(\frac{k}{n} - \frac{k-1}{n} \right) \\
&= \sum_{k=1}^{n} \frac{k-1}{n} \cdot \frac{1}{n} = \frac{1}{n^2} \sum_{k=1}^{n} (k-1).
\end{aligned}
$$

The sum of the first n integers is given by the well-known formula $\sum_{k=1}^{n} k = \frac{n(n+1)}{2}$, a result that can be readily established using induction on n. Thus

$$L(f;\mathcal{P}_n) = \frac{1}{n^2}\left[\frac{(n-1)n}{2}\right] = \frac{1}{2}\left(1 - \frac{1}{n}\right).$$

Similarly, for the upper Darboux sum we have

$$U(f;\mathcal{P}_n) = \sum_{k=1}^{n} M_k(f)\cdot(x_k - x_{k-1}) = \sum_{k=1}^{n}\frac{k}{n}\left(\frac{k}{n} - \frac{k-1}{n}\right) = \frac{1}{n^2}\sum_{k=1}^{n}k.$$

And since $\sum_{k=1}^{n} k = \frac{n(n+1)}{2}$, we have

$$U(f;\mathcal{P}_n) = \frac{1}{n^2}\left[\frac{n(n+1)}{2}\right] = \frac{1}{2}\left(1 + \frac{1}{n}\right).$$

It is not too hard to see that

$$L(f;\mathcal{P}_n) \leqslant \frac{1}{2} \leqslant U(f;\mathcal{P}_n), \quad \text{for all } n.$$

Also notice that $U(f;\mathcal{P}_n) - L(f;\mathcal{P}_n) = 1/n$ can be made as small as we like as n becomes as large as we like. So in the limit of n becoming very large, that is as $n \to \infty$, one has

$$\lim_{n\to\infty} L(f;\mathcal{P}_n) = \frac{1}{2} = \lim_{n\to\infty} U(f;\mathcal{P}_n),$$

and we conclude that $\int_0^1 f(x)\,dx = \frac{1}{2}$. ▶

What the foregoing example shows is that even when the integrability of a function is known in advance, finding the value for the Riemann integral is still a rather lengthy process. In Chapter 3 the problem of evaluating Riemann integrals will be taken up once more where it will be shown how the process can be significantly sped up and simplified.

§ The Definite Integral and Area

Currently the only way we have of finding the value for a definite integral is to use upper and lower Darboux sums. However, from the geometric interpretation of the Riemann integral as corresponding to the area beneath the curve, the x-axis, and the two bounding vertical lines given by the limits of integration, under very special circumstances it may be possible to find its value by calculating an appropriate area using known area formulae. We now consider two examples where the values for the definite integral are found in this way.

Example 1.4 Evaluate $\int_{-1}^{1}(x+1)\,dx$.

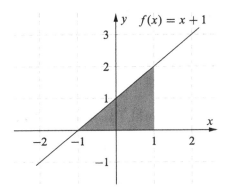

Figure 1.4. Sketch of the required area, which is triangular.

Solution A sketch of the area beneath the graph of the function $f(x) = x + 1$ (a straight line) and the x-axis between the vertical lines $x = -1$ and $x = 1$ is shown in Figure 1.4.

As the enclosed area is that of a triangle of base 2 units and height 2 units, the value of the definite integral is

$$\int_{-1}^{1} (x + 1)\, dx = \frac{1}{2} \cdot 2 \cdot 2 = 2.$$ ▶

Example 1.5 Evaluate $\displaystyle\int_{0}^{3} \sqrt{9 - x^2}\, dx$.

Solution Figure 1.5 a sketch of the area beneath the graph of the function $f(x) = \sqrt{9 - x^2}$ (a semicircle lying above the x-axis, centre at the origin, and of radius 3 units) and the x-axis between the vertical lines $x = 0$ and $x = 1$ is shown.

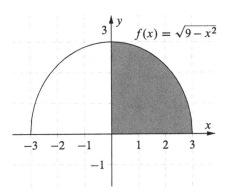

Figure 1.5. Sketch of the required area, which is a quarter of a circle.

As the enclosed area is a quarter of the area of a circle of radius 3 units, the value of the definite integral is

$$\int_0^3 \sqrt{9 - x^2}\, dx = \frac{1}{4} \cdot \pi \cdot 3^2 = \frac{9\pi}{4}. \qquad \blacktriangleright$$

§ The Fundamental Theorems of Calculus

From Example 1.3 we see that the business of evaluating definite integrals from the definition for the Riemann integral using Darboux sums, even for the simplest of functions, is at best, difficult, while at worst is all but impossible. If this were the only way to calculate definite integrals for functions that are Riemann integrable, then progress would be slow and would place real limits on the usefulness and usability of the definite integral. Fortunately there is another, far simpler way to calculate definite integrals for those functions that are integrable. It is done using what is the most fundamental and perhaps most important theorem in all of the calculus – the *fundamental theorem of calculus*.

By now you should be very familiar with the process of differentiation. It is a local process in that the value of the derivative at a point depends only on the values of the function in a small interval about that point. Integration, on the other hand, is a global process in the sense that the integral of an integrable function depends on the values of the function on the entire interval. Each of these processes, at least initially, appears to be very different, with no apparent connection between them. From the geometric point of view, differentiation corresponds to finding gradients of tangents to curves, while integration corresponds to finding areas under these curves. Why each of these different geometric notions should be intimately related is not at all obvious.

We now present a result (we actually will break it up and present two results) that indicates a strong connection between integration and differentiation. Recall that if $f : [a, b] \to \mathbb{R}$ is differentiable, then one obtains a new function $f' : [a, b] \to \mathbb{R}$, called the derivative of f. Likewise, if $f : [a, b] \to \mathbb{R}$ is integrable in the Riemann sense, then one obtains a new function $F : [a, b] \to \mathbb{R}$, which we call a *primitive* or *antiderivative* of f, defined by

$$F(x) = \int_a^x f(t)\, dt, \quad \text{for all } x \in [a, b].$$

Geometrically, if f is positive, the primitive function F can be interpreted as an 'area function' for the graph of the function $y = f(x)$ bounded by the curve and the x-axis on the interval $[a, x]$ (see Figure 1.6).

We are now in a position to present the central result of the calculus known as the *fundamental theorem of calculus*. It states that differentiation and

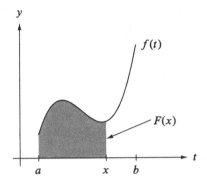

Figure 1.6. The area function $F(x) = \int_a^x f(t)\,dt$.

integration are (nearly) inverse operations of each other. Roughly speaking, the fundamental theorem of calculus says that differentiation undoes what integration does to a function. That is, if one integrates a function and differentiates the result, then one gets back the original function. Similarly, integration undoes what differentiation does to a function up to an arbitrary additive constant. That is, if one differentiates a function and then integrates the result, to within an arbitrary additive constant, one again gets back the original function. We now state the theorem in two parts and give some of their consequences.

Theorem 1.3 (The first fundamental theorem of calculus). *If f is a continuous function defined on the interval $[a, b]$ and therefore integrable, then the function $F : [a, b] \to \mathbb{R}$ defined by*

$$F(x) = \int_a^x f(t)\,dt,$$

is continuous on the closed interval $[a, b]$, is differentiable on the open interval (a, b), and has a derivative F' given by

$$\boxed{F'(x) = \frac{d}{dx} \int_a^x f(t)\,dt = f(x)}$$

for all $x \in (a, b)$.

As the proof requires ideas outside the scope of this text it will be omitted.

From the first fundamental theorem of calculus we see that every continuous function f on $[a, b]$ has an antiderivative F. This is demonstrated by the fact that F satisfies the equation $F' = f$ on (a, b). Secondly, since any two antiderivatives of a function f differ by at most a constant, as every antiderivative of f takes on the form of F plus an additivity constant, one sees that the processes of integration and antidifferentiation are essentially the same thing; this suggests that integration and differentiation are inverse operations. Thus if one takes a function f, integrates it, and then differentiates the result, one gets back the original function f. Differentiation therefore undoes what integration does to the function f. Whether or not the converse is true will be taken up shortly. The first fundamental theorem of calculus therefore suggests there may be an alternative way of calculating the area under a curve given by the graph of f. Instead of integrating via applying Darboux sums, one may instead try to integrate a function via antidifferentiation.

It is clear from the first fundamental theorem of calculus that if f is a continuous function on the interval $[a, b]$ (and hence Riemann integrable) and has a primitive given by F, then one must have

$$F(x) = \int_a^x f(t)\,dt + F(a), \quad \text{for all } x \in [a, b].$$

F is called the *indefinite integral* of f. If we have no particular interest in the interval $[a, x]$ and merely want to indicate that F is an indefinite integral of f, then we write

$$\int f(x)\,dx = F(x) + C.$$

Here C denotes an arbitrary constant and indicates that a primitive function of f is unique only up to an arbitrary additive constant. The indefinite integral on an interval is a collection of all its primitives F on that interval whose derivatives are the given function f; it is the reason why the indefinite article 'a' is used when referring to the primitive of an indefinite integral rather than the definite article 'the': to do otherwise would suggest the primitive is something that can be uniquely specified, which it cannot be.

Example 1.6 If $g(x) = \int_{\frac{\pi}{3}}^x \sin^5\left(\frac{\pi t^2}{2}\right)\,dt$, find the value of $g'(1)$.

Solution Applying the first fundamental theorem of calculus to the function g gives

$$g'(x) = \frac{d}{dx}\int_{\frac{\pi}{3}}^x \sin^5\left(\frac{\pi t^2}{2}\right)\,dt = \sin^5\left(\frac{\pi x^2}{2}\right).$$

And when $x = 1$, we have

$$g'(1) = \sin^5\left(\frac{\pi}{2}\right) = 1. \qquad \blacktriangleright$$

If the first fundamental theorem of calculus says that the processes of integration and antidifferentiation are essentially the same, the second fundamental theorem of calculus (see Theorem 1.4 (highlighted)) gives a fast way of calculating integrals by exploiting this similarity.

Theorem 1.4 (The second fundamental theorem of calculus). *Suppose that f is a continuous function on the interval* $[a, b]$ *and therefore integrable. If F is a primitive of f on* (a, b), *then*

$$\int_a^b f(t)\, dt = F(b) - F(a)$$

Proof To prove the result we make use of the following property for the definite integral $\int_a^a f(t)\, dt = 0$, a property to be considered in Chapter 4 (see page 40).

Suppose that f is continuous on $[a, b]$ and that F is a primitive of f on (a, b). From the first fundamental theorem of calculus, we have

$$\frac{dF}{dx} = \frac{d}{dx} \int_a^x f(t)\, dt.$$

Rearranging, we have

$$\frac{d}{dx} \left[F(x) - \int_a^x f(t)\, dt \right] = 0.$$

So there is a constant C such that

$$F(x) - \int_a^x f(t)\, dt = C,$$

for all $x \in [a, b]$. Writing this as

$$F(x) = \int_a^x f(t)\, dt + C,$$

and setting $x = a$ we have

$$F(a) = \int_a^a f(t)\, dt + C = C,$$

since $\int_a^a f(t)\, dt = 0$. Thus

$$\int_a^x f(t)\, dt = F(x) - F(a),$$

which on setting x equal to b becomes

$$\int_a^b f(t)\, dt = F(b) - F(a),$$

as required to prove. ∎

The second fundamental theorem of calculus is a truly remarkable result. It says that in order to calculate the area beneath the graph of a continuous function all one needs to know is the value of any primitive function of f at its end-points. It recasts the problem of evaluating definite integrals to that of finding an explicit primitive of f, and while an explicit primitive cannot always be found, as we shall see soon it can be found for a wide class of relatively elementary integrands f.

The second part of the fundamental theorem of calculus provides the most widely used method for evaluating Riemann integrals. As a basic tool it allows one to find the integral of a function f if we can find a primitive function F such that $F' = f$. Of course, using the method relies on being able to conjure up a function whose derivative is the given function, something that at least in the case of simple functions, you are by now already familiar with doing. Compared to the rules of differentiation that provide a mechanical algorithm for the calculation of the derivative of a function, there is no systematic procedure for finding primitives, which means that the task of finding an integral of a function is, in general, not the easiest of things to do.

Finally, we return to the question of whether integration and differentiation are inverse operations of each other. The first fundamental theorem shows that if one integrates a function f and differentiates the result one obtains f again. As for the converse, the result is not true, for if a function f is differentiated and the result is integrated, one does not obtain f. Instead one obtains f to within an arbitrary additivity constant.

Taken together the two fundamental theorems of calculus show that differentiation and integration are intimately related to each other. They provide the link between the differential calculus (the calculus of change) and the integral calculus (the calculus of accumulation). The consequences of the fundamental theorem of calculus are, on the one hand, deeply profound while on the other are extremely versatile and useful.

Exercises for Chapter 1

✠ **Warm-ups**

1. Compare the following two expressions:

$$\frac{d}{dx}\left[\int_a^x f(t)\,dt\right] \quad \text{and} \quad \int_a^x \frac{d}{dt}[f(t)]\,dt.$$

Are they the same? Carefully explain your answer.

✠ Practice questions

2. Consider the function $f(x) = x^2$.

   (a) Show that f is increasing on the interval $[0, 1]$.
   (b) By considering a partition on $[0, 1]$ with n equally sized sub-intervals of length $1/n$, that is $\mathcal{P}_n = \{0, \frac{1}{n}, \frac{2}{n}, \ldots, 1\}$, calculate lower and upper Darboux sums for f.

   To do this, the formula $\displaystyle\sum_{k=1}^{n} k^2 = \frac{n(n+1)(2n+1)}{6}$ will be needed.

   (c) By taking the limit as $n \to \infty$, calculate $\displaystyle\int_0^1 f(x)\,dx$.

3. Consider the function $f(x) = x^3$.

   (a) Show that f is increasing on the interval $[0, 1]$.
   (b) By considering a partition on $[0, 1]$ with n equally sized sub-intervals of length $1/n$, that is $\mathcal{P}_n = \{0, \frac{1}{n}, \frac{2}{n}, \ldots, 1\}$, calculate lower and upper Darboux sums for f.

   To do this, the formula $\displaystyle\sum_{k=1}^{n} k^3 = \frac{n^2(n+1)^2}{4}$ will be needed.

   (c) By taking the limit as $n \to \infty$, calculate $\displaystyle\int_0^1 f(x)\,dx$.

4. For each given definite integral $\int_a^b f(x)\,dx$, (i) sketch the area between the graph of the function $y = f(x)$, the x-axis, and the vertical lines $x = a$ and $x = b$, and (ii) using known area formulae, evaluate each integral:

   (a) $\displaystyle\int_0^3 x\,dx$ 　　　　　　　　　　 (b) $\displaystyle\int_0^2 (x+1)\,dx$

   (c) $\displaystyle\int_{-1}^1 (1-x)\,dx$ 　　　　　　 (d) $\displaystyle\int_0^2 \sqrt{4-x^2}\,dx$

   (e) $\displaystyle\int_{-3}^3 \sqrt{9-x^2}\,dx$ 　　　　 (f) $\displaystyle\int_{-2}^2 |x|\,dx$

5. Using the first fundamental theorem of calculus, evaluate the following expressions:

   (a) $\displaystyle\frac{d}{dx}\int_2^x t\sin t\,dt$ 　　　　　 (b) $\displaystyle\frac{d}{dx}\int_0^x e^{t^2}\,dt$

   (c) $\displaystyle\frac{d}{dx}\int_{\frac{\pi}{3}}^x \cos(t^2)\,dt$ 　　　 (d) $\displaystyle\frac{d}{dx}\int_1^x (t^2 - \cos t)\,dt$

6. Evaluate $\displaystyle\int_1^3 f'(x)\,dx$ if $f(1) = 5$ and $f(3) = 11$.

7. If $\displaystyle f(x) = x \int_0^x \sin(t^3)\,dt$, show that $xf'(x) - f(x) = x^2 \sin(x^3)$ for $x \neq 0$.

8. If $\displaystyle F(x) = \int_1^x f(t)\,dt$ and $\displaystyle f(t) = \int_1^t \sqrt{1 + u^2}\,du$, find the value of $F''(1)$.

9. A special function known as the *error function* erf(x) is defined in terms of a definite integral as follows:

$$\mathrm{erf}(x) = \frac{2}{\sqrt{\pi}} \int_0^x e^{-u^2}\,du, \quad \text{for all } x.$$

(a) Using the first fundamental theorem of calculus, find $\dfrac{d}{dx}\big(\mathrm{erf}(x)\big)$.

(b) Show that the error function is an increasing function for all $x \in \mathbb{R}$.

✠ Extension questions and Challenge problems

10. In this question the definite integral $\displaystyle\int_0^{\frac{\pi}{2}} \sin x\,dx$ will be evaluated directly using Darboux sums.

Suppose that $f : [0, \frac{\pi}{2}] \to \mathbb{R}$ is defined by $f(x) = \sin x$.

(a) Show that f is increasing on the interval $[0, \frac{\pi}{2}]$.

(b) Show that the lower and upper Darboux sums for f with respect to the partition $\mathscr{P}_n = \left\{0, \frac{\pi}{2n}, \frac{2\pi}{2n}, \ldots, \frac{(n-1)\pi}{2n}, \frac{n\pi}{2n}\right\}$ of $[0, \frac{\pi}{2}]$ are given, respectively, by

$$L(f; \mathscr{P}_n) = \frac{\pi}{2n} \sum_{k=1}^n \sin\left(\frac{(k-1)\pi}{2n}\right), \quad \text{and}$$

$$U(f; \mathscr{P}_n) = \frac{\pi}{2n} \sum_{k=1}^n \sin\left(\frac{k\pi}{2n}\right).$$

(c) The sums given in (b) can be found and expressed in simple form. Let

$$S_n(\alpha) = \sin\alpha + \sin 2\alpha + \cdots + \sin(n-1)\alpha + \sin n\alpha.$$

By multiplying $S_n(\alpha)$ by the term $2\sin\frac{\alpha}{2}$ and summing the resulting telescoping series, show that

$$S_n(\alpha) = \sum_{k=1}^{n} \sin k\alpha = \frac{\sin\left[\left(\dfrac{n+1}{4n}\right)\pi\right]\sin\left(\dfrac{\pi}{4}\right)}{\sin\left(\dfrac{\pi}{4n}\right)}.$$

(d) Hence deduce that

$$U(f;\mathscr{P}_n) = \frac{\pi}{2n}\frac{\sin\left[\left(\dfrac{n+1}{4n}\right)\pi\right]\sin\left(\dfrac{\pi}{4}\right)}{\sin\left(\dfrac{\pi}{4n}\right)}, \quad\text{and}$$

$$L(f;\mathscr{P}_n) = U(f;\mathscr{P}_n) - \frac{\pi}{2n}\sin\left(\frac{n\pi}{2n}\right).$$

(f) As $n \to \infty$, recognising that

$$\lim_{n\to\infty} U(f;\mathscr{P}_n) = 2\sin\left(\frac{\pi}{4}\right)\lim_{n\to\infty}\sin\left[\left(1+\frac{1}{n}\right)\frac{\pi}{4}\right]$$

$$\times \lim_{n\to\infty}\frac{\dfrac{\pi}{4n}}{\sin\left(\dfrac{\pi}{4n}\right)},$$

if you are given that $\displaystyle\lim_{n\to\infty}\frac{k/n}{\sin(k/n)} = 1$ for all $k \neq 0$, use this to show that $\displaystyle\lim_{n\to\infty} U(f;\mathscr{P}_n) = 1$.

(f) Hence deduce that $\displaystyle\lim_{n\to\infty} L(f;\mathscr{P}_n) = 1$.

(g) Explain why f is integrable on the interval $[0, \frac{\pi}{2}]$ and hence deduce that $\displaystyle\int_0^{\frac{\pi}{2}} \sin x \, dx = 1$.

Endnotes

1. The integral is so named after the nineteenth-century German mathematician Georg Friedrich Bernhard Riemann (1826–1866) who was the first to give a rigorous formulation of the integral in 1854 in his habilitation.
2. So named after the French mathematician Jean-Gaston Darboux (1842–1917), who introduced the idea in 1875.
3. In a strict technical sense the maximum value of f should be replaced with the *supremum* of f on the interval (the least upper bound for f on the interval), while the minimum value for f should be replaced with the *infimum* of f on the interval (the greatest lower bound of f on the interval).

2

Basic Properties of the Definite Integral: Part I

'Ignorance of Axioms', the Lecturer continued, 'is a great drawback in life. It wastes so much time to have to say them over and over again.'
— Lewis Carroll, *Sylvie and Bruno Concluded*

In this chapter we give some basic properties of the definite integral. Each property to be given should seem obvious if the definite integral is interpreted geometrically as an area under a curve.

The properties for the definite integral we give apply to any bounded, piecewise continuous function on the closed interval $[a, b]$. According to Theorem 1.1 on page 8 of Chapter 1, such functions are integrable in the Riemann sense with the definite integral existing.

Property 1 – Additive Property
The additive property states that the definite integral of the sum of two integrable functions is equal to the sum of the definite integrals of these functions.

$$\int_a^b [f(x) + g(x)]\, dx = \int_a^b f(x)\, dx + \int_a^b g(x)\, dx$$

As the proof of this property depends on the definition of the integral, it will not be given here.

Property 2 – Homogeneous Property
The homogeneous property states that if all the function values are multiplied by a constant k and then integrated, this is equal to the definite integral of the original function multiplied by the constant k, namely

$$\int_a^b k \cdot f(x)\, dx = k \int_a^b f(x)\, dx$$

Once again the proof of this property depends on the definition of the definite integral and is therefore omitted.

Property 3 – Linearity Property

The first two properties for the definite integral can be combined into a single formula known as the linearity property. If $k_1, k_2 \in \mathbb{R}$, then

$$\int_a^b [k_1 f(x) + k_2 g(x)]\, dx = k_1 \int_a^b f(x)\, dx + k_2 \int_a^b g(x)\, dx$$

This property shows us that the definite integral of a sum is equal to the sum of the definite integrals.

Example 2.1 If $\displaystyle\int_0^8 f(x)\, dx = 11$ and $\displaystyle\int_0^8 g(x)\, dx = 7$, find the value of $\displaystyle\int_0^8 [2f(x) + 5g(x)]\, dx$.

Solution From the linearity property we can write

$$\int_0^8 [2f(x) + 5g(x)]\, dx = 2 \int_0^8 f(x)\, dx + 5 \int_0^8 g(x)\, dx$$
$$= 2 \cdot 11 + 5 \cdot 7 = 57. \qquad \blacktriangleright$$

Example 2.2 If f is an integrable function, use the linearity property to show that

$$\int_a^b [f(x) - g(x)]\, dx = \int_a^b f(x)\, dx - \int_a^b g(x)\, dx.$$

Solution Writing $f(x) - g(x) = f(x) + (-1)g(x)$, from the linearity property (here $k_1 = 1, k_2 = -1$) we have

$$\int_a^b [f(x) - g(x)]\, dx = \int_a^b [f(x) + (-1)g(x)]\, dx$$
$$= \int_a^b f(x)\, dx + (-1) \int_a^b g(x)\, dx$$
$$= \int_a^b f(x)\, dx - \int_a^b g(x)\, dx,$$

as required to show. $\qquad \blacktriangleright$

Property 4 – Signed Property

If a function f is positive throughout the interval $[a, b]$, then its definite integral will also be positive. That is, if $f(x) > 0$ for all $x \in [a, b]$ then

$$\int_a^b f(x)\, dx > 0$$

The proof of this property again depends on the definition of the definite integral and is omitted.

Similarly, if a function g is negative throughout the interval $[a, b]$, then its definite integral will also be negative. That is, if $g(x) < 0$ for all $x \in [a, b]$ then

$$\int_a^b g(x)\, dx < 0$$

Proof Let $f(x) > 0$ for all $x \in [a, b]$ and define $g(x) = -f(x)$. Thus $g(x) < 0$ for all $x \in [a, b]$ since $f(x) > 0$ for all $x \in [a, b]$. Now for the definite integral of g one has

$$\int_a^b g(x)\, dx = -\int_a^b f(x)\, dx < 0,$$

since by the signed property $\int_a^b f(x)\, dx > 0$, as required to prove. ∎

The signed property can be interpreted geometrically. For the case where $f(x) > 0$ for all $x \in [a, b]$ the graph of the curve of f lies above the x-axis, meaning the area of the region bounded by the curve and the x-axis between $x = a$ and $x = b$ is positive (see Figure 2.1). The definite integral, representing the area under the curve, will also be positive.

Example 2.3 Show that the definite integral $\int_{\frac{1}{2}}^1 e^{-x} \ln x \, dx$ is negative.

Solution Noting that $e^{-x} > 0$ for $x \in (1/2, 1)$ while $\ln x < 0$ for $x \in (1/2, 1)$, for the product we have $e^{-x} \ln x < 0$ for $x \in (1/2, 1)$. So from the signed property for the definite integral one has

$$\int_{\frac{1}{2}}^1 e^{-x} \ln x \, dx < 0.$$

Thus the definite integral is negative, as required to show. ▶

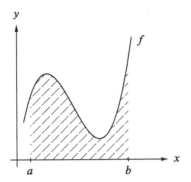

Figure 2.1. Graphical illustration of the signed property for the definite integral.

Exercises for Chapter 2

✠ **Practice questions**

1. If $\int_{-1}^{2} f(x)\,dx = 7$ and $\int_{-1}^{2} g(x)\,dx = 3$, find the value of the following:

(a) $\int_{-1}^{2} 5f(x)\,dx$ 　　　　(b) $\int_{-1}^{2} \left(f(x) - g(x)\right) dx$

(c) $\int_{-1}^{2} \left(f(x) + 3g(x)\right) dx$

2. Without attempting to evaluate the integral, determine if the following definite integrals are positive or negative:

(a) $\int_{0}^{2} x^5(1 - x^2)^2\,dx$ 　　　　(b) $\int_{0}^{1} x(x - 1)(x - 2)\,dx$

(c) $\int_{0}^{\pi} \sin^3 x\,dx$ 　　　　(d) $\int_{0}^{3} [(x - 1)(x - 2) - 2]\,dx$

(e) $\int_{1}^{2} \dfrac{dx}{x(x - 3) - 2}$ 　　　　(f) $\int_{\frac{1}{e}}^{1} \dfrac{dx}{\ln x - 2}$

(g) $\int_{-\frac{\pi}{2}}^{\frac{\pi}{2}} \cos^5 x\,dx$ 　　　　(h) $\int_{\frac{1}{2}}^{1} \ln x\,dx$

3. The graph for the function $f(x) = x - 1$ is given below.

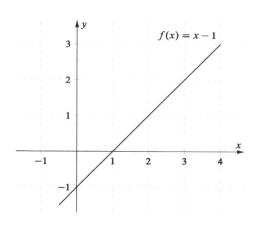

Find

(a) $\displaystyle\int_{1}^{3} 3f(x)\,dx$ (b) $\displaystyle\int_{2}^{3} 2f(x)\,dx$ (c) $\displaystyle\int_{2}^{4} f(x)\,dx$

4. If $\displaystyle\int_{1}^{b} f(x)\,dx = 1 - \frac{1}{b}$ where $b > 1$, find the value of the following integrals:

(a) $\displaystyle\int_{1}^{3} [2f(x) + 1]\,dx$ (b) $\displaystyle\int_{1}^{6} [6f(x) - 4]\,dx$

(c) $\displaystyle\int_{1}^{4} f(x)\,dx + \int_{1}^{5} f(x)\,dx + \int_{1}^{6} f(x)\,dx$

3

Some Basic Standard Forms

No one can be a good mathematician who cannot differentiate $\arccos\left(\dfrac{a\cos x + b}{a + b\cos x}\right)$, or show a little ingenuity in devices of integration.

— *G. H. Hardy*

Integration is hard. Very hard. Compared to finding the derivative of a function the process of integration is a complex and subtle affair. As was shown in Chapter 1, every continuous function f on a closed interval $[a, x]$ has a primitive. Our job in this chapter is to establish some basic primitives for commonly encountered functions that can be written in terms of familiar functions, such as polynomials, roots, and the trigonometric, logarithmic, and exponential functions.

Table 3.1 lists a number of basic elementary primitives. The list can be obtained by simply differentiating each of the various functions and noting the result. Many more primitives for other commonly encountered functions will be given as we progress through the text, so the list given here should be thought of as the simplest and most basic. Each of the eleven integrals listed in the table needs to be known and represents the basic building blocks to be drawn upon when we later attempt to find more difficult integrals by reducing them to a known, standard form. For brevity, the arbitrary constant of integration in the table has been omitted.

We now give a number of examples that make use of the integrals found in Table 3.1. At times, in some of the integrals to be considered, the integrand will need to be either expanded or factorised, or may first require a change in form by exploiting some particular property of the function under consideration. Extensive use of the linearity property for the integral will also be made. In the first three examples to be considered, application of the first of the

Table 3.1. *Table of standard integrals. For brevity the arbitrary constant of integration has been omitted.*

Table of Standard integrals

1. $\displaystyle\int x^n \, dx = \frac{1}{n+1}x^{n+1}, \quad n \neq -1; \ x \neq 0 \text{ if } n < 0$

2. $\displaystyle\int \frac{dx}{x} = \ln x, \quad x > 0$

3. $\displaystyle\int e^{ax} \, dx = \frac{1}{a}e^{ax}, \quad a \neq 0$

4. $\displaystyle\int \cos ax \, dx = \frac{1}{a}\sin ax, \quad a \neq 0$

5. $\displaystyle\int \sin ax \, dx = -\frac{1}{a}\cos ax, \quad a \neq 0$

6. $\displaystyle\int \sec^2 ax \, dx = \frac{1}{a}\tan ax, \quad a \neq 0$

7. $\displaystyle\int \sec ax \tan ax \, dx = \frac{1}{a}\sec ax, \quad a \neq 0$

8. $\displaystyle\int \frac{dx}{a^2 + x^2} = \frac{1}{a}\tan^{-1}\left(\frac{x}{a}\right), \quad a \neq 0$

9. $\displaystyle\int \frac{dx}{\sqrt{a^2 - x^2}} = \sin^{-1}\left(\frac{x}{a}\right), \quad a > 0, \ |x| < a$

10. $\displaystyle\int \frac{dx}{\sqrt{x^2 - a^2}} = \ln\left|x + \sqrt{x^2 - a^2}\right|, \quad 0 < a < x$

11. $\displaystyle\int \frac{dx}{\sqrt{x^2 + a^2}} = \ln\left|x + \sqrt{x^2 + a^2}\right|$

integrals appearing in the table will be made. Known as the *index law* it applies to a simple monomial term in x raised to a power $n \neq -1$.

Example 3.1 Find $\displaystyle\int 4x^4 \, dx$.

Solution From the linearity property for the integral we can write

$$\int 4x^4 \, dx = 4\int x^4 \, dx.$$

Now from the index law a primitive is

$$\int 4x^4 \, dx = 4 \cdot \frac{1}{4+1}x^{4+1} + C = 4 \cdot \frac{1}{5}x^5 + C = \frac{4}{5}x^5 + C.$$

As a simple check of the answer found, you should notice that

$$\frac{d}{dx}\left(\frac{4}{5}x^5 + C\right) = 4x^4,$$

the integrand of the integral, as expected.   ▶

Example 3.2 Find $\int x^2 \sqrt{x}\, dx$.

Solution We begin by first rewriting the integral as

$$\int x^2 \sqrt{x}\, dx = \int x^{5/2}\, dx.$$

Now on applying the index law, a primitive can be found. Here

$$\int x^2 \sqrt{x}\, dx = \frac{2}{7}x^{7/2} + C.$$ ▶

Example 3.3 Find $\int (4x^3 + 2x + 5)\, dx$.

Solution We begin by applying the linearity law first. Doing so yields

$$\int (4x^3 + 2x + 5)\, dx = 4\int x^3\, dx + 2\int x\, dx + 5\int dx.$$

Now applying the index law to each of the integrals gives

$$\int (4x^3 + 2x + 5)\, dx = x^4 + x^2 + 5x + C.$$ ▶

In future examples the linearity for the integral will be implicitly assumed, allowing us to write down the primitive in a single step without having to split up the integral.

Example 3.4 Find $\int (2x + 3)^2\, dx$.

Solution Expanding the term appearing in the integrand, we have

$$\int (2x + 3)^2\, dx = \int (4x^2 + 12x + 9)\, dx.$$

Now applying the index law (where the linearity of the integral has been implicitly assumed) we have

$$\int (2x + 3)^2\, dx = \frac{4}{3}x^3 + 6x^2 + 9x + C.$$ ▶

Example 3.5 Evaluate $\int_0^1 (\sqrt[4]{x} - \sqrt{x})\, dx$.

Solution Rewriting the integral as

$$\int_0^1 \left(\sqrt[4]{x} - \sqrt{x} \right) dx = \int_0^1 \left(x^{1/4} - x^{1/2} \right) dx,$$

on applying the index law to find a primitive for the integral together with the second fundamental theorem of calculus, we have

$$\int_0^1 \left(\sqrt[4]{x} - \sqrt{x} \right) dx = \left[\frac{4}{5} x^{5/4} - \frac{2}{3} x^{3/2} \right]_0^1 = \frac{4}{5} - \frac{2}{3} = \frac{2}{15}. \quad \blacktriangleright$$

Example 3.6 Find $\displaystyle\int \frac{4x^4 + 1}{x} \, dx$.

Solution The rational function in the integrand can be simplified to

$$\int \frac{4x^4 + 1}{x} \, dx = \int \left(4x^3 + \frac{1}{x} \right) dx.$$

To find the primitive the index law is applied to the first term, while the second entry in the table of standard integrals is applied to the second term. Thus

$$\int \frac{4x^4 + 1}{x} \, dx = x^4 + \ln x + C. \quad \blacktriangleright$$

Example 3.7 Find $\displaystyle\int \left(e^{5x} - \sin(5x) + \frac{1}{\sqrt[5]{x}} \right) dx$.

Solution Rewrite the integral as

$$\int \left(e^{5x} - \sin(5x) + \frac{1}{\sqrt[5]{x}} \right) dx = \int \left(e^{5x} - \sin(5x) + x^{-1/5} \right) dx.$$

From the first, third, and fifth entries appearing in the table of standard integrals, a primitive can be found. The result is

$$\int \left(e^{5x} - \sin(5x) + \frac{1}{\sqrt[5]{x}} \right) dx = \frac{1}{5} e^{5x} + \frac{1}{5} \cos(5x) + \frac{5}{4} x^{4/5} + C. \quad \blacktriangleright$$

In the next example properties of the sine function will be used to help put the integrand into a standard form before a primitive can be found.

Example 3.8 Find $\displaystyle\int \sin(7x - 2) \, dx$.

Solution Making use of the difference identity for the sine function, namely

$$\sin(A - B) = \sin A \cos B - \cos A \sin B,$$

the sine term can be written as

$$\sin(7x - 2) = \sin 7x \cos 2 - \cos 7x \sin 2.$$

Thus

$$\int \sin(7x - 2)\, dx = \int (\sin 7x \cos 2 - \cos 7x \sin 2)\ dx$$

$$= \cos 2 \int \sin 7x\, dx - \sin 2 \int \cos 7x\, dx$$

$$= -\frac{1}{7} \cos 2 \cos 7x - \frac{1}{7} \sin 2 \sin 7x + C.$$

Recognising that $\cos 7x \cos 2 + \sin 7x \sin 2 = \cos(7x - 2)$, the primitive may be rewritten as

$$\int \sin(7x - 2)\, dx = -\frac{1}{7} \cos(7x - 2) + C. \qquad \blacktriangleright$$

Example 3.8 suggests the following generalisations. If $a, b \in \mathbb{R}, a \neq 0$ then

$$\boxed{\int \cos(ax + b)\, dx = \frac{1}{a} \sin(ax + b) + C}$$

and

$$\boxed{\int \sin(ax + b)\, dx = -\frac{1}{a} \cos(ax + b) + C}$$

These results will be proved in the exercise set at the end of the chapter. A similar result for the exponential function with an argument that is linear can also be readily proved (see Exercise 6).

We now provide a few more examples that make use of some of the other standard integrals listed in Table 3.1.

Example 3.9 Find $\displaystyle\int \frac{dx}{11 + x^2}$.

Solution Writing the integral as

$$\int \frac{dx}{11 + x^2} = \int \frac{dx}{(\sqrt{11})^2 + x^2},$$

from the eighth entry in the table of standard integrals one immediately has

$$\int \frac{dx}{11 + x^2} = \frac{1}{\sqrt{11}} \tan^{-1}\left(\frac{x}{\sqrt{11}}\right) + C. \qquad \blacktriangleright$$

Example 3.10 Find $\int \dfrac{dx}{1 + 16x^2}$.

Solution Factoring out a factor of 16 from the denominator one has

$$\int \frac{dx}{1 + 16x^2} = \frac{1}{16} \int \frac{dx}{\frac{1}{16} + x^2} = \frac{1}{16} \int \frac{dx}{\left(\frac{1}{4}\right)^2 + x^2},$$

which again from the eighth entry in the table of standard integrals gives

$$\int \frac{dx}{1 + 16x^2} = \frac{1}{16} \cdot 4 \tan^{-1}(4x) + C = \frac{1}{4} \tan^{-1}(4x) + C. \qquad \blacktriangleright$$

Having a number of known primitives at hand allows a particularly profitable interaction between Darboux sums considered in Chapter 1 and antidifferentiation to be found.

Example 3.11 Consider the function $f(x) = \dfrac{1}{\sqrt{1 - x^2}}$.

(a) Show that f is increasing on the interval $[0, \frac{1}{2}]$.
(b) Find the upper Darboux sum for f with respect to the partition

$$\mathcal{P}_n = \left\{ 0, \frac{1}{2n}, \frac{2}{2n}, \frac{3}{2n}, \ldots, \frac{n}{2n} \right\} \text{ of } [0, \tfrac{1}{2}].$$

(c) Hence show that

$$\lim_{n \to \infty} \left(\frac{1}{\sqrt{4n^2 - 1^2}} + \frac{1}{\sqrt{4n^2 - 2^2}} + \cdots + \frac{1}{\sqrt{4n^2 - n^2}} \right) = \frac{\pi}{6}.$$

Solution (a) As

$$f'(x) = \frac{x}{(1 - x^2)^{\frac{3}{2}}} > 0 \text{ for } x \in [0, \tfrac{1}{2}],$$

we see f is an increasing function on the interval $[0, \frac{1}{2}]$, as required to show.

(b) Since f is increasing on $[0, \frac{1}{2}]$, the maximum value of f on any particular sub-interval is given by its end-point. So for the kth sub-interval in $\mathcal{P}_n$, since the end-point is $x_k = \dfrac{k}{2n}$, we have

$$M_k(f) = f(x_k) = f\left(\frac{k}{2n} \right) = \frac{1}{\sqrt{1 - \frac{k^2}{4n^2}}} = \frac{2n}{\sqrt{4n^2 - k^2}}.$$

So for the upper Darboux sum on the partition $\mathcal{P}_n$ we have

$$U(f;\mathcal{P}_n) = \sum_{k=1}^{n} M_k(f) \cdot (x_k - x_{k-1})$$

$$= \sum_{k=1}^{n} \frac{2n}{\sqrt{4n^2 - k^2}} \cdot \left(\frac{k}{2n} - \frac{k-1}{2n} \right)$$

$$= \sum_{k=1}^{n} \frac{1}{2n} \cdot \frac{2n}{\sqrt{4n^2 - k^2}} = \sum_{k=1}^{n} \frac{1}{\sqrt{4n^2 - k^2}}.$$

(c) Since f is bounded and continuous on $[0, \frac{1}{2}]$ it is integrable on this interval and

$$\lim_{n \to \infty} U(f;\mathcal{P}_n) = \int_0^{\frac{1}{2}} \frac{dx}{\sqrt{1 - x^2}}.$$

As a primitive for the integral is known, which corresponds to the inverse sine function, from the second fundamental theorem of calculus we have

$$\int_0^{\frac{1}{2}} \frac{dx}{\sqrt{1 - x^2}} = \left[\sin^{-1} x \right]_0^{\frac{1}{2}} = \frac{\pi}{6}.$$

Hence

$$\lim_{n \to \infty} \sum_{k=1}^{n} \frac{1}{\sqrt{4n^2 - k^2}} = \lim_{n \to \infty} \left(\frac{1}{\sqrt{4n^2 - 1^2}} + \frac{1}{\sqrt{4n^2 - 2^2}} + \cdots \right.$$

$$\left. \cdots + \frac{1}{\sqrt{4n^2 - n^2}} \right) = \frac{\pi}{6},$$

as required to show. ▶

Exercises for Chapter 3

✠ **Warm-ups**

1. Find $\int \cos(ax)\, dx$ if (a) $a = 0$, and (b) $a \neq 0$.

✠ **Practice questions**

2. Using the index law, find the following:

(a) $\displaystyle\int x^3\, dx$

(b) $\displaystyle\int x\sqrt{x}\, dx$

(c) $\int \dfrac{dx}{3x^5}$

(d) $\int (3x^2 - 2x + 1)\, dx$

(e) $\int \dfrac{x^4 + 3x^2 + 1}{\sqrt{x}}\, dx$

(f) $\int (2x - 1)^2\, dx$

(g) $\int x(x - 6)\, dx$

(h) $\int \sqrt[3]{x}(x + 1)\, dx$

(i) $\int (x - 6)(x - 3)\, dx$

(j) $\int \dfrac{x^5 - 1}{x^2}\, dx$

(k) $\int (x + 3)^3\, dx$

(l) $\int \left(\sqrt{x} + \dfrac{1}{\sqrt{x}} \right) dx$

3. Evaluate the following:

(a) $\int_0^1 \sqrt{x}(x^2 - 4)\, dx$

(b) $\int_{-1}^1 (x^4 - 2x^2 + 1)\, dx$

(c) $\int_{-2}^2 x(x^2 - 1)\, dx$

(d) $\int_4^9 \dfrac{1 + x}{\sqrt{x}}\, dx$

(e) $\int_0^1 (\sqrt[6]{x} - \sqrt[4]{x})\, dx$

(f) $\int_{-1}^2 (2 - x^2)^2\, dx$

4. Find the following:

(a) $\int \dfrac{dx}{2x}$

(b) $\int \dfrac{3x^3 + 1}{x}\, dx$

(c) $\int (4e^{4x} - 2\sin 2x)\, dx$

(d) $\int \dfrac{dx}{3 + x^2}\, dx$

(e) $\int \sin(3x + 2)\, dx$

(f) $\int \dfrac{dx}{1 + 4x^2}\, dx$

(g) $\int \left[\sec^2 \left(\dfrac{x}{2} \right) + \cos \left(\dfrac{x}{2} \right) \right] dx$

(h) $\int \dfrac{dx}{\sqrt{1 - 4x^2}}$

(i) $\int (3 \tan x \sec x + \sin 3x)\, dx$

(j) $\int \dfrac{\sqrt{1 - 9x^2} + 1 + 9x^2}{(1 + 9x^2)\sqrt{1 - 9x^2}}\, dx$

5. Evaluate the following:

(a) $\displaystyle\int_0^1 (x+1)^2(x^2+1)\,dx$

(b) $\displaystyle\int_0^\pi \cos 3x\,dx$

(c) $\displaystyle\int_{\frac{1}{e}}^e \frac{dx}{3x}$

(d) $\displaystyle\int_0^{\frac{\pi}{4}} \cos(2x+5)\,dx$

(e) $\displaystyle\int_0^{\frac{1}{2}} \frac{dx}{1+4x^2}$

(f) $\displaystyle\int_0^{\frac{1}{2}} \frac{1+\pi+4\pi x^2}{1+4x^2}\,dx$

(g) $\displaystyle\int_0^{\frac{1}{3}} \frac{dx}{\sqrt{1-9x^2}}$

(h) $\displaystyle\int_0^{\frac{1}{\sqrt{3}}} \frac{dx}{1+9x^2}$

6. If $a, b \in \mathbb{R}$ such that $a \neq 0$, prove the following extensions for arguments that are linear:

(a) $\displaystyle\int e^{ax+b}\,dx = \frac{1}{a}e^{ax+b} + C$

(b) $\displaystyle\int \cos(ax+b)\,dx = \frac{1}{a}\sin(ax+b) + C$

(c) $\displaystyle\int \sin(ax+b)\,dx = -\frac{1}{a}\cos(ax+b) + C$

7. (a) Show that $\displaystyle\frac{d}{dx}\left(\frac{1}{(3x+1)^3}\right) = -\frac{9}{(3x+1)^4}$.

(b) Hence evaluate $\displaystyle\int_0^1 \frac{dx}{(3x+1)^4}$.

8. (a) Show that

$$\frac{d}{dx}\left[\tan^{-1}\left(\frac{\sqrt{\tan x}-\sqrt{\cot x}}{\sqrt{2}}\right)\right] = \frac{1}{\sqrt{2}}\left(\sqrt{\tan x}+\sqrt{\cot x}\right).$$

(b) Hence find $\displaystyle\int \left(\sqrt{\tan x}+\sqrt{\cot x}\right)dx$.

9. (a) Using the binomial theorem or otherwise, expand $(x-1)^6$.

(b) Hence evaluate $\displaystyle\int_1^2 (x-1)^6\,dx$.

10. By making use of trigonometric identities, find

(a) $\displaystyle \int \frac{\sin x}{1 - \sin^2 x}\, dx$ (b) $\displaystyle \int \frac{\cos x}{1 - \cos^2 x}\, dx$

11. Evaluate $\displaystyle \int_1^x t^3\, dt$ and verify that $\displaystyle \frac{d}{dx} \int_1^x t^3\, dt = x^3$.

12. Let $\displaystyle f(x) = \int_{\frac{\pi}{6}}^x (e^t + \cos t)\, dt$.

   (a) Verify by direct integration and differentiation that $f'(x) = e^x + \cos x$.

   (b) Does the result found in (a) agree with the result that would be found by applying the first fundamental theorem of calculus to the integral directly?

13. If $u_0 = 1$ and $\displaystyle u_n = \int_0^{u_{n-1}} (1 - x)^{4-n}\, dx$ for $n = 1, 2, 3$, find the value of u_3.

14. Suppose that $f(x) = \sin x$.

   (a) Show that f is increasing on the interval $[0, \frac{\pi}{2}]$.

   (b) Find the upper Darboux sum for f with respect to the partition

$$\mathcal{P}_n = \left\{ 0, \frac{\pi}{2n}, \frac{2\pi}{2n}, \ldots, \frac{n\pi}{2n} \right\} \text{ of } [0, \tfrac{\pi}{2}].$$

   (c) Hence show that

$$\lim_{n \to \infty} \frac{1}{n} \left[\sin\left(\frac{\pi}{2n} \right) + \sin\left(\frac{2\pi}{2n} \right) + \cdots + \sin\left(\frac{n\pi}{2n} \right) \right] = \frac{2}{\pi}.$$

15. Suppose that $\displaystyle f(x) = \frac{1}{1 + x^2}$.

   (a) Show that f is increasing on the interval $[0, 1]$.

   (b) Find the upper Darboux sum for f with respect to the partition

$$\mathcal{P}_n = \left\{ 0, \frac{1}{n}, \frac{2}{n}, \ldots, \frac{n-1}{n}, \frac{n}{n} \right\} \text{ of } [0, 1].$$

   (c) Hence show that

$$\lim_{n \to \infty} \left(\frac{n}{n^2 + 1^2} + \frac{n}{n^2 + 2^2} + \cdots + \frac{n}{n^2 + n^2} \right) = \frac{\pi}{4}.$$

16. Suppose that $f(x) = \dfrac{1}{x}$.

(a) Show that f is decreasing on the interval $[1, 2]$.

(b) Find the lower Darboux sum for f with respect to the partition

$$\mathcal{P}_n = \left\{ \frac{n}{n}, \frac{n+1}{n}, \frac{n+2}{n}, \ldots, \frac{2n}{n} \right\} \text{ of } [1, 2].$$

(c) Hence show that

$$\int_1^2 \frac{dx}{x} = \lim_{n \to \infty} \left(\frac{1}{n+1} + \frac{1}{n+2} + \cdots + \frac{1}{2n-1} + \frac{1}{2n} \right).$$

(d) Show that for all integers $n \geqslant 1$

$$1 - \frac{1}{2} + \frac{1}{3} - \frac{1}{4} + \cdots + \frac{1}{2n-1} - \frac{1}{2n} = \frac{1}{n+1} + \frac{1}{n+2}$$

$$+ \cdots + \frac{1}{2n-1} + \frac{1}{2n}.$$

(e) Hence deduce that the infinite series $1 - \dfrac{1}{2} + \dfrac{1}{3} - \dfrac{1}{4} + \cdots$ has a sum equal to $\ln 2$.

✠ Extension questions and Challenge problems

17. Suppose f is a continuous function for all real x that satisfies the equation

$$\int_0^x f(t)\, dt = -\int_1^x t^4 f(t)\, dt + \frac{x^{12}}{6} + \frac{x^{16}}{8} + k,$$

where k is a constant. Find an explicit formula for $f(x)$ and find the value of the constant k.

18. A special function known as the *digamma function* $\psi(x)$ is defined by

$$\psi(x) = -\gamma + \int_0^1 \frac{1 - t^{x-1}}{1 - t}\, dt.$$

Here γ is a mathematical constant known as the *Euler–Mascheroni* constant ($\gamma = 0.577\,215\,664\ldots$). Show that the digamma function satisfies the following functional relation:

$$\psi(x + 1) = \psi(x) + \frac{1}{x}, \quad x \neq 0, -1, -2, \ldots$$

4

Basic Properties of the Definite Integral: Part II

I could never resist the challenge of a definite integral.

— G. H. Hardy

Now that we know how to find a number of basic integrals for several commonly encountered functions, in this chapter we continue our discussion of basic properties for the definite integral.

Following on from where we left off in Chapter 2, the first property given here is labelled Property 5.

Property 5 – Comparison Property I

If one function is larger than or equal to another function throughout the interval $[a, b]$, then its definite integral will also be larger. That is, if $f(x) \geqslant g(x)$ for all $x \in [a, b]$ then

$$\int_a^b f(x)\, dx \geqslant \int_a^b g(x)\, dx$$

This property shows us that integration preserves inequalities between functions.

Proof By the assumption $f(x) \geqslant g(x)$, we have $f(x) - g(x) \geqslant 0$. So from the signed property, we have

$$\int_a^b [f(x) - g(x)]\, dx \geqslant 0.$$

From the linearity property we can write this as

$$\int_a^b f(x)\, dx - \int_a^b g(x)\, dx \geqslant 0 \quad \text{or} \quad \int_a^b f(x)\, dx \geqslant \int_a^b g(x)\, dx,$$

as required to prove. ∎

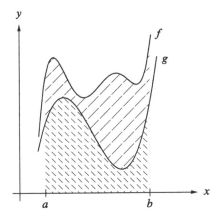

Figure 4.1. Graphical illustration of the comparison property for the definite integral.

Once again this property can be interpreted geometrically. If $f(x) > g(x) > 0$, then the property says that the graph of the curve of g lies beneath the graph of the curve of f, meaning the area of the region bounded by the graph of the curve g and the x-axis between $x = a$ and $x = b$ will be less than the corresponding area of the region bounded by the graph of the curve f and the x-axis between $x = a$ and $x = b$ (see Figure 4.1).

Example 4.1 Given that $\dfrac{\sin x}{x} > \cos x$ for $(0, \frac{\pi}{2})$, use this to show that

$$\int_{\frac{\pi}{6}}^{\frac{\pi}{2}} \frac{\sin x}{x}\,dx > \frac{1}{2}.$$

Solution Let $f(x) = \dfrac{\sin x}{x}$ and $g(x) = \cos x$. Since $\dfrac{\sin x}{x} > \cos x$ for $0 < x < \dfrac{\pi}{2}$, from the first comparison property for definite integrals we have

$$\int_{\frac{\pi}{6}}^{\frac{\pi}{2}} \frac{\sin x}{x}\,dx > \int_{\frac{\pi}{6}}^{\frac{\pi}{2}} \cos x\,dx = \sin x \Big|_{\frac{\pi}{6}}^{\frac{\pi}{2}} = 1 - \frac{1}{2} = \frac{1}{2},$$

as required to show.   ▶

Property 6 – Comparison Property II
The first comparison property can be extended to a comparison property between three (or more) functions. If $g(x) \leqslant f(x) \leqslant h(x)$ for all $x \in [a, b]$,

then

$$\int_a^b g(x)\,dx \leqslant \int_a^b f(x)\,dx \leqslant \int_a^b h(x)\,dx$$

Proof By the assumption, as $f(x) \geqslant g(x)$, from the first comparison property we have

$$\int_a^b f(x)\,dx \geqslant \int_a^b g(x)\,dx.$$

Also from the assumption, since $h(x) \geqslant f(x)$, again from the first comparison property we have

$$\int_a^b h(x)\,dx \geqslant \int_a^b f(x)\,dx.$$

On combining these two results the desired result follows.   ∎

Example 4.2 Show that $\dfrac{\ln 2}{\sqrt{3}} \leqslant \displaystyle\int_{\frac{\pi}{6}}^{\frac{\pi}{3}} \dfrac{\tan x}{x} \leqslant \sqrt{3}\ln 2.$

Solution Since $\tan x$ is an increasing function on the interval $[\frac{\pi}{6}, \frac{\pi}{4}]$ we have

$$\tan\left(\frac{\pi}{6}\right) \leqslant \tan x \leqslant \tan\left(\frac{\pi}{3}\right) \quad\text{or}\quad \frac{1}{\sqrt{3}} \leqslant \tan x \leqslant \sqrt{3}.$$

As $x > 0$ for all $x \in [\pi/6, \pi/3]$, on dividing the inequality by x we have

$$\frac{1}{\sqrt{3}} \cdot \frac{1}{x} \leqslant \frac{\tan x}{x} \leqslant \sqrt{3} \cdot \frac{1}{x}, \quad \text{for} \quad \frac{\pi}{6} \leqslant x \leqslant \frac{\pi}{3}.$$

Now let $s(x) = \dfrac{k}{x}$ where $k = \sqrt{3}$ or $\dfrac{1}{\sqrt{3}}$. Note that

$$\int_{\frac{\pi}{6}}^{\frac{\pi}{3}} s(x)\,dx = k\int_{\frac{\pi}{6}}^{\frac{\pi}{3}} \frac{dx}{x} = k\,[\ln x]_{\frac{\pi}{6}}^{\frac{\pi}{3}} = k\left[\ln\left(\frac{\pi}{3}\right) - \ln\left(\frac{\pi}{6}\right)\right] = k\ln 2.$$

Setting $g(x) = \dfrac{1}{\sqrt{3}} \cdot \dfrac{1}{x}$ and $h(x) = \sqrt{3} \cdot \dfrac{1}{x}$, from the second comparison property, namely

$$\int_a^b g(x)\,dx \leqslant \int_a^b f(x)\,dx \leqslant \int_a^b h(x)\,dx,$$

we have

$$\frac{1}{\sqrt{3}} \int_{\frac{\pi}{6}}^{\frac{\pi}{3}} \frac{dx}{x} \leqslant \int_{\frac{\pi}{6}}^{\frac{\pi}{3}} \frac{\tan x}{x}\,dx \leqslant \sqrt{3}\int_{\frac{\pi}{6}}^{\frac{\pi}{3}} \frac{dx}{x},$$

or

$$\frac{\ln 2}{\sqrt{3}} \leqslant \int_{\frac{\pi}{6}}^{\frac{\pi}{3}} \frac{\tan x}{x} \, dx \leqslant \sqrt{3} \ln 2,$$

as required to show. ▶

Property 7 – Absolute Value Property
The absolute value property states that if f is integrable then $|f|$ is also integrable, and the absolute value of the definite integral of the function f is less than or equal to the definite integral of the absolute value of this function, namely

$$\left| \int_a^b f(x) \, dx \right| \leqslant \int_a^b |f(x)| \, dx$$

Proof From the properties of the absolute value we have $-|f(x)| \leqslant f(x) \leqslant |f(x)|$ for all $x \in [a, b]$. Since f is integrable, $|f|$ is also integrable as it is bounded and piecewise continuous. Therefore, on applying the second comparison property to the inequality one has

$$-\int_a^b |f(x)| \, dx \leqslant \int_a^b f(x) \, dx \leqslant \int_a^b |f(x)| \, dx.$$

Note the negative sign in the left integral can be moved outside the integral sign using the homogeneous property. Also, from properties of the absolute value, since $-b \leqslant a \leqslant b$ can be written as $|a| \leqslant b$, then

$$\left| \int_a^b f(x) \, dx \right| \leqslant \int_a^b |f(x)| \, dx,$$

as required to prove. ■

Example 4.3 Show that $\int_0^\pi \sqrt{x} \sin x \, dx \leqslant \frac{2}{3} \pi^{\frac{3}{2}}$.

Solution First we note that as $\sqrt{x} \sin x > 0$ for $x \in [0, \pi]$, the first of the comparison properties for definite integrals gives $\int_0^\pi \sqrt{x} \sin x \, dx > 0$. Thus

$$\int_0^\pi \sqrt{x} \sin x \, dx = \left| \int_0^\pi \sqrt{x} \sin x \, dx \right|.$$

Next, on applying the absolute value property to the integral we have

$$\left| \int_0^\pi \sqrt{x} \sin x \, dx \right| \leqslant \int_0^\pi |\sqrt{x} \sin x| \, dx = \int_0^\pi \sqrt{x} |\sin x| \, dx.$$

Now since $|\sin x| \leqslant 1$ for all x, one has $\sqrt{x}\,|\sin x| \leqslant \sqrt{x}$ for $x \in [0, \pi]$. So from the first comparison property one has

$$\left| \int_0^\pi \sqrt{x} \sin x\, dx \right| \leqslant \int_0^\pi \sqrt{x}\, dx = \left[\frac{2}{3} x^{\frac{3}{2}} \right]_0^\pi = \frac{2}{3} \pi^{\frac{3}{2}}.$$

Thus $\int_0^\pi \sqrt{x} \sin x \leqslant \frac{2}{3} \pi^{\frac{3}{2}}$, as required to show.   ▶

Property 8 – Interchanging the Limits of Integration Property

This property states that if the limits of integration in the definite integral are interchanged then the sign of the definite integral also changes.

$$\boxed{\int_a^b f(x)\, dx = - \int_b^a f(x)\, dx}$$

Proof While the proof of this property depends on the formal definition for the integral, we can at least give a justification that shows the result is true. If f is continuous the integral $\int f(x)\, dx$ exists and has a primitive given by $F(x)$. From the second fundamental theorem of calculus we have

$$\int_a^b f(x)\, dx = F(b) - F(a),$$

and

$$-\int_b^a f(x)\, dx = -\big(F(a) - F(b)\big) = F(b) - F(a),$$

clearly showing the two results are the same.   ■

Property 9 – Equal Limits of Integration Property

The equal limits of integration property states that if the lower and upper limits of integration of the definite integral are the same, the definite integral will be equal to zero.

$$\boxed{\int_a^a f(x)\, dx = 0}$$

Proof If the limits of integration are interchanged, from Property 8 we have

$$\int_a^a f(x)\, dx = - \int_a^a f(x)\, dx,$$

or

$$2 \int_a^a f(x)\, dx = 0,$$

from which the assertion follows.   ■

The next result is an important property that relates definite integrals over different intervals.

Property 10 – Additivity with Respect to the Interval of Integration Property

If $c \in \mathbb{R}$ is any number that is either inside or outside the interval $[a, b]$, then

$$\int_a^b f(x)\,dx = \int_a^c f(x)\,dx + \int_c^b f(x)\,dx$$

This property tells one how to integrate over adjacent intervals $[a, c]$ and $[c, b]$, where c need not be between a and b.

Proof While the proof of this property depends on the formal definition for the integral, we can at least give a justification that shows the result is true. If f is continuous the integral $\int f(x)\,dx$ exists and has a primitive given by $F(x)$. From the second fundamental theorem of calculus we have

$$\int_a^b f(x)\,dx = F(b) - F(a).$$

Also

$$\int_a^c f(x)\,dx + \int_c^b f(x)\,dx = \big(F(c) - F(a)\big) + \big(F(b) - F(c)\big)$$

$$= F(b) - F(a),$$

and we see the two results are the same. ∎

Geometrically, for $a < c < b$ this property can be understood as follows. If we consider that the region Ω between the curve $y = f(x)$ and the x-axis from $x = a$ to $x = b$ splits into two non-overlapping regions between the curve and the x-axis, the first, Ω_1, from $x = a$ to $x = c$, and the second, Ω_2, from $x = c$ to $x = b$, then the (signed) area of Ω is equal to the sum of the (signed) areas of Ω_1 and Ω_2. This is shown in Figure 4.2.

Example 4.4 If $\int_0^5 f(x)\,dx = 16$ and $\int_0^4 f(x)\,dx = 12$, find the value of $\int_4^5 f(x)\,dx$.

Solution From the additivity property with respect to the interval of integration we can write

$$\int_0^5 f(x)\,dx = \int_0^4 f(x)\,dx + \int_4^5 f(x)\,dx.$$

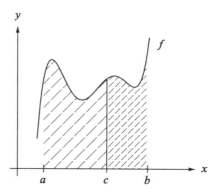

Figure 4.2. Graphical illustration of the additivity property with respect to the interval of integration for the definite integral.

So

$$\int_4^5 f(x)\,dx = \int_0^5 f(x)\,dx - \int_0^4 f(x)\,dx = 16 - 12 = 4. \qquad \blacktriangleright$$

Example 4.5 Consider the piecewise continuous function

$$f(x) = \begin{cases} \dfrac{x}{3} + 1, & 0 \leqslant x \leqslant 3 \\ 5 - x, & 3 \leqslant x \leqslant 5. \end{cases}$$

Evaluate $\displaystyle\int_0^5 f(x)\,dx$.

Solution From the additivity property with respect to the interval of integration we can write

$$\int_0^5 f(x)\,dx = \int_0^3 f(x)\,dx + \int_3^5 f(x)\,dx$$

$$= \int_0^3 \left(\frac{x}{3} + 1\right) dx + \int_3^5 (5 - x)\,dx$$

$$= \left[\frac{x^2}{6} + x\right]_0^3 + \left[5x - \frac{x^2}{2}\right]_3^5$$

$$= \left(\frac{3}{2} + 3\right) + \left(25 - \frac{25}{2}\right) - \left(15 - \frac{9}{2}\right) = \frac{13}{2}. \qquad \blacktriangleright$$

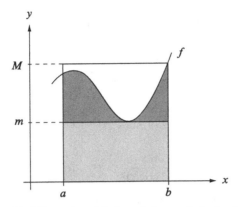

Figure 4.3. Graphical illustration of the bounded property for the definite integral.

Property 11 – Bounded Property

If m and M are real numbers such that $m \leqslant f(x) \leqslant M$ for all $x \in [a, b]$, then

$$m(b - a) \leqslant \int_a^b f(x)\, dx \leqslant M(b - a)$$

This property tells us that if m and M are the smallest and largest values of the function f in the interval $[a, b]$, respectively, then the definite integral of f is bounded between the product of these two numbers with the size of the interval width $(b - a)$.

Proof Applying the second of the comparison properties to the assumption, namely to $m \leqslant f(x) \leqslant M$ for all $x \in [a, b]$, one has

$$m \int_a^b dx \leqslant \int_a^b f(x)\, dx \leqslant M \int_a^b dx$$

$$\Rightarrow m(b - a) \leqslant \int_a^b f(x)\, dx \leqslant M(b - a),$$

as required to prove. ∎

Let us interpret the property geometrically. If $f(x) > 0$ then the property says that, since the graph of the curve of f lies between the lines $y = m$ and $y = M$ on the interval $[a, b]$, the area of the region between the curve $y = f(x)$ and the x-axis is greater than the area of the rectangle of height m but less than the area of the rectangle of height M (see Figure 4.3).

Example 4.6 Suppose that $\int_{\frac{\pi}{3}}^{\frac{\pi}{2}} \dfrac{dx}{2 + 3\cos x}.$

(a) Show that $f(x) = \dfrac{1}{2 + 3\cos x}$ is an increasing function on the interval $[\frac{\pi}{3}, \frac{\pi}{2}]$.

(b) Hence deduce that $\dfrac{\pi}{21} \leqslant \displaystyle\int_{\frac{\pi}{3}}^{\frac{\pi}{2}} \dfrac{dx}{2 + 3\cos x} \leqslant \dfrac{\pi}{12}$.

Solution (a) Since $f'(x) = \dfrac{3\sin x}{(2 + 3\cos x)^2}$, as $\sin x > 0$ for all $x \in [\frac{\pi}{3}, \frac{\pi}{2}]$, we have

$$f'(x) = \frac{3\sin x}{(2 + 3\cos x)^2} > 0, \quad \text{for } \frac{\pi}{3} \leqslant x \leqslant \frac{\pi}{2}.$$

Thus f is an increasing function on the interval $[\frac{\pi}{3}, \frac{\pi}{2}]$, as required to show.

(b) As f is an increasing function on $[\frac{\pi}{3}, \frac{\pi}{2}]$ the least value of f occurs at the end-point, $x = \pi/3$. Thus

$$m = f\left(\frac{\pi}{3}\right) = \frac{1}{2 + 3\cos(\frac{\pi}{3})} = \frac{2}{7},$$

while the greatest value of f occurs at the opposite end-point of $x = \pi/2$. Thus

$$M = f\left(\frac{\pi}{2}\right) = \frac{1}{2 + 3\cos(\frac{\pi}{2})} = \frac{1}{2}.$$

So from the bounded property for definite integrals, namely

$$m(b - a) \leqslant \int_a^b f(x)\,dx \leqslant M(b - a),$$

we have

$$\frac{2}{7}\left(\frac{\pi}{2} - \frac{\pi}{3}\right) \leqslant \int_{\frac{\pi}{3}}^{\frac{\pi}{2}} \frac{dx}{2 + 3\cos x} \leqslant \frac{1}{2}\left(\frac{\pi}{2} - \frac{\pi}{3}\right),$$

or

$$\frac{\pi}{21} \leqslant \int_{\frac{\pi}{3}}^{\frac{\pi}{2}} \frac{dx}{2 + 3\cos x} \leqslant \frac{\pi}{12},$$

as required to show.   ▶

As a final comment, it should be clear the variable used in a definite integral is of no consequence. For example, the following definite integrals are all equivalent:

$$\int_a^b f(x)\,dx = \int_a^b f(u)\,du = \int_a^b f(t)\,dt = \int_a^b f(y)\,dy = \cdots.$$

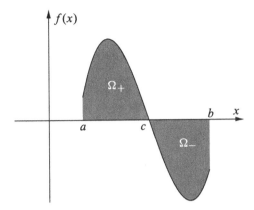

Figure 4.4. The definite integral and signed area.

As such, the variable that appears in a definite integral is often referred to as a *dummy* variable of integration since it plays no part in the final answer found.

§ Signed Area

The linearity property of the integral together with the additivity with respect to the interval of integration property allows one to interpret more general definite integrals in terms of areas. If a function f is integrable on an interval $[a, b]$, the definite integral

$$\int_a^b f(x)\, dx,$$

is interpreted as the area under the graph of f and the x-axis from a to b if $f(x) \geqslant 0$ for all $x \in [a, b]$. An obvious question to ask is how to interpret the definite integral if the sign of f switches at a finite number of points within the interval. For example, suppose $a < c < b$ such that $f(x) \geqslant 0$ when $a \leqslant x \leqslant c$ and $f(x) \leqslant 0$ when $c \leqslant x \leqslant b$ with $f(c) = 0$ as is shown in Figure 4.4. In this case, how is the definite integral on the interval $[a, b]$ to be interpreted?

From the geometric interpretation of the definite integral, the definite integral

$$\int_a^c f(x)\, dx = \Omega_+,$$

gives the area of the region Ω_+ beneath the graph of the function f and the x-axis bounded between the lines $x = a$ and $x = c$. Also, as $f(x) \leqslant 0$ on $[c, b]$, then $-f(x) \geqslant 0$ on this interval, and so from the geometric interpretation for

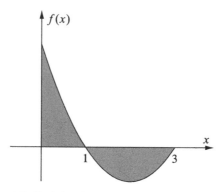

Figure 4.5. Shaded region corresponding to the signed area.

the definite integral we have

$$\int_c^b (-f(x))\, dx = -\int_c^b f(x)\, dx.$$

The definite integral thus gives the area of the region above the x-axis and below the graph of $-f$ bounded by the lines $x = c$ and $x = b$. But this is just the reflection about the x-axis of the region below the x-axis and above the graph of f bounded between the lines $x = c$ and $x = b$. Denoting the area of this region by Ω_-, we have

$$\int_c^b f(x)\, dx = -\Omega_-,$$

and it follows that

$$\int_a^b f(x)\, dx = \Omega_+ - \Omega_-.$$

This quantity is known as the *signed area* under the graph of f from a to b.

Example 4.7 Calculate the area of the region bounded by the function $f(x) = x^2 - 4x + 3$, the x-axis, and the lines $x = 0$ and $x = 3$.

Solution Since $f(x) = (x - 1)(x - 3)$ we see the graph of the function cuts the x-axis at $x = 1$ and $x = 3$. The signed area of the required region is shown shaded in Figure 4.5.

The area $\mathcal{A}$ of the required region therefore corresponds to the following signed area:

$$\mathcal{A} = \int_0^1 (x^2 - 4x + 3)\, dx - \int_1^3 (x^2 - 4x + 3)\, dx.$$

Evaluating each of these integrals, we have

$$
\mathcal{A} = \left[\frac{x^3}{3} - 2x^2 + 3x\right]_0^1 - \left[\frac{x^3}{3} - 2x^2 + 3x\right]_1^3
$$

$$
= \left[\frac{1}{3} - 2 + 3\right] - \left[\left(\frac{3^3}{3} - 2 \cdot 3^2 + 3 \cdot 3\right) - \left(\frac{1}{3} - 2 + 3\right)\right] = \frac{8}{3}. \quad \blacktriangleright
$$

Exercises for Chapter 4

✠ **Practice questions**

1. (a) If $\displaystyle\int_0^3 f(x)\,dx = 3$, $\displaystyle\int_{-1}^0 f(x)\,dx = -7$, and $\displaystyle\int_3^5 f(x)\,dx = 1$, find
   the value of $\displaystyle\int_{-1}^5 f(x)\,dx$.

   (b) If $\displaystyle\int_0^1 g(x)\,dx = -89$ and $\displaystyle\int_0^2 g(x)\,dx = 87$, find $\displaystyle\int_1^2 g(x)\,dx$.

2. Suppose $F : [0, 12] \to \mathbb{R}$ is defined by $F(x) = \displaystyle\int_0^x f(t)\,dt$ where f is a
   piecewise continuous function given by

   $$
   f(t) = \begin{cases} 4 - \dfrac{4t}{3}, & 0 \leqslant t \leqslant 6 \\ t - 10, & 6 \leqslant t \leqslant 12. \end{cases}
   $$

   (a) Calculate the value of $F(6)$ and $F(12)$.
   (b) Find all values of x for which $F(x) = 0$.
   (c) Find all stationary points for F and determine their nature.

3. (a) Calculate the area of the region bounded by the function $f(x) = x^2 - 5x + 6$, the x-axis, and the lines $x = 1$ and $x = 3$.
   (b) Calculate the area of the region bounded by the function $f(x) = x^3 - 6x^2 + 2x + 2$, the x-axis, and the lines $x = 0$ and $x = 2$.
   (c) Calculate the area of the region bounded by the function $f(x) = x^2 - 6x + 8$, the x-axis, and the lines $x = 0$ and $x = 5$.

4. Given that $1 - x^2 \leqslant e^{-x^2}$ for $0 \leqslant x \leqslant 1$, use this result to show that
   $\displaystyle\int_0^1 e^{-x^2}\,dx \geqslant \frac{2}{3}$.

5. Suppose f is a function that is continuous everywhere such that $f(1) = 8$
   and $\displaystyle\int_0^1 f(t)\,dt = 3$.

Let $g(x) = \dfrac{1}{2} \displaystyle\int_0^x (x-t)^2 f(t)\,dt$.

(a) Show that $g'(x) = x \displaystyle\int_0^x f(t)\,dt - \int_0^x tf(t)\,dt$.

(b) Hence find values for $g''(1)$ and $g'''(1)$.

6. *Jordan's inequality* states that $\dfrac{2}{\pi}x \leqslant \sin x \leqslant x$ for $[0, \frac{\pi}{2}]$.

Use this inequality to show that $1 < \displaystyle\int_0^{\frac{\pi}{2}} \dfrac{\sin x}{x}\,dx < \dfrac{\pi}{2}$.

7. Each of the following statements is either true or false. Answer each statement with either TRUE or FALSE and give a brief reason for your answer.

(a) $\displaystyle\int_{-2}^2 e^{-x^2}\,dx = 0$

(b) $\displaystyle\int_0^\pi \cos^4 x\,dx > 0$

(c) $\displaystyle\int_0^\pi (\sin^6 x - \cos^6 x)\,dx = 0$

(d) $\displaystyle\int_0^1 \dfrac{dx}{1+x^3} > \int_0^1 \dfrac{dx}{1+x^5}$

8. (a) For $0 < x < 1$, starting with $x(1+x) > 0$, show that

$$\dfrac{1}{\sqrt{1+x}} < \dfrac{1}{\sqrt{1-x^2}}.$$

(b) Use the inequality given in (a) to show that $\displaystyle\int_0^1 \dfrac{dx}{\sqrt{1+x}} < \dfrac{\pi}{2}$.

9. If $\displaystyle\int_0^1 f(x)\,dx = 8$, $\displaystyle\int_{-1}^1 f(x)\,dx = 3$, and $\displaystyle\int_0^5 f(x)\,dx = 13$, find the value of $\displaystyle\int_{-1}^5 f(x)\,dx$.

10. If f is a bounded continuous function such that $|f(x)| \leqslant M$ for all x in the interval $[a,b]$, show that $\left| \displaystyle\int_a^b f(x)\,dx \right| \leqslant M(b-a)$. Here M is a positive constant.

11. For $0 < x < 1$, starting with $x(1-x) > 0$, by applying the first comparison property to a suitable inequality, show that $\ln(1+x) < \tan^{-1}(1+x^2)$ for $0 < x < 1$.

12. If n is a positive integer such that $n \geqslant 2$, show that $\dfrac{\pi}{4} \leqslant \displaystyle\int_0^1 \dfrac{dx}{1+x^n} < 1$.

13. Find all continuous functions $\varphi(x)$ that satisfy $\varphi(x) = x + \int_0^{\frac{1}{2}} \varphi(u)\, du$.

14. (a) Evaluate $\int_{-2}^2 \varphi(x)\, dx$ if $\varphi(x) = \max\{1, x^2\}$. Here $\max\{f, g\}$ denotes the maximum of the two functions f and g, which is the function whose value at any x is the larger of $f(x)$ and $g(x)$.

(b) Evaluate $\int_{-1}^3 x\varphi(x)\, dx$ if $\varphi(x) = \min\{1, x^2\}$. Here $\min\{f, g\}$ denotes the minimum of the two functions f and g, which is the function whose value at any x is the smaller of $f(x)$ and $g(x)$.

15. Evaluate $\displaystyle\int_0^1 F(\sin \pi x)\, dx$ if $F(u) = \begin{cases} 1, & u \geqslant \frac{1}{2} \\ 0, & u < \frac{1}{2} \end{cases}$.

16. Using properties for the definite integral, prove the following bounds for the given definite integral:

(a) $\displaystyle\int_0^{\pi/2} x \sin x\, dx \leqslant \frac{\pi^2}{8}$

(b) $\displaystyle\left| \int_0^{2\pi} f(x) \sin(2x)\, dx \right| \leqslant \int_0^{2\pi} |f(x)|\, dx$

17. The graph of the derivative of f, that is f', is shown in the figure below.

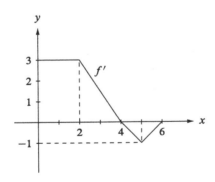

If $f(0) = 5$, find the value of $f(6)$.

18. Suppose that $f : [0, \pi/2] \to \mathbb{R}$ is given by $f(x) = \dfrac{\sec x + \tan x}{2 \sec x + 3 \tan x}$.

(a) Show that f is a decreasing function for $0 < x < \dfrac{\pi}{2}$.

(b) Hence show that $\dfrac{\pi(\sqrt{2} - 1)}{4} < \displaystyle\int_0^{\frac{\pi}{4}} \dfrac{\sec x + \tan x}{2 \sec x + 3 \tan x}\, dx < \dfrac{\pi}{8}$.

19. Using properties for the definite integral, establish the following integral bounds:

(a) $\dfrac{1}{3\sqrt{2}} \leqslant \displaystyle\int_0^1 \dfrac{x^2}{\sqrt{1+x^2}}\,dx \leqslant \dfrac{1}{3}$

(b) $1 \leqslant \displaystyle\int_0^{\frac{\pi}{2}} \sqrt{\sin x}\,dx \leqslant \dfrac{\pi}{2}$

(c) $\dfrac{3}{8} \leqslant \displaystyle\int_0^{\frac{1}{2}} \sqrt{\dfrac{1-x}{1+x}}\,dx \leqslant \dfrac{\sqrt{3}}{4}$

(d) $\dfrac{1}{10\sqrt{2}} \leqslant \displaystyle\int_0^1 \dfrac{x^9}{\sqrt{1+x}}\,dx \leqslant \dfrac{1}{10}$

(e) $\dfrac{1}{4}\left(\dfrac{1}{e} - \dfrac{1}{e^2}\right) \leqslant \displaystyle\int_e^{e^2} \dfrac{dx}{x^2 \ln^2 x}\,dx \leqslant \dfrac{1}{e} - \dfrac{1}{e^2}$

20. (a) Show that $\displaystyle\int_1^2 \dfrac{dx}{\sqrt{1+x^3}} < 2 - \sqrt{2}$ and $\displaystyle\int_1^2 \dfrac{dx}{x\sqrt{1+x}} < 2 - \sqrt{2}$.

(b) Hence deduce that $2 - \sqrt{2} > \displaystyle\int_1^2 \dfrac{dx}{\sqrt{1+x^3}} > \displaystyle\int_1^2 \dfrac{dx}{x\sqrt{1+x}}$.

✠ Extension questions and Challenge problems

21. If $I = \displaystyle\int_0^{4\pi} e^t(\sin^6 t + \cos^6 t)\,dt$ and $J = \displaystyle\int_0^{\pi} e^t(\sin^6 t + \cos^6 t)\,dt$, find the value of I/J.

22. Assume the following result: $\int_a^b f(x)\,dx = \int_a^c f(x)\,dx + \int_c^b f(x)\,dx$, where f is a continuous function on some interval I such that $a < b$ and c is any point *inside* the interval (a, b), that is, $c \in (a, b)$.

Show that this result is true for any choice in $c \in I$ by extending the above result to any choice in c that is either inside, on, or outside of the interval (a, b). Here one needs to consider four separate cases: $c = a, c = b, c < a$, and $b < c$.

23. In this question increasingly 'sharper' polynomial bounds (inequalities) will be found for the sine and cosine functions on the interval $0 \leqslant x \leqslant 1$ by repeated application of the second of the comparison properties.

(a) Start with the well-known result of $0 \leqslant \cos t \leqslant 1$ for $t \in [0, x]$ such that $x \in [0, 1]$.

By applying the second of the comparison properties for definite integrals to the above inequality, show that

$$0 \leqslant \sin x \leqslant x, \quad x \in [0, 1].$$

(b) From the inequality in (a), writing this as $0 \leqslant \sin t \leqslant t$ for $t \in [0, x]$ such that $x \in [0, 1]$, on applying once more the second of the comparison properties for definite integrals to the inequality, show that

$$1 - \frac{x^2}{2!} \leqslant \cos x \leqslant 1, \quad x \in [0, 1].$$

(c) From the inequality in (b), writing this as $1 - \frac{t^2}{2!} \leqslant \cos t \leqslant 1$ for $t \in [0, x]$ such that $x \in [0, 1]$, on applying for the third time the second of the comparison properties for definite integrals to the inequality, show that

$$x - \frac{x^3}{3!} \leqslant \sin x \leqslant x, \quad x \in [0, 1].$$

(d) From the inequality in (c), writing this as $t - \frac{t^3}{3!} \leqslant \sin t \leqslant t$ for $t \in [0, x]$ such that $x \in [0, 1]$, on applying for the fourth time the second of the comparison properties for definite integrals to the inequality, show that

$$1 - \frac{x^2}{2!} \leqslant \cos x \leqslant 1 - \frac{x^2}{2!} + \frac{x^4}{4!}, \quad x \in [0, 1].$$

Note that we may continue to repeat the above process as many times as we wish. In doing so, increasingly sharper and sharper polynomial bounds for the sine and cosine functions on the interval $[0, 1]$ will be found.

24. Let f and g be two continuous bounded functions on some interval $[a, b]$ such that $\int_a^b |f(x) - g(x)| \, dx = 0$. Show that $\int_a^b |f(x) - g(x)|^2 \, dx = 0$.

25. Find all continuous positive functions f that are bounded on the interval $0 \leqslant x \leqslant 1$ such that

$$\int_0^1 f(x) \, dx = 1, \quad \int_0^1 x f(x) \, dx = \alpha, \quad \int_0^1 x^2 f(x) \, dx = \alpha^2.$$

Here $\alpha \in \mathbb{R}$.

26. Suppose that f and g are two continuous bounded functions on $[a, b]$.

(a) Prove the *Cauchy–Schwarz inequality*[1]

$$\left(\int_a^b f(x)g(x) \, dx \right)^2 \leqslant \int_a^b f^2(x) \, dx \cdot \int_a^b g^2(x) \, dx.$$

Hint: Consider the function $h(x) = (f(x) + \lambda g(x))^2$ where $\lambda \in \mathbb{R}$.

(b) What is the condition for equality to hold?

(c) Hence show that $\left(\displaystyle\int_0^1 f(x)\,dx\right)^2 \leqslant \displaystyle\int_0^1 f^2(x)\,dx.$

27. In this question the following inequality involving the tangent function will be proved:

$$x \tan x > \frac{4x - \pi}{\pi - 2x}, \qquad \text{for } x \in (0, \tfrac{\pi}{2}).$$

We will have a need for this inequality in a later chapter.

(a) Show the inequality is obviously true for $0 < x \leqslant \frac{\pi}{4}$. So we need only prove the inequality is true for $x \in (\frac{\pi}{4}, \frac{\pi}{2})$.

(b) By performing the change of variable $x \mapsto \frac{\pi}{2} - x$, show that under this change the inequality becomes

$$\tan x < \frac{x(\pi - 2x)}{\pi - 4x}, \qquad \text{for } x \in (0, \tfrac{\pi}{4}).$$

(c) Let $g(x) = \dfrac{x(\pi - 2x)}{\pi - 4x}.$

  (i) Show that $g(0) = 0$.

  (ii) Show that $g'(x) - g^2(x) - 1 = \dfrac{x(\pi - 2x)(2x^2 - \pi x + 4)}{(\pi - 4x)^2}.$

  (iii) Let $h(x) = 2x^2 - \pi x + 4$. By showing that h has an absolute minimum at $x = \frac{\pi}{4}$, use this to conclude that $h(x) > 0$ for all $x \in (0, \frac{\pi}{4})$.

  (iv) Hence deduce that $g'(x) > 1 + g^2(x)$ for $x \in (0, \frac{\pi}{4})$.

(d) Show that $\displaystyle\int_0^x \frac{g'(t)}{1 + g^2(t)}\,dt = \tan^{-1}(g(x)).$

(e) By applying the inequality found in (c)(iv) to the integral given in (d), using the first comparison property for the definite integral, show that $\tan^{-1}(g(x)) > x$ for $x \in (0, \frac{\pi}{4})$.

(f) Hence deduce that $x \tan x > \dfrac{4x - \pi}{\pi - 2x}$ for $x \in (0, \frac{\pi}{2})$.

Endnote

1. The inequality is named after the French mathematician Augustin-Louis Cauchy (1789–1857) and the German mathematician Karl Hermann Amandus Schwarz (1843–1921).

5

Standard Forms

As an inverse operation, [integration] requires a great deal of pattern-recognition skill and experience.
— Jonathan Borwein and Keith Devlin, *The Computer as Crucible:*
An Introduction to Experimental Mathematics

In this chapter we will begin to extend those ideas first introduced in Chapter 3 for finding primitives of functions that while initially may not be in standard form can be brought into such a form through either manipulation of the integrand or by making use of standard identities for the function under consideration. In all cases we only consider integrals whose primitive can be found in terms of familiar functions such as polynomials, roots, the trigonometric functions and their inverses, and the logarithmic and exponential functions. Function which can be written in this way are said to be *elementary*. An *elementary function* is one built up from polynomial and rational functions,[1] exponential, logarithmic, and the trigonometric functions and their inverses,[2] and obtained by addition, multiplication, division, root extraction, and the operation of repeated composition. Finding an elementary primitive means the integral has been found in *elementary terms*. For most elementary functions, it is rare to find a primitive in terms of elementary functions. For example, there is no elementary function F such that $F'(x) = e^{x^2}$ for all x.[3]

In this chapter we will confine our attention to those cases where elementary primitives for the elementary functions can be found. And while the examples and problems considered here may give the impression an elementary primitive for any elementary function that is continuous and bounded on some closed interval can always be found, it is worth keeping in mind that in general this is very far from being the case.

The general approach to finding an integral in elementary terms largely consists of trying to reduce the integral to a well-known form for which a primitive

53

is known. The list of primitives that we take as being known are those given in Table 3.1 on page 26 of Chapter 3. As primitives for a larger class of functions are slowly found, these will be added to our ever-expanding list of 'standard' integrals.

We commence by extending the list of standard integrals we gave in Table 3.1. For the second of the integrals appearing in the table, namely

$$\int \frac{dx}{x} = \ln x + C,$$

it is defined for all positive values of x, but can be readily extended to all real x other than $x = 0$ as follows. Suppose that x is negative. Then $-x$ will be positive and so $\ln(-x)$ is defined. Also, as

$$\frac{d}{dx} \ln(-x) = \frac{-1}{-x} = \frac{1}{x},$$

so that when x is negative

$$\int \frac{dx}{x} = \ln(-x), \quad x < 0.$$

One can of course combine the two results and write

$$\int \frac{dx}{x} = \ln(\pm x) = \ln |x|, \quad x \neq 0,$$

a result that holds for all real x other than $x = 0$, with the ambiguous sign in the argument of the natural logarithmic function chosen so that $\pm x$ is positive.

An important extension of the logarithmic result is the following. Recalling

$$\frac{d}{dx} \ln(\pm f(x)) = \frac{f'(x)}{f(x)},$$

it is immediate that

$$\boxed{\int \frac{f'(x)}{f(x)} \, dx = \ln |f(x)| + C}$$

The definite integral for the tangent and cotangent functions can now be readily found. For the tangent function we have

$$\int \tan x \, dx = \int \frac{\sin x}{\cos x} \, dx = - \int \frac{-\sin x}{\cos x} \, dx,$$

or

$$\boxed{\int \tan x \, dx = -\ln |\cos x| + C}$$

Similarly, for the cotangent function we have

$$\int \cot x \, dx = \int \frac{\cos x}{\sin x} \, dx,$$

or

$$\int \cot x \, dx = \ln |\sin x| + C$$

We now give a number of examples where either manipulation of the integrand or making use of known identities can reduce an integral into standard form before being readily found.

Example 5.1 Find $\int \frac{x}{x-3} \, dx$.

Solution Notice in the integrand that the numerator is almost identical to the denominator, except the constant term of negative three in the numerator is missing. We can therefore add a term of three followed by subtracting it out. Doing so produces

$$\int \frac{x}{x-3} \, dx = \int \frac{(x-3)+3}{x-3} \, dx = \int \frac{x-3}{x-3} \, dx + \int \frac{3}{x-3} \, dx$$

$$= \int dx + 3 \int \frac{dx}{x-3} = x + 3 \ln |x-3| + C. \qquad \blacktriangleright$$

Example 5.2 Find $\int \frac{x^2}{1+x^2} \, dx$.

Solution Again note that the numerator of the integrand is almost the same as the form of the denominator, except for the missing constant term of one. Manipulating the integrand we have

$$\int \frac{x^2}{1+x^2} \, dx = \int \frac{(1+x^2)-1}{1+x^2} \, dx = \int dx - \int \frac{dx}{1+x^2}$$

$$= x - \tan^{-1} x + C. \qquad \blacktriangleright$$

Example 5.3 Find $\int a^x \, dx$ where a is a constant such that $a > 0$ and $a \neq 1$.

Solution We begin by noting that as the variable x appears in the index, the normal power rule does not apply. From properties of the exponential function one can write

$$a^x = e^{\ln(a^x)} = e^{x \ln a}.$$

So

$$\int a^x \, dx = \int e^{x \ln a} \, dx = \frac{1}{\ln a} e^{x \ln a} + C = \frac{a^x}{\ln a} + C. \qquad \blacktriangleright$$

Integrals containing trigonometric functions have plenty of scope for simplification, given the many identities that exist for these functions. To demonstrate this, we consider three examples.

Example 5.4 Find $\int \dfrac{\sin^2 x}{1 + \cos x} \, dx$.

Solution Recalling $\sin^2 x = 1 - \cos^2 x$, we are able to rewrite the integrand as

$$\int \frac{\sin^2 x}{1 + \cos x} \, dx = \int \frac{1 - \cos^2 x}{1 + \cos x} \, dx = \int \frac{(1 - \cos x)(1 + \cos x)}{1 + \cos x} \, dx$$

$$= \int (1 - \cos x) \, dx = x - \sin x + C. \qquad \blacktriangleright$$

Example 5.5 Find $\int \dfrac{(\sin x + \cos x)^3}{1 + \sin 2x} \, dx$.

Solution In the denominator of the integrand, rewriting $\sin^2 x + \cos^2 x$ for 1 and $2 \sin x \cos x$ for $\sin 2x$, we have

$$\int \frac{(\sin x + \cos x)^3}{1 + \sin 2x} \, dx = \int \frac{(\sin x + \cos x)^3}{(\sin^2 x + \cos^2 x) + 2 \sin x \cos x} \, dx$$

$$= \int \frac{(\sin x + \cos x)^3}{(\sin x + \cos x)^2} \, dx = \int (\sin x + \cos x) \, dx$$

$$= -\cos x + \sin x + C. \qquad \blacktriangleright$$

For our final example involving only trigonometric functions in the integrand, the angle sum identity for the tangent function will be used in a clever way.

Example 5.6 Find $\int \tan 2x \tan 3x \tan 5x \, dx$.

Solution From the angle sum identity for the tangent function, we can write

$$\tan 5x = \tan(2x + 3x) = \frac{\tan 2x + \tan 3x}{1 - \tan 2x \tan 3x},$$

giving

$$\tan 5x - \tan 2x \tan 3x \tan 5x = \tan 2x + \tan 3x,$$

or

$$\tan 2x \tan 3x \tan 5x = \tan 5x - \tan 3x - \tan 2x.$$

Thus

$$\int \tan 2x \tan 3x \tan 5x \, dx = \int (\tan 5x - \tan 3x - \tan 2x) \, dx$$

$$= \int \tan 5x \, dx - \int \tan 3x \, dx - \int \tan 2x \, dx$$

$$= -\frac{1}{5} \ln |\cos 5x| + \frac{1}{3} \ln |\cos 3x|$$

$$+ \frac{1}{2} \ln |\cos 2x| + C. \qquad \blacktriangleright$$

As our final example we show how it is possible to find an integral through the clever manipulation of the integrand.

Example 5.7 Find $\displaystyle \int \frac{e^x + \sin x + 4 \cos x + 3}{e^x + 5 \sin x + 3 \cos x + 6} \, dx$.

Solution Let $I = \displaystyle \int \frac{e^x + \sin x + 4 \cos x + 3}{e^x + 5 \sin x + 3 \cos x + 6} \, dx$.

The presence of an exponential, and sine and cosine terms in both the denominator and numerator of the integrand leads one to suspect it may be possible to write the numerator of the integrand as the derivative of the denominator. Attempting to do so, we have

$$I = \int \frac{e^x + 5 \cos x - 3 \sin x + (-\cos x + 4 \sin x + 3)}{e^x + 5 \sin x + 3 \cos x + 6} \, dx$$

$$= \int \frac{e^x + 5 \cos x - 3 \sin x}{e^x + 5 \sin x + 3 \cos x + 6} \, dx + \int \frac{4 \sin x - \cos x + 3}{e^x + 5 \sin x + 3 \cos x + 6} \, dx$$

$$= \ln |e^x + 5 \sin x + 3 \cos x + 6| + \int \frac{4 \sin x - \cos x + 3}{e^x + 5 \sin x + 3 \cos x + 6} \, dx.$$

Initially, this does not have seem to have taken us very far, but do not despair just yet. Next, returning to the original integral, if we rewrite the numerator so the term from the denominator appears, one has

$$I = \int \frac{e^x + 5\sin x + 3\cos x + 6 + (-4\sin x + \cos x - 3)}{e^x + 5\sin x + 3\cos x + 6} \, dx$$

$$= \int dx - \int \frac{4\sin x - \cos x + 3}{e^x + 5\sin x + 3\cos x + 6} \, dx$$

$$= x - \int \frac{4\sin x - \cos x + 3}{e^x + 5\sin x + 3\cos x + 6} \, dx.$$

Adding the two results for I together, the integrals appearing in each expression cancel out and we are left with

$$I + I = 2I = x + \ln |e^x + 5\sin x + 3\cos x + 6| + C,$$

or

$$\int \frac{e^x + \sin x + 4\cos x + 3}{e^x + 5\sin x + 3\cos x + 6} \, dx = \frac{1}{2}\ln |e^x + 5\sin x + 3\cos x + 6|$$

$$+ \frac{x}{2} + C. \qquad \blacktriangleright$$

Exercises for Chapter 5

✠ **Warm-ups**

1. Find $\displaystyle\int x\sin x\cos x\tan x\cosec x\sec x\cot x\,dx$.

✠ **Practice questions**

2. By either manipulating the integrand or employing identities, find the following integrals.

(a) $\displaystyle\int \frac{x}{x+5}\,dx$

(b) $\displaystyle\int \frac{2x^2}{x^2+4}\,dx$

(c) $\displaystyle\int \sqrt{2\sqrt{3\sqrt{x}}}\,dx$

(d) $\displaystyle\int \frac{\sin x - \cos x}{\cos x + \sin x}\,dx$

(e) $\displaystyle\int \frac{x-2}{x^2+x-6}\,dx$

(f) $\displaystyle\int \left(\tan^{-1} x + \tan^{-1}\frac{1}{x}\right)dx, \ x > 0$

(g) $\displaystyle\int \sin^3 x\,dx$

(h) $\displaystyle\int \tan(x-5)\tan(x+5)\tan 2x\,dx$

(i) $\displaystyle\int 2^{\ln x}\,dx$

(j) $\displaystyle\int \cos^2\left[\tan^{-1}\left(\sin(\cot^{-1}x)\right)\right]\,dx$

(k) $\displaystyle\int \frac{x^6-1}{x-1}\,dx$

(l) $\displaystyle\int \frac{\sin 2x + \sin 4x - \sin 6x}{\cos 2x + \cos 4x + \cos 6x + 1}\,dx$

(m) $\displaystyle\int \sin 3x \cos 2x\,dx$

(n) $\displaystyle\int \sin^6 x \cos^6 x\,dx$

(o) $\displaystyle\int \frac{x^3 - 3x^2 + x - 3}{x^2+1}\,dx$

(p) $\displaystyle\int \frac{\sin(x-2)}{\sin x}\,dx$

(q) $\displaystyle\int (\cot x - \tan x)^2\,dx$

(r) $\displaystyle\int \sqrt{\sqrt{\sqrt{x}}}\left(x + \frac{1}{x}\right)dx$

(s) $\displaystyle\int \frac{\tan x - \tan 2}{1 + \tan 2 \tan x}\,dx$

(t) $\displaystyle\int \frac{x^6 - 1}{x^4 + x^3 - x - 1}\,dx$

3. Evaluate

(a) $\displaystyle\int_0^1 x^\pi\,dx$

(b) $\displaystyle\int_0^1 \pi^x\,dx$

(c) $\displaystyle\int_0^{\frac{\pi}{8}} \frac{\tan x + 1}{1 - \tan x}\,dx$

(d) $\displaystyle\int_0^{\frac{\pi}{12}} \sec 3x(\sec 3x + \tan 3x)\,dx$

(e) $\displaystyle\int_0^1 \frac{x^3}{x-2}\,dx$

(f) $\displaystyle\int_0^{\ln 2} \frac{e^x + e^{2x} + e^{3x}}{e^{5x}}\,dx$

(g) $\displaystyle\int_0^2 \frac{x^2-1}{x^2+4}\,dx$

(h) $\displaystyle\int_1^2 \frac{dx}{\sqrt{x+1} + \sqrt{x-1}}\,dx$

4. Find

(a) $\displaystyle\int \sec x \operatorname{cosec} x \cot x\,dx$

(b) $\displaystyle\int \sin x \cos x \tan x\,dx$

5. (a) By finding $\dfrac{d}{dx}\left[\tan^{-1}\left(\sqrt{2+x^2}\right)\right]$, use this to evaluate

$$\int_0^1 \frac{x}{(3+x^2)\sqrt{2+x^2}}\,dx.$$

(b) By finding $\dfrac{d}{dx}\left[\tan^{-1}\left(\dfrac{x}{\sqrt{2+x^2}}\right)\right]$, use this to evaluate

$$\int_0^1 \frac{dx}{(1+x^2)\sqrt{2+x^2}}\,dx.$$

6. (a) If $a > 0$ such that $a \neq 1$, find $\displaystyle\int a^x e^x\,dx$.

   (b) If $a > 0$ such that $a \neq \dfrac{1}{e}$, find $\displaystyle\int a^{\ln x}\,dx$.

7. (a) Show that $\cos\left(x + \dfrac{\pi}{4}\right) = \dfrac{1}{\sqrt{2}}(\cos x - \sin x)$.

   (b) Hence find $\displaystyle\int \frac{2}{(\cos x - \sin x)^2}\,dx$.

8. (a) Show that $\sec^2\left(\dfrac{x}{2} - \dfrac{\pi}{4}\right) = \dfrac{2}{1 + \sin x}$.

   (b) Hence find $\displaystyle\int \frac{dx}{1 + \sin x}$.

9. Suppose that $I = \displaystyle\int_0^{\frac{\pi}{2}} \frac{\sin x}{3\sin x + \cos x}\,dx$ and $J = \displaystyle\int_0^{\frac{\pi}{2}} \frac{\cos x}{3\sin x + \cos x}\,dx$.

   (a) Show that $3I + J = \dfrac{\pi}{2}$ and $3J - I = \ln 3$.

   (b) Hence find values for I and J.

10. In this question a primitive for the secant function will be found. Later, a different method for finding the integral for the secant function will be given (see page 112 of Chapter 8).

    (a) By writing $\sec x$ as the reciprocal of $\cos x$ before multiplying throughout by $\cos x / \cos x$, show that the integral for secant can be rewritten as

    $$\int \sec x\,dx = \int \frac{\cos x}{(1 - \sin x)(1 + \sin x)}\,dx.$$

    (b) Verify that

    $$\frac{\cos x}{(1 - \sin x)(1 + \sin x)} = \frac{\cos x}{2(1 - \sin x)} + \frac{\cos x}{2(1 + \sin x)}.$$

    (c) Hence show that $\displaystyle\int \sec x\,dx = \frac{1}{2}\ln\left|\frac{1 + \sin x}{1 - \sin x}\right| + C.$

(d) Show that the primitive for secant given in (c) can be rewritten in the equivalent form of

$$\int \sec x \, dx = \ln |\sec x + \tan x| + C.$$

11. Find $\dfrac{d}{dx}\left[\displaystyle\int_{\sqrt{x}}^{x^2} \tan 6t \, dt\right].$

12. Suppose that $F(x) = \displaystyle\int_{x}^{1} t(t^2 + 1) \, dt$. Find all stationary points for F and determine their nature.

13. (a) Verify that the function $f(x) = e^x - 1$ satisfies the following *integral equation*[4]

$$f(x) - x - \int_{0}^{x} f(t) \, dt = 0.$$

(b) Verify that the function $\varphi(x) = \cos x + \sin x$ satisfies the following *integro-differential equation*[5]

$$\varphi'(x) + \int_{\frac{\pi}{4}}^{x} \varphi(t) \, dt = 0.$$

14. If n is a positive integer, find

(a) $\displaystyle\int \frac{x^{2n}}{1 - x^2} \, dx$ (b) $\displaystyle\int \frac{x^{2n}}{1 + x^2} \, dx$

15. The formula for the arc length ℓ of a simple curve lying in the plane described by the equation $y = y(x)$ between the points $x = a$ and $x = b$ ($b > a$) is given by

$$\ell = \int_{a}^{b} \sqrt{1 + \left(\frac{dy}{dx}\right)^2} \, dx.$$

Using this formula, find the arc length of the curve given by $y = x^2 - \dfrac{\ln x}{8}$ between the points $x = 1$ and $x = e$.

16. Solve for $f(x)$ and $g(x)$ the system given by

$$f(x) = 1 + \int_{0}^{x} g(t) \, dt \quad \text{and} \quad g(x) = x(x - 1) + \int_{-1}^{1} f(t) \, dt.$$

17. If $I = \displaystyle\int_0^\pi (\sin^2\theta + \cos^4\theta)\,d\theta$ and $J = \displaystyle\int_0^\pi (\cos^2\theta + \sin^4\theta)\,d\theta$, find the value of $I - J$.

18. (a) Let $f(x) = 1 + x + \dfrac{x^2}{2!} + \cdots + \dfrac{x^n}{n!}$, $n \in \mathbb{N}$. Show that

$$f(x) - f'(x) = \frac{x^n}{n!}.$$

(b) Hence show that

$$\int \frac{x^n}{1 + x + \frac{x^2}{2!} + \cdots + \frac{x^n}{n!}}\,dx = n!\ln\left|1 + x + \frac{x^2}{2!} + \cdots + \frac{x^n}{n!}\right|$$

$$- n!\,x + C.$$

19. Consider the following integral $\displaystyle\int \frac{2n!\sin x + x^n}{e^x + \sin x + \cos x + P_n(x)}\,dx$, where $n \in \mathbb{N}$ and $P_n(x)$ is the polynomial given by

$$P_n(x) = 1 + \frac{x}{1!} + \frac{x^2}{2!} + \cdots + \frac{x^{n-1}}{(n-1)!} + \frac{x^n}{n!}.$$

(a) Show that $P_n'(x) = P_n(x) - \dfrac{x^n}{n!}$.

(b) Let $f(x) = e^x + \cos x + \sin x + P_n(x)$. Show that

$$\int \frac{2n!\sin x + x^n}{e^x + \sin x + \cos x + P_n(x)}\,dx = n!\int \frac{f(x) - f'(x)}{f(x)}\,dx.$$

(c) Hence find the integral.

20. Consider the integral $\displaystyle\int \frac{e^{2x} - e^x + 1}{(e^x\cos x - \sin x)(e^x\sin x + \cos x)}\,dx$.

(a) Let $f(x) = e^x\cos x - \sin x$ and $g(x) = e^x\sin x + \cos x$. Show that

$$f'(x)g(x) - f(x)g'(x) = e^{2x} - e^x + 1.$$

(b) From (a) we note the integral can be rewritten as

$$\int \frac{f'(x)g(x) - f(x)g'(x)}{f(x)g(x)}\,dx = \int \left(\frac{f'(x)}{f(x)} - \frac{g'(x)}{g(x)}\right)\,dx.$$

Use this to find the integral.

✠ **Extension questions and Challenge problems**

21. If $n \in \mathbb{N}$, find $\displaystyle\int \frac{2016 - x^n}{1 - x}\,dx$.

22. Find

(a) $\displaystyle\int \frac{\sqrt[4]{x^{10} + x^8 + 1}}{x^6} (3x^{10} + 2x^8 - 2)\, dx$

(b) $\displaystyle\int \frac{\sin x + 4\sin 3x + 6\sin 5x + 3\sin 7x}{\sin 2x + 3\sin 4x + 3\sin 6x}\, dx$

23. Assuming that it is permitted to interchange the infinite sum and the integral, find $\displaystyle\sum_{k=0}^{\infty} \int_0^{\frac{\pi}{3}} \sin^{2k} x\, dx$.

24. If $\displaystyle\gamma = \int_0^{\ln 2} \frac{2e^{3x} + e^{2x} - 1}{e^{3x} + e^{2x} - e^x + 1}\, dx$ find the value of e^{γ}.

Endnotes

1. Polynomials and rational functions are examples of *algebraic* functions.
2. The exponential, logarithmic, the trigonometric functions and their inverses are examples of *transcendental* functions.
3. The fact that $\exp(x^2)$ has no primitive in terms of the elementary functions is a very famous example of such a case. It was first proved to be so by the French mathematician Joseph Liouville (1809–1882) in his paper 'Suite du émoire sur la classification des Transcendantes, et sur l'impossibilité d'exprimer les racines de certaines équations en fonction finie explicite des coefficients', published in 1838. For a relatively modern description and proof of this result see pages 59 and 60 of G. H. Hardy's *The Integration of Functions of a Single Variable*, 2nd edition (Cambridge University Press, Cambridge, 1916).
4. An *integral equation* is an equation that involves integrals of a function.
5. An *integro-differential equation* is an equation that involves both integrals and derivatives of a function.

6

Integration by Substitution

I almost wish I hadn't gone down the rabbit-hole...
— Lewis Carroll, *Alice in Wonderland*

The substitution method is by far the most important of all the methods that can be used to find an integral. In a large number of cases, when the form of the integral is not directly obvious, it can enable one to reduce an integral to standard form. The idea behind the method is that, by employing a substitution, one is led to a simpler integral that can be readily found. It achieves this by using the function of a function rule for differentiation in reverse. Of course, not all substitutions lead to a simpler integral, and in many cases no such substitution may exist that leads to a simpler integral. Knowing when to use a substitution and what form the substitution should be is not always immediately obvious, and is a skill best acquired through long and careful practice. We start with relatively simple and obvious substitutions to find an integral before moving onto substitutions that are slightly more sophisticated and often require more insight to be spotted before being used.

Theorem 6.1 (Integration by substitution). *If $f(x)$ and $g'(x)$ are continuous, then*

$$\boxed{\int_{g(a)}^{g(b)} f(u)\, du = \int_a^b f(g(x)) \cdot g'(x)\, dx}$$

Proof If F is a primitive of f, then the left side is

$$\int_{g(a)}^{g(b)} f(u)\, du = F(u)\Big|_{g(a)}^{g(b)} = F(g(b)) - F(g(a)).$$

On the other hand, from the function of a function rule, we have

$$\frac{d}{dx}[F(g(x))] = F'(g(x)) \cdot g'(x) = f(g(x)) \cdot g'(x),$$

as F is a primitive of f, that is $F' = f$. So we see $F(g(x))$ is a primitive of $f(g(x)) \cdot g'(x)$ and the right side is

$$\int_a^b f(g(x)) \cdot g'(x)\, dx = F(g(x))\Big|_a^b = F(g(b)) - F(g(a)),$$

as required. ∎

The simplest use of the substitution method depends on recognising that the integrand is of the form $f(g(x)) \cdot g'(x)$.

Example 6.1 Evaluate $\displaystyle\int_0^{\frac{\pi}{2}} \cos x \sin^3 x\, dx$.

Solution Let $f(x) = x^3$ and $g(x) = \sin x$. So $f(g(x)) = \sin^3 x$ and $g'(x) = \cos x$. Also $g(0) = 0$ and $g\left(\frac{\pi}{2}\right) = 1$. So from the substitution formula

$$\int_a^b f(g(x)) \cdot g'(x)\, dx = \int_{g(a)}^{g(b)} f(u)\, du,$$

we have

$$\int_0^{\frac{\pi}{2}} \cos x \sin^3 x\, dx = \int_0^1 u^3\, du = \left[\frac{u^4}{4}\right]_0^1 = \frac{1}{4}. \qquad \blacktriangleright$$

Having to work formally with the substitution formula, as we have done in our first example, takes considerable time and effort. Fortunately the method can be streamlined in the following manner. First, note that u can be substituted for $g(x)$ and du for $g'(x)\, dx$ followed by a change in the limits of integration. The substitutions are therefore performed directly on the original integrand and account for the name of the method. As the variable changes, the method is sometimes called the *change of variable* method. We now redo the previous example using this more streamlined approach.

Example 6.2 Evaluate $\displaystyle\int_0^{\frac{\pi}{2}} \cos x \sin^3 x\, dx$.

Solution Let $u = \sin x$, $du = \cos x\, dx$ while for the limits of integration we have

$$\text{when } x = 0, \qquad u = 0$$

$$\text{and when } x = \frac{\pi}{2}, \qquad u = 1.$$

So we have

$$\int_0^{\frac{\pi}{2}} \cos x \sin^3 x\, dx = \int_0^1 u^3\, du = \left[\frac{u^4}{4}\right]_0^1 = \frac{1}{4},$$

as before. $\blacktriangleright$

How to Integrate It

As a second example, consider the following.

Example 6.3 Evaluate $\displaystyle\int_0^1 x\sqrt{1-x^2}\,dx$.

Solution Let $u = 1 - x^2$, $du = -2x\,dx$ so that $-\frac{1}{2}du = x\,dx$. For the limits of integration we have

$$\text{when } x = 0, \qquad u = 1$$
$$\text{and when } x = 1, \qquad u = 0.$$

Thus

$$\int_0^1 x\sqrt{1-x^2}\,dx = -\frac{1}{2}\int_1^0 \sqrt{u}\,du = \frac{1}{2}\int_0^1 \sqrt{u}\,du = \frac{1}{2}\left[\frac{2}{3}u^{\frac{3}{2}}\right]_0^1 = \frac{1}{3}. \quad \blacktriangleright$$

If the definite integral $\int_a^b f(x)\,dx$ can be found for all a and b, then the indefinite integral $\int f(x)\,dx$ can certainly be found on some interval of x for which the function f is bounded and continuous; this means the change of variable method can be equally applied to the indefinite case. We now consider a variety of indefinite integrals that can be found using the substitution method.

Example 6.4 Find $\displaystyle\int x^2(1+x)^{100}\,dx$.

Solution Let $u = 1 + x$ or $x = u - 1$ giving $du = dx$. Thus

$$\int x^2(1+x)^{100}\,dx = \int (u-1)^2 u^{100}\,du$$

$$= \int (u^2 - 2u + 1)u^{100}\,du$$

$$= \int \left(u^{102} - 2u^{101} + u^{100}\right)du$$

$$= \frac{1}{103}u^{103} - \frac{1}{51}u^{102} + \frac{1}{101}u^{101} + C$$

$$= \frac{1}{103}(1+x)^{103} - \frac{1}{51}(1+x)^{102}$$

$$+ \frac{1}{101}(1+x)^{101} + C,$$

since $u = 1 + x$. $\qquad\qquad\qquad\qquad\qquad\qquad\blacktriangleright$

IMPORTANT: In applying the substitution method to indefinite integrals, one should not forget to substitute back the original variable you started out with.

Example 6.5 Find $\int \dfrac{x^3}{\sqrt{x^4 + 4}} \, dx$.

Solution Let $u = x^4 + 4$. So $du = 4x^3 \, dx$ or $\frac{1}{4} du = x^3 \, dx$. Thus

$$\int \frac{x^3}{\sqrt{x^4 + 1}} \, dx = \frac{1}{4} \int \frac{du}{\sqrt{u}} = \frac{1}{4} \int u^{-\frac{1}{2}} \, du$$

$$= \frac{1}{4} \cdot 2u^{\frac{1}{2}} + C = \frac{1}{2} \sqrt{x^4 + 1} + C. \qquad \blacktriangleright$$

Example 6.6 Find $\int \dfrac{x^2 + 1}{x^3 + 3x + 8} \, dx$.

Solution Let $u = x^3 + 3x + 8$. So $du = (3x^2 + 3) \, dx$ or $\frac{1}{3} du = (x^2 + 1) \, dx$. Thus

$$\int \frac{x^2 + 1}{x^3 + 3x + 8} \, dx = \frac{1}{3} \int \frac{du}{u} = \frac{1}{3} \ln |u| + C$$

$$= \frac{1}{3} \ln \left| x^3 + 3x + 8 \right| + C. \qquad \blacktriangleright$$

Of course, in the previous example no substitution was necessary. Simple manipulation of the numerator appearing in the integrand into the exact form for the derivative of the denominator and recognising it as of the logarithmic form, namely $\int \frac{f'(x)}{f(x)} dx$, are all that is needed in this case.

Often an integral can be performed by inspection. In those cases, the substitution is simple enough it can be done mentally in one's head as the next example illustrates.

Example 6.7 Find $\int xe^{x^2} \, dx$.

Solution Immediately writing down the answer for the integral one has

$$\int xe^{x^2} \, dx = \frac{1}{2} e^{x^2} + C.$$

The mental substitution used here was $u = x^2$ giving $\frac{1}{2} du = x \, dx$. $\qquad \blacktriangleright$

We complete this section by considering several more examples that make use of a few different types of substitutions and approaches.

Example 6.8 Find $\displaystyle\int \frac{dx}{x \ln x}$.

Solution Let $u = \ln x$, so $du = \dfrac{dx}{x}$. Thus

$$\int \frac{dx}{x \ln x} = \int \frac{du}{u} = \ln|u| + C = \ln|\ln x| + C.$$

▶

Example 6.9 Find $\displaystyle\int \frac{x^3}{x^8 + 9}\, dx$.

Solution First, we begin by writing

$$\int \frac{x^3}{x^8 + 9}\, dx = \int \frac{x^3}{(x^4)^2 + 9}\, dx.$$

Now let $u = x^4$, so $du = 4x^3\, dx$ or $\frac{1}{4} du = x^3\, dx$. Thus

$$\int \frac{x^3}{x^8 + 9}\, dx = \frac{1}{4} \int \frac{du}{u^2 + 9} = \frac{1}{4} \cdot \frac{1}{3} \tan^{-1}\left(\frac{u}{3}\right) + C$$

$$= \frac{1}{12} \tan^{-1}\left(\frac{x^4}{3}\right) + C.$$

▶

Example 6.10 Find $\displaystyle\int \frac{dx}{1 + \sqrt{x}}$.

Solution Let $x = u^2$ (here $x > 0$), $dx = 2u\, du$. Thus

$$\int \frac{dx}{1 + \sqrt{x}} = 2 \int \frac{u}{1 + u}\, du = 2 \int \frac{(1 + u) - 1}{1 + u}\, du$$

$$= 2 \int du - 2 \int \frac{du}{1 + u} = 2u - 2\ln|1 + u| + C$$

$$= 2\sqrt{x} - 2\ln\left|1 + \sqrt{x}\right| + C.$$

▶

Example 6.11 Find $\displaystyle\int \frac{e^x}{\sqrt{1 - e^{2x}}}\, dx$.

Solution Begin by writing

$$\int \frac{e^x}{\sqrt{1 - e^{2x}}}\, dx = \int \frac{e^x}{\sqrt{1 - (e^x)^2}}\, dx.$$

Now let $u = e^x$, $du = e^x \, dx$. Thus

$$\int \frac{e^x}{\sqrt{1 - e^{2x}}} \, dx = \int \frac{du}{\sqrt{1 - u^2}} = \sin^{-1} u + C$$

$$= \sin^{-1}(e^x) + C. \qquad \blacktriangleright$$

Example 6.12 Find $\displaystyle\int \frac{dx}{1 + \sin^2 x} \, dx$.

Solution Before a substitution is made we begin by manipulating the integrand into a form where the choice of the substitution to be made will become obvious.

$$\int \frac{dx}{1 + \sin^2 x} = \int \frac{dx}{(\cos^2 x + \sin^2 x) + \sin^2 x} = \int \frac{dx}{\cos^2 x + 2 \sin^2 x}$$

$$= \int \frac{dx}{\cos^2 x (1 + 2 \tan^2 x)} = \int \frac{\sec^2 x}{1 + 2 \tan^2 x} \, dx.$$

The 'obvious' substitution to make is therefore $u = \sqrt{2} \tan x$. So $du = \sqrt{2} \sec^2 x \, dx$ and we have

$$\int \frac{dx}{1 + \sin^2 x} = \frac{1}{\sqrt{2}} \int \frac{du}{1 + u^2} = \frac{1}{\sqrt{2}} \tan^{-1} u + C$$

$$= \frac{1}{\sqrt{2}} \tan^{-1} \left(\sqrt{2} \tan x \right) + C. \qquad \blacktriangleright$$

Exercises for Chapter 6

✠ **Warm-ups**

1. Find the following integrals by *inspection*. That is, try to find the integrals mentally in your head by simply looking at them.

(a) $\displaystyle\int \cos(2x) \, dx$

(b) $\displaystyle\int \sec^2 x \tan x \, dx$

(c) $\displaystyle\int 2x(1 + x^2)^3 \, dx$

(d) $\displaystyle\int \cos x e^{\sin x} \, dx$

(e) $\displaystyle\int \cos^5 x \sin x \, dx$

(f) $\displaystyle\int \frac{\sin \sqrt{x}}{\sqrt{x}} \, dx$

✠ Practice questions

2. By using an appropriate substitution find the following integrals.

(a) $\displaystyle\int \sqrt{5x+7}\,dx$ (b) $\displaystyle\int x^2 e^{-x^3}\,dx$ (c) $\displaystyle\int \frac{\ln x}{x}\,dx$

(d) $\displaystyle\int \frac{\tan^{-1}x}{1+x^2}\,dx$ (e) $\displaystyle\int \frac{2x^3}{\sqrt{1+x^4}}\,dx$ (f) $\displaystyle\int \frac{dx}{x\ln^2 x}$

(g) $\displaystyle\int \frac{e^{\sqrt{x}}}{\sqrt{x}}\,dx$ (h) $\displaystyle\int \frac{x^4}{\sqrt{1-x^5}}\,dx$ (i) $\displaystyle\int \frac{\sin x}{\cos^4 x}\,dx$

(j) $\displaystyle\int \frac{\sin x}{2-\cos x}\,dx$ (k) $\displaystyle\int \frac{e^x - e^{2x}}{1+e^{2x}}\,dx$ (l) $\displaystyle\int \frac{\sqrt{x}}{1-x}\,dx$

(m) $\displaystyle\int \frac{x^5}{\sqrt{25-x^2}}\,dx$ (n) $\displaystyle\int \frac{\sqrt[6]{x}-1}{\sqrt[3]{x^2}-\sqrt{x}}\,dx$ (o) $\displaystyle\int \sqrt{\frac{x^3-3}{x^{11}}}\,dx$

(p) $\displaystyle\int \frac{e^{\tan x}}{\cos^2 x}\,dx$ (q) $\displaystyle\int x\sin(x^2)e^{\cos(x^2)}\,dx$

(r) $\displaystyle\int \frac{dx}{3^x - 8}$ (s) $\displaystyle\int \frac{\sin x + \cos x}{\sqrt[3]{\sin x - \cos x}}\,dx$

(t) $\displaystyle\int [\ln 2]^{\ln x}\,dx$ (u) $\displaystyle\int (1+\sin 2x)^3 \cos 2x\,dx$

3. Evaluate the following integrals.

(a) $\displaystyle\int_0^6 x\sqrt{36-x^2}\,dx$ (b) $\displaystyle\int_0^1 \frac{x}{\sqrt{2-x^2}}\,dx$

(c) $\displaystyle\int_1^{\sqrt{2}} \sqrt{x^2+x^4}\,dx$ (d) $\displaystyle\int_0^e e^{e^x + x}\,dx$

(e) $\displaystyle\int_1^{\ln 2} \frac{e^{2x}}{e^x+1}\,dx$ (f) $\displaystyle\int_0^{\frac{\pi}{2}} \sin^2 x \cos^2 x\,dx$

(g) $\displaystyle\int_1^2 \frac{(2x-1)e^{x^2}}{e^x}\,dx$ (h) $\displaystyle\int_0^1 (1-\sqrt{x})^{p-1}\,dx,\ p>0$

4. (a) If $\int_0^3 f(x)\,dx = 6$ find $\int_0^1 f(3x)\,dx$.

(b) If $\int_0^2 g(x)\,dx = 5$ find $\int_0^1 [g(x) + g(x + 1)]\,dx$.

5. Find the value of k if $\int_0^k \dfrac{\cos x}{1 + \sin^2 x}\,dx = \dfrac{\pi}{4}$.

6. (a) Let f be a continuous function satisfying

$$f(x) = f\left(\frac{100}{x}\right) \quad \text{for all } x > 0.$$

If $\int_1^{10} \dfrac{f(x)}{x}\,dx = 5$, find the value of $\int_1^{100} \dfrac{f(x)}{x}\,dx$.

(b) Let f be a continuous function satisfying

$$f\left(\frac{1}{x}\right) + x^2 f(x) = 0 \quad \text{for all } x \neq 0.$$

Find the value of $\int_{\sin \theta}^{\operatorname{cosec} \theta} f(x)\,dx$. Here θ is a constant.

7. If $a, b, x > 0$, by using the substitution $u = \ln x$, find $\displaystyle\int \frac{\ln(ax)}{x \ln(bx)}\,dx$.

8. Suppose that $I_n = \displaystyle\int_0^1 (1 - \sqrt{x})^n\,dx$. Here $n \in \mathbb{N}$.

(a) By using an appropriate substitution, show that $I_n = \dfrac{2}{(n + 1)(n + 2)}$.

(b) By applying the binomial theorem to the integrand of I_n before integrating, show that $I_n = \displaystyle\sum_{k=0}^{n} \binom{n}{k} \frac{(-1)^k 2}{k + 2}$.

(c) Hence find a value for the sum $\displaystyle\sum_{k=0}^{100} \binom{100}{k} \frac{(-1)^k}{k + 2}$ as a simple fraction.

9. If $f(x) = \displaystyle\int_0^x \left(\cos^2(2x - 2t) - \frac{1}{4}\right) dt$ such that $[0, \frac{\pi}{2}]$, find the maximum and minimum value(s) of $f(x)$.

10. If $\int_0^1 f(x)\,dx = 5$, $\int_{-1}^1 f(x)\,dx = 3$, $\int_0^2 f(x)\,dx = 8$, and $\int_0^4 f(x)\,dx = 11$, evaluate

(a) $\displaystyle\int_0^2 f(2x)\,dx$ (b) $\displaystyle\int_0^\pi f(\cos x)\sin x\,dx$

(c) $\displaystyle\int_2^3 xf(8-x^2)\,dx$

11. Let $0 < a, b < \dfrac{\pi}{2}$.

(a) Find $I = \displaystyle\int_a^b \dfrac{\cos x}{1+\sin x}\,dx$ in terms of a and b.

(b) Find $J = \displaystyle\int_a^b \dfrac{\sin x}{1+\cos x}\,dx$ in terms of a and b.

(c) Show that if $a + b = \dfrac{\pi}{2}$, then $I = J$.

12. Find $\displaystyle\int \dfrac{1+x-x^2}{(1-x^2)^{\frac{3}{2}}}\,dx$.

13. (a) Using a suitable substitution, show that

$$\int_a^1 \frac{dx}{1+x^2} = \int_1^{\frac{1}{a}} \frac{dx}{1+x^2}, \quad a > 0.$$

(b) Hence show that $\tan^{-1} a + \tan^{-1}\left(\dfrac{1}{a}\right) = \dfrac{\pi}{2}$, $a > 0$.

14. Find the value of a such that the identity,

$$\ln(\ln(\ln x)) = \int_a^x \frac{dt}{t\,\ln t\,\ln(\ln t)},$$

holds. Here $x > a$.

15. By using an appropriate substitution find the following integrals.

(a) $\displaystyle\int \dfrac{x+1}{x(\ln x + 1)}\,dx$

(b) $\displaystyle\int e^{\cot x + 2\ln(\operatorname{cosec} x)}\,dx$

(c) $\displaystyle\int \sin^{12}(7x)\cos^3(7x)\,dx$

(d) $\displaystyle\int \sec x \tan x \sqrt{1 + \sec x}\,dx$

16. If a and b are real numbers, show that

$$\int_0^1 x^a(1-x)^b\,dx = \int_0^1 x^b(1-x)^a\,dx.$$

17. If $0 < a < b$, use the substitution $x = \dfrac{ab}{u}$ to evaluate $\displaystyle\int_a^b \dfrac{e^{\frac{x}{a}} - e^{\frac{b}{x}}}{x}\, dx$.

18. (a) Find $\displaystyle\int \dfrac{(x+1)\sqrt{x + \ln x}}{x \ln^2 x + 2x^2 \ln x + x^3}\, dx$.

    (b) Hence show that

$$\int \dfrac{\ln^2 x + 2x \ln x + x^2 + (x+1)\sqrt{x + \ln x}}{x \ln^2 x + 2x^2 \ln x + x^3}\, dx = \ln x$$

$$-\dfrac{2}{\sqrt{x + \ln x}} + C.$$

19. By 'rationalising the numerator' first, find $\displaystyle\int \sqrt{\dfrac{1-x}{1+x}}\, dx$.

20. If y is a function of x such that $y(x - y)^2 = x$, by letting $t = x - y$, find

$$\int \dfrac{dx}{x - 3y}.$$

21. Consider the integral $\displaystyle\int_1^\alpha \dfrac{dt}{1+t^2}$. Here $\alpha = \dfrac{1-x}{\sqrt{1-x^2}}$ such that $|x| < 1$.

    (a) Find the value of the integral by directly evaluating it.

    (b) Find the value of the integral after the substitution $t = \dfrac{1-u}{\sqrt{1-u^2}}$ has been made.

    (c) Hence deduce that $\tan^{-1}\left(\dfrac{1-x}{\sqrt{1-x^2}}\right) = \dfrac{\pi}{4} - \dfrac{1}{2}\sin^{-1} x$ for $|x| < 1$.

22. Suppose $\displaystyle\int \dfrac{x^2 + 1}{x^4 - x^2 + 1}\, dx$.

    (a) By rearranging the integrand, show that

$$\int \dfrac{x^2+1}{x^4 - x^2 + 1}\, dx = \int \dfrac{\frac{1}{x^2} + 1}{\left(x - \frac{1}{x}\right)^2 + 1}\, dx.$$

    (b) Hence find the integral by using the substitution $u = x - \dfrac{1}{x}$.

23. Evaluate $\displaystyle\int_0^1 x(x+1)^m\, dx$ for all the different cases that arise according to the value of $m \in \mathbb{R}$.

24. By making use of the substitution $u = e^x + 2e^{-x}$ find

$$\int \dfrac{e^{3x} - 2e^x}{4 + 8e^{2x} + e^{4x}}\, dx.$$

25. Suppose $\displaystyle\int \frac{2x^3 - 1}{x^6 + 2x^3 + 9x^2 + 1}\, dx.$

(a) By rearranging the integrand, show that

$$\int \frac{2x^3 - 1}{x^6 + 2x^3 + 9x^2 + 1}\, dx = \int \frac{2x - \frac{1}{x^2}}{\left(x^2 + \frac{1}{x}\right)^2 + 9}\, dx.$$

(b) Hence find the integral by using the substitution $u = x^2 + \dfrac{1}{x}$.

26. (a) Find $\displaystyle\int \sqrt{\frac{x + 2}{x}}\, dx$ **and** $\displaystyle\int \sqrt{\frac{x}{x + 2}}\, dx.$

(b) Hence find $\displaystyle\int \frac{dx}{\sqrt{x(x + 2)}}.$

27. Find $\displaystyle\int \frac{dx}{\left(\sqrt{\sin x} + \sqrt{\cos x}\right)^4}.$

28. Suppose $\displaystyle\int_0^1 \frac{(x - 1)^3}{(x + 1)^4}\, dx.$

(a) By using the substitution $u = x + 1$, evaluate the integral.

(b) By using the substitution $u = \dfrac{x - 1}{x + 1}$, evaluate the integral.

(c) Do your answers found in (a) and (b) agree?

29. By using a substitution of $x = 1/u$, **evaluate**

(a) $\displaystyle\int_{\frac{1}{\pi}}^{\pi} \frac{\tan^{-1} x}{x}\, dx$ \qquad (b) $\displaystyle\int_{\frac{1}{2}}^{2} \left(1 + \frac{1}{x^2}\right) \tan^{-1} x\, dx$

30. (a) Evaluate $\displaystyle\int_{\frac{1}{\sqrt{3}}}^{\sqrt{3}} \frac{\ln x}{1 + x^2}\, dx.$

(b) Hence show that $\displaystyle\int_1^3 \frac{\ln x}{3 + x^2}\, dx = \frac{\pi\sqrt{3}\ln 3}{36}.$

31. Suppose $I = \displaystyle\int_{\sqrt{2}-1}^{\sqrt{2}+1} \frac{dx}{(1 + x^2)(1 + 5^{\ln x})}.$

(a) Using the substitution $x = 1/u$, show that

$$I = \int_{\sqrt{2}-1}^{\sqrt{2}+1} \frac{dx}{(1 + x^2)(1 + 5^{-\ln x})}.$$

Note here the dummy variable has been changed back to x.

(b) By adding the integral in (a) to the original integral, deduce that
$I = \pi/8$.

32. Consider the integral $\displaystyle\int \frac{2\sin^{-1} x + 2\cos^{-1} x}{\sqrt{1 - x^2}}\, dx$.

(a) By using appropriate substitutions, find

   (i) $\displaystyle\int \frac{2\sin^{-1} x}{\sqrt{1 - x^2}}\, dx$ (ii) $\displaystyle\int \frac{2\cos^{-1} x}{\sqrt{1 - x^2}}\, dx$

(b) Hence deduce that

$$\int \frac{2\sin^{-1} x + 2\cos^{-1} x}{\sqrt{1 - x^2}}\, dx = (\sin^{-1} x)^2 - (\cos^{-1} x)^2 + C_1.$$

(c) By recalling $\sin^{-1} x + \cos^{-1} x = \dfrac{\pi}{2}$, show that

$$\int \frac{2\sin^{-1} x + 2\cos^{-1} x}{\sqrt{1 - x^2}}\, dx = \pi \sin^{-1} x + C_2.$$

(d) Using (b) and (c), show that

$$(\sin^{-1} x)^2 - (\cos^{-1} x)^2 = \pi\left(\sin^{-1} x - \frac{\pi}{4}\right).$$

33. Two special functions known as the *error function* erf(x) and the *cumulative distribution function for the standard normal distribution* $\Phi(x)$ are defined in terms of definite integrals as follows:

$$\text{erf}(x) = \frac{2}{\sqrt{\pi}} \int_0^x e^{-u^2}\, du \quad \text{and} \quad \Phi(x) = \frac{1}{2} + \frac{1}{\sqrt{2\pi}} \int_0^x e^{-t^2/2}\, dt.$$

By using a suitable substitution, show that the two functions are related to each other by showing that erf(x) = $2\Phi(\sqrt{2}x) - 1$.

34. Consider the definite integral $\displaystyle\int_0^1 \frac{\tan^{-1} x}{1 + x}\, dx$.

(a) By writing down the angle sum identity formula for $\tan(\alpha + \beta)$, show that

$$\tan^{-1} \alpha - \tan^{-1} \beta = \tan^{-1}\left(\frac{\alpha - \beta}{1 + \alpha\beta}\right).$$

(b) By making use of the result in (a), together with the substitution $u = (1 - x)/(1 + x)$, show that

$$\int_0^1 \frac{\tan^{-1} x}{1 + x}\, dx = \frac{\pi}{8} \ln 2.$$

35. Let $I = \int_0^1 \frac{dx}{\sqrt{1 + x^4}}$ and $J = \int_0^1 \frac{dx}{\sqrt{1 - x^4}}$. By making a substitution of $x = \frac{1 - u^2}{1 + u^2}$ in J, show that $\frac{I}{J} = \frac{1}{\sqrt{2}}$.

36. Consider the integral $\int \frac{2x^3 - 3x^2 + 1}{x^6 - 2x^3 + x^2 - 2x + 2} \, dx$.

(a) After factoring, show that the integral can be rewritten as

$$\int \frac{2x^3 - 3x^2 + 1}{x^6 - 2x^3 + x^2 - 2x + 2} \, dx = \int \frac{2x + 1}{(x^2 + x + 1)^2 + 1} \, dx.$$

(b) Hence find the integral.

37. Suppose $H_n = \int_0^1 \frac{1 - (1 - x)^n}{x} \, dx$ where $n \in \mathbb{N}$.

(a) Using an appropriate substitution, show that $H_n = \sum_{k=1}^n \frac{1}{k}$. The numbers H_n are known as the *harmonic numbers*.

(b) Hence write down values for the first five harmonic numbers.

(c) By expanding the term $(1 - x)^n$ using the binomial theorem before integrating, deduce that

$$H_n = \sum_{k=1}^n \binom{n}{k} \frac{(-1)^{k+1}}{k} = \sum_{k=1}^n \frac{1}{k}.$$

38. Let $I = \int_0^1 \left((1 - x^a)^{1/a} - x \right)^2 \, dx$ where $a > 0$.

(a) By using the substitution $u = (1 - x^a)^{1/a}$, show that

$$I = \int_0^1 \frac{u^a (1 - u^a)^{1/a}}{u(1 - u^a)} \left(u - (1 - u^a)^{1/a} \right)^2 \, du.$$

(b) By adding the two forms for the integral I together (the original form together with the one found in (a)) show that

$$I = \frac{1}{2} \int_0^1 \left(\frac{x^a (1 - x^a)^{1/a}}{x(1 - x^a)} + 1 \right) \left(x - (1 - x^a)^{1/a} \right)^2 \, dx.$$

(c) Now use the substitution $u = (1 - x^a)^{1/a} - x$ in the integral given in (b), and deduce that $I = \frac{1}{3}$ for all $a > 0$.

39. In this question a method for finding the following two integrals

$$I = \int \frac{1+x^4}{1-x^4} \frac{dx}{\sqrt{1+x^4}} \quad \text{and} \quad J = \int \frac{x^2}{1-x^4} \frac{dx}{\sqrt{1+x^4}},$$

will be developed.

(a) If $A = \int \frac{1+x^2}{1-x^2} \frac{dx}{\sqrt{1+x^4}}$ and $B = \int \frac{1-x^2}{1+x^2} \frac{dx}{\sqrt{1+x^4}}$, show that

$I = \frac{1}{2}(A+B)$ and $J = \frac{1}{4}(A-B)$.

(b) The integrals A and B can be evaluated using the substitution $u = \sqrt{1+x^4}/x$. By using this substitution, show that

$$A = \frac{1}{2\sqrt{2}} \ln \left| \frac{\sqrt{1+x^4} + \sqrt{2}x}{\sqrt{1+x^4} - \sqrt{2}x} \right| + C \text{ and } B = -\frac{1}{\sqrt{2}} \tan^{-1} \left(\frac{\sqrt{1+x^4}}{\sqrt{2}x} \right) + C.$$

(c) Hence find I and J.

✠ Extension questions and Challenge problems

40. If $n \in \mathbb{N}$, show that $\displaystyle\int \frac{dx}{\sqrt[n]{x} - 1} = n \ln \left| 1 - \sqrt[n]{x} \right| + n \sum_{k=1}^{n-1} \frac{x^{\frac{k}{n}}}{k} + C.$

41. Evaluate $\displaystyle\int_3^6 \left[\sqrt{x + \sqrt{12x - 36}} + \sqrt{x - \sqrt{12x - 36}} \right] dx.$

42. Find $\displaystyle\int \left(x^n + x^{2n} + x^{3n} \right) \sqrt[n]{2x^{2n} + 3x^n + 6} \, dx$ if $n \in \mathbb{N}$ and $x > 0$.

43. By using a suitable substitution, show that

$$\int_0^a \frac{\ln(x+a)}{x^2 + a^2} \, dx = \frac{\pi}{8a} \ln(2a^2), \ a > 0.$$

44. A so-called *Euler substitution* is a substitution method used for finding integrals of the form

$$\int \mathcal{R} \left(x, \sqrt{ax^2 + bx + c} \right) dx,$$

where $\mathcal{R}$ is a rational function of x and $\sqrt{ax^2 + bx + c}$. In such cases the integrand may be changed into a rational function using one of three broadly similar substitution types. In what is known as the *third* type of substitution, if the polynomial $ax^2 + bx + c$, where $a, b, c \in \mathbb{R}$, has real roots α and β, one may choose a substitution t of the form

$$\sqrt{ax^2 + bx + c} = \sqrt{a(x - \alpha)(x - \beta)} = (x - \alpha)t = (x - \beta)t.$$

Note here that either of the real roots may be chosen in the substitution. Using an Euler substitution of the third type, find $\displaystyle\int \frac{dx}{\sqrt{2 - x - x^2}}$.

45. Let $\displaystyle I = \int_0^1 \left(\frac{2}{x + 1} - x - 1\right)^n dx$ where n is a positive even integer.

(a) By expanding the integrand using the binomial theorem, show that when evaluated in this way the integral can be expressed as

$$I = \sum_{k=0}^{n} \binom{n}{k} \frac{(-1)^k 2^{n-k}}{2k - n + 1} \left(2^{2k-n+1} - 1\right).$$

(b) By using the substitution $u = \dfrac{2}{x + 1} - 1$, show that the integral can be rewritten as

$$I = \int_0^1 \frac{2}{(u + 1)^2} \left(\frac{2}{u + 1} - u - 1\right)^n du.$$

(c) By adding the two forms for the integral I together (the original form together with the one found in (b)), show that

$$I = \frac{1}{2} \int_0^1 \left(\frac{2}{(x + 1)^2} + 1\right) \left(\frac{2}{x + 1} - x - 1\right)^n dx.$$

(d) Now using the substitution $u = \dfrac{2}{x + 1} - x - 1$ in the integral given in (c), show that

$$\int_0^1 \left(\frac{2}{x + 1} - x - 1\right)^n dx = \frac{1}{n + 1}.$$

(e) Hence deduce a simple expression for the sum given in (a).

46. Let $\ln^{[0]}(x) = x$ and $\ln^{[n]}(x) = \ln\left(\ln^{[n-1]}(x)\right)$, $n \in \mathbb{N}$. Using induction, prove that

$$\int \frac{dx}{\prod_{k=0}^n \ln^{[k]}(x)} = \ln^{[n+1]}(x) + C.$$

7

Integration by Parts

'Begin at the beginning,' the King said, very gravely, 'and go on till you come to the end: then stop.'

— Lewis Carroll, *Alice in Wonderland*

After integration by substitution, the next most important technique is what is known as *integration by parts*. There is no counterpart for the product rule for the derivative when it comes to integration. Integration by parts, however, is a consequence of the product rule for differentiation. We first state and give a proof for this very important result before giving a number of examples that make use of the rule.

Theorem 7.1 (Integration by parts). *If $f'(x)$ and $g'(x)$ are continuous functions then*

$$\int f(x)g'(x)\,dx = f(x) \cdot g(x) - \int f'(x)g(x)\,dx$$

Proof From the product rule for the derivative we have

$$(f(x)g(x))' = f'(x)g(x) + f(x)g'(x).$$

Rewriting this as

$$f(x)g'(x) = (f(x)g(x))' - f'(x)g(x),$$

one has

$$\int f(x)g'(x)\,dx = \int (f(x)g(x))'\,dx - \int f'(x)g(x)\,dx$$

$$= f(x) \cdot g(x) - \int f'(x)g(x)\,dx,$$

which completes the proof. ∎

For a definite integral on the interval $[a, b]$ the corresponding result will be

$$\int_a^b f(x)g'(x)\, dx = f(x) \cdot g(x)\Big|_a^b - \int_a^b f'(x)g(x)\, dx$$

It is possible to write the rule more compactly. If $u(x)$ (for $f(x)$) and $v(x)$ (for $g(x)$) are two continuously differentiable functions of x, we can write $du = u'(x)\, dx$ and $dv = v'(x)\, dx$, and the rule for integration by parts can be written more compactly as

$$\int u\, dv = uv - \int v\, du$$

At first sight, using integration by parts may not seem very useful as it appears that one integral on the left side has simply been replaced with another integral on the right side. While this is true, the real power of the rule comes about from this exact change. In many instances the integral on the right side will be more amenable to integration than the integral one started out with. In many commonly encountered cases this is exactly what one finds. Also, to apply the rule, the function $g'(x)$ to be integrated needs to be relatively simple to begin with. A common example of a situation is in the integration where the integrand consists of a product between a polynomial and either a sine or cosine function or the exponential function.

To apply integration by parts, the overall strategy is to divide the integrand into a product between two functions before carefully choosing which to integrate and which to differentiate, the goal being to end up with an integral that is simpler and can be more readily found. Knowing what choice to make comes with practice. In general, when deciding on possible choices for u and dv, one tries to choose $u = f(x)$ as the function that becomes simpler when differentiated, provided a primitive for dv can still be readily found. We now consider a number of examples to show how the method of integration by parts works in practice.

Example 7.1 Using integration by parts, find $\int x \cos x\, dx$.

Solution For our first few examples we will slowly work our way through the process of carefully applying the rule for integration by parts. As one becomes more familiar with the workings of the rule, the formula can be dispensed with altogether as one integrates one part and differentiates the other to arrive at the expression for the new integral.

By differentiating x we will be left with a constant so the following choices for u and dv are made:

$$u = x \qquad dv = \cos x \, dx$$

$$du = dx \qquad v = \int \cos x \, dx = \sin x$$

Note here that we can take any primitive function for v we like (usually the simplest), so in general the arbitrary constant of integration is set equal to zero (though see the example on page 84 where a nonzero constant is selected).

Integrating by parts we have

$$\int x \cos x \, dx = x \sin x - \int \sin x \, dx.$$

Notice the choice made for u and v has led to a far simpler integral compared to the original. Integrating we have

$$\int x \cos x \, dx = x \sin x + \cos x + C. \qquad \blacktriangleright$$

What would have happened if instead we had chosen $\cos x$ for u and x for v? Well, let us see. In this case one would have

$$u = \cos x \qquad dv = x \, dx$$

$$du = -\sin x \, dx \qquad v = \int x \, dx = \frac{1}{2} x^2,$$

and integrating by parts gives

$$\int x \cos x \, dx = -x \sin x + \frac{1}{2} \int x^2 \sin x \, dx.$$

While the above result is certainly true, as we have ended up with an integral that is more difficult to find compared to the original one, it has not taken us any closer to a solution. This is often what will happen if an inappropriate choice for u and dv is made.

Example 7.2 Using integration by parts, find $\displaystyle\int x^2 e^{-x} \, dx$.

Solution As we are about to see, integration by parts may be applied more than once to find an integral.

We set

$$u = x^2 \qquad dv = e^{-x}\, dx$$

$$du = 2x\, dx \qquad v = \int e^{-x}\, dx = -e^{-x}.$$

Integrating by parts leads to

$$\int x^2 e^{-x}\, dx = -x^2 e^{-x} - \int 2x \cdot (-e^{-x})\, dx = -x^2 e^{-x} + 2 \int x e^{-x}\, dx.$$

To evaluate the integral $\int x e^{-x}\, dx$ that has appeared, integration by parts can be applied again. Here we set

$$u = x \qquad dv = e^{-x}\, dx$$

$$du = dx \qquad v = \int e^{-x}\, dx = -e^{-x}.$$

Integrating by parts leads to

$$\int x e^{-x}\, dx = -x e^{-x} - \int -e^{-x}\, dx$$

$$= -x e^{-x} + \int e^{-x}\, dx = -x e^{-x} - e^{-x} + C.$$

So for the original integral one has

$$\int x^2 e^{-x}\, dx = -x^2 e^{-x} + 2\left(-x e^{-x} - e^{-x}\right) + C$$

$$= -x^2 e^{-x} - 2x e^{-x} - 2e^{-x} + C$$

$$= -(x^2 + 2x + 2)e^{-x} + C. \qquad \blacktriangleright$$

This example shows that there is no reason why integration by parts cannot be used multiple times to find an integral. To help streamline the computations involved when such a situation arises, shortly we will introduce a method referred to as *tabular integration by parts*. We now give an example where the method of integration by parts is applied to a definite integral.

Example 7.3 Using integration by parts, evaluate $\displaystyle\int_0^{\frac{\pi}{2}} x \sin 2x\, dx.$

Solution Set

$$u = x \qquad dv = \sin 2x\, dx$$

$$du = dx \qquad v = \int \sin 2x\, dx = -\frac{1}{2}\cos 2x.$$

Integrating by parts leads to

$$\int_0^{\frac{\pi}{2}} x \sin 2x \, dx = \left[-\frac{x}{2} \cos 2x \right]_0^{\frac{\pi}{2}} + \frac{1}{2} \int_0^{\frac{\pi}{2}} \cos 2x \, dx$$

$$= \frac{\pi}{4} + \frac{1}{2} \left[\frac{1}{2} \sin 2x \right]_0^{\frac{\pi}{2}} = \frac{\pi}{4}.$$ ▶

The first three examples show that integration by parts is particularly suited to removing powers of x in products involving either the sine or cosine function or the exponential function. In the next two examples we show how integration by parts is also particularly suited to removing logarithms.

Example 7.4 Using integration by parts, find $\int x^3 \ln x \, dx$.

Solution Set

$$u = \ln x \qquad dv = x^3 \, dx$$

$$du = \frac{1}{x} dx \qquad v = \int x^3 \, dx = \frac{x^4}{4}.$$

Integrating by parts leads to

$$\int x^3 \ln x \, dx = \frac{1}{4} x^4 \ln x - \frac{1}{4} \int x^4 \cdot \frac{1}{x} dx$$

$$= \frac{1}{4} x^4 \ln x - \frac{1}{4} \int x^3 \, dx = \frac{1}{4} x^4 \ln x - \frac{1}{12} x^4 + C$$

$$= x^4 \left(\frac{3 \ln x - 1}{12} \right) + C.$$ ▶

Example 7.5 Using integration by parts, find $\int 2x \ln(x + 12) \, dx$.

Solution Set

$$u = \ln(x + 12) \qquad dv = 2x \, dx$$

$$du = \frac{1}{x + 12} dx \qquad v = \int 2x \, dx = x^2.$$

Integrating by parts leads to

$$\int 2x \ln(x+12)\,dx = x^2 \ln(x+12) - \int x^2 \cdot \frac{1}{x+12}\,dx$$

$$= x^2 \ln(x+12) - \int \frac{(x^2-144)+144}{x+12}\,dx$$

$$= x^2 \ln(x+12) - 144\int \frac{dx}{x+12} - \int \frac{x^2-144}{x+12}\,dx$$

$$= x^2 \ln(x+12) - 144\ln|x+12| - \int \frac{(x-12)(x+12)}{x+12}\,dx$$

$$= (x^2-144)\ln(x+12) - \int (x-12)\,dx$$

$$= (x^2-144)\ln(x+12) - \frac{x^2}{2} + 12x + C. \qquad \blacktriangleright$$

§ The Intermediate Constant of Integration in v

So far when performing integration by parts, you may have noticed that in all cases we have set the arbitrary constant of integration in the function integrated leading to $v(x)$ equal to zero. While a value of zero is invariably chosen on the assumption that it is the simplest, this need not be the case. As the constant of integration is arbitrary, any value one may care to consider can be selected. Often a judicious choice of this value can lead to a significant simplification in the integral to be found.

As an example, consider the last example we gave. Instead of choosing zero for the constant of integration in v, consider what would happen if a value of -144 is selected instead. Why a choice of -144 for the constant is made in the first place should become obvious in a moment's time. In this case for v we would have

$$v = \int 2x\,dx = x^2 - 144,$$

and on applying integration by parts to the integral we now have

$$\int 2x \ln(x+12)\,dx = (x^2-144)\ln(x+12) - \int \frac{x^2-144}{x+12}\,dx$$

$$= (x^2-144)\ln(x+12) - \int \frac{(x-12)(x+12)}{x+12}\,dx,$$

which immediately reduces to

$$\int 2x \ln(x+12)\,dx = (x^2-144)\ln(x+12) - \frac{x^2}{2} + 12x + C,$$

as expected, but with considerably less effort compared to when the 'automatic' choice of zero for the arbitrary constant in v was made.

Integration by parts can also be used to integrate certain trigonometric functions, such as the square of the sine or cosine functions, without the need to recall any double-angle identities, as the following example shows.

Example 7.6 Using integration by parts, find $\int \sin^2 x \, dx$.

Solution We begin by writing the integrand as a product as follows:

$$\int \sin^2 x \, dx = \int \sin x \cdot \sin x \, dx.$$

Set

$$u = \sin x \qquad dv = \sin x \, dx$$

$$du = \cos x \, dx \qquad v = \int \sin x \, dx = -\cos x.$$

So by parts one has

$$\int \sin^2 x \, dx = -\sin x \cos x + \int \cos^2 x \, dx$$

$$= -\sin x \cos x + \int (1 - \sin^2 x) \, dx$$

$$= -\sin x \cos x + \int dx - \int \sin^2 x \, dx$$

$$= -\sin x \cos x + x - \int \sin^2 x \, dx.$$

At this point we notice the integral we started out with has reappeared. The integral can now be finished off using algebra alone. Doing so yields

$$2 \int \sin^2 x \, dx = -\sin x \cos x + x + C$$

$$\Rightarrow \int \sin^2 x \, dx = -\frac{1}{2} \sin x \cos x + \frac{x}{2} + C. \qquad \blacktriangleright$$

What the previous example shows is a useful trick. Often when performing integration by parts the integral you started out with reappears. The next example gives a common situation for this kind of reappearance.

Example 7.7 Using integration by parts, find $\int e^{2x} \sin 3x \, dx$.

Solution To speed up the process, we now proceed by applying the rule for integration by parts *directly* without first writing down what u and dv are (this

step can now be done mentally in one's head). For this example, it is not important which function one chooses to initially integrate and which function one
chooses to differentiate. We will chose to integrate the exponential function
first. Doing so yields

$$\int e^{2x} \sin 3x \, dx = \frac{1}{2} e^{2x} \sin 3x - \int \frac{1}{2} e^{2x} \cdot 3 \cos 3x \, dx \quad \text{(by parts)}$$

$$= \frac{1}{2} e^{2x} \sin 3x - \frac{3}{2} \int e^{2x} \cos 3x \, dx$$

$$= \frac{1}{2} e^{2x} \sin 3x - \frac{3}{2} \left(\frac{1}{2} e^{2x} \cos 3x \right.$$

$$\left. - \int \frac{1}{2} e^{2x} \cdot -3 \sin 3x \, dx \right) \quad \text{(by parts again)}$$

$$= \frac{1}{2} e^{2x} \sin 3x - \frac{3}{4} e^{2x} \cos 3x - \frac{9}{4} \int e^{2x} \sin 3x \, dx.$$

As the integral we started out with has reappeared, move it to the left side:

$$\frac{13}{4} \int e^{2x} \sin 3x \, dx = \frac{1}{2} e^{2x} \sin 3x - \frac{3}{4} e^{2x} \cos 3x + C,$$

or

$$\int e^{2x} \sin 3x \, dx = \frac{e^{2x}}{13} (2 \sin 3x - 3 \cos 3x) + C. \qquad \blacktriangleright$$

Example 7.8 Using integration by parts, find $\int \sec^3 x \, dx$.

Solution

$$\int \sec^3 x \, dx = \int \sec^2 \cdot \sec x \, dx$$

$$= \tan x \sec x - \int \tan x \cdot \sec x \tan x \, dx \quad \text{(by parts)}$$

$$= \tan x \sec x - \int \sec x \tan^2 x \, dx$$

$$= \tan x \sec x - \int \sec x (\sec^2 x - 1) \, dx$$

$$= \tan x \sec x + \int \sec x \, dx - \int \sec^3 x \, dx.$$

Recalling

$$\int \sec x \, dx = \ln|\sec x + \tan x| + C,$$

one has

$$\int \sec^3 x \, dx = \tan x \sec x + \ln|\sec x + \tan x| - \int \sec^3 x \, dx.$$

As the integral we started out with has reappeared, moving it to the left side and completing the problem using algebra alone give

$$\int \sec^3 x \, dx = \frac{1}{2} \tan x \sec x + \frac{1}{2} \ln|\sec x + \tan x| + C. \qquad \blacktriangleright$$

One can use the method of integration by parts to find the indefinite integral for many of the inverse elementary transcendental functions such as the logarithm and the inverse trigonometric functions. While this may appear strange at first, since the integrand appearing in such integrals is not a product, the integrand can always by multiplied by unity, thereby turning it into a product between two functions. Indefinite integrals for the remainder of the elementary transcendental functions that have not yet been given can therefore be found in this way. We give two examples that demonstrate the method.

Example 7.9 Using integration by parts, find $\int \ln x \, dx$.

Solution Since we can write

$$\int \ln x \, dx = \int 1 \cdot \ln x \, dx,$$

integrating by parts, one has

$$\int \ln x \, dx = x \ln x - \int x \cdot \frac{1}{x} dx = x \ln x - \int dx = x \ln x - x + C. \quad \blacktriangleright$$

Example 7.10 Using integration by parts, find $\int \sin^{-1} x \, dx$.

Solution Again, since we can write

$$\int \sin^{-1} x \, dx = \int 1 \cdot \sin^{-1} x \, dx,$$

integrating by parts, one has

$$\int \sin^{-1} x \, dx = x \sin^{-1} x - \int x \cdot \frac{1}{\sqrt{1 - x^2}} dx.$$

The second integral can be found using a substitution of $u = 1 - x^2$. Since $du = -2x\, dx$, for this integral one has

$$\int \frac{x}{\sqrt{1 - x^2}}\, dx = -\frac{1}{2} \int \frac{du}{\sqrt{u}} = -\sqrt{u} + C = -\sqrt{1 - x^2} + C.$$

So finally for the integral of the inverse sine function we have

$$\int \sin^{-1} x\, dx = x \sin^{-1} x + \sqrt{1 - x^2} + C. \qquad \blacktriangleright$$

§ Tabular Integration by Parts

When the method of integration by parts needs to be applied more than once, the work and effort involved can become quite tedious. As a way of organising and arranging everything when integration by parts is performed multiple times, a method known as *tabular integration by parts*[1] can be used. As we will see, it is not a new technique; rather it is a way of organising in a systematic fashion the mechanics of applying integration by parts multiple times.

As its name suggests, a table is used. It consists of two columns that we will label D for 'differentiate' (the function to be differentiated) and I for 'integrate' (the function to be integrated). The entries in the first row are the functions selected to be differentiated u, and integrated dv. Each successive entry in the D column is the derivative of its previous column entry, while each successive entry in the I column is the integral of its previous column entry. While the table can be terminated at any time, the real usefulness of the tabular method is brought to the fore when the table is terminated when one of the four following cases arise.

(a) The entry in the first column D is zero.
(b) The functions found in the first row, to within a constant multiple, reappear in a subsequent row.
(c) The product of the two functions in any given row becomes an easy integral to find.
(d) The rows continue without end in which case a series will be developed.

When the table is terminated, the integral is written down as the sum of the products of the function in the nth row of the first column (column D) with the function in the $(n + 1)$th row of the second column (column I), with the sign of the products alternating after starting with a plus sign. To help remember the technique, the appropriate products between the functions are formed by drawing a diagonal arrow in the table while its corresponding sign is placed just above each arrow. We now give a number of examples showing the tabular method in action.

Example 7.11 Find $\int x^2 e^{2x}\, dx$.

Solution Choosing $u = x^2$ and $dv = e^{2x}\, dx$, and constructing the table of derivatives and integrals we have

D(differentiate)	I(integrate)
x^2	e^{2x}
$2x$	$\frac{1}{2}e^{2x}$
2	$\frac{1}{4}e^{2x}$
0	$\frac{1}{8}e^{2x}$

Note the table is terminated as a zero appears in the first column. The integral is now equal to the sum of the appropriate products indicated along the diagonal arrows. Thus

$$\int x^2 e^{2x}\, dx = x^2 \cdot \frac{1}{2}e^{2x} - 2x \cdot \frac{1}{2}e^{2x} + 2 \cdot \frac{1}{8}e^{2x} + C = \frac{e^{2x}}{4}(2x^2 - 2x + 1) + C.$$

▶

To see how the tabular form is nothing more than integration by parts performed multiple times, if the table in the previous example were to be terminated after the first row, one would have

$$\int x^2 e^{2x}\, dx,$$

that is, one has yet to perform any integration by parts. Terminating the table after the second row, we would have one product only and an integral of the product between the two functions in the second row, namely

$$\int x^2 e^{2x}\, dx = x^2 \cdot \frac{1}{2}e^{2x} - \int 2x \cdot \frac{1}{2}e^{2x}\, dx = \frac{x^2}{2}e^{2x} - \int x e^{2x}\, dx,$$

a result corresponding to a single application of integration by parts. Finally, terminating the table after the third row, a result corresponding to two applications of integration by parts, one has the following sum between two products and an integral of a product between the two functions in the third row of

$$\int x^2 e^{2x}\, dx = x^2 \cdot \frac{1}{2}e^{2x} - 2x \cdot \frac{1}{4}e^{2x} + \int 2 \cdot \frac{1}{4}e^{2x}\, dx$$

$$= \frac{x^2}{2}e^{2x} - \frac{x}{2}e^{2x} + \frac{1}{2}\int e^{2x}\, dx.$$

The tabular method used in Example 7.11 was thus a convenient way of streamlining the process of performing integration by parts three times.

The next example demonstrates the real efficiency of the tabular method for finding an integral that would otherwise have to be found by performing integration by parts no less than four times.

Example 7.12 Find $\int (3x^3 + 2x - 1)e^{-x}\, dx$.

Solution Choosing to differentiate $3x^3 + 2x - 1$ and integrate e^{-x}, and constructing the table of derivatives and integrals we have

D	I
$3x^3 + 2x - 1$	e^{-x}
$9x^2 + 2$	$-e^{-x}$
$18x$	e^{-x}
18	$-e^{-x}$
0	e^{-x}

Thus

$$\int (3x^3 + 2x - 1)e^{-x}\, dx = -e^{-x} \cdot (3x^3 + 2x - 1) - e^{-x} \cdot (9x^2 + 2)$$

$$- e^{-x} \cdot 18x - e^{-x} \cdot 18 + C$$

$$= -e^{-x}(3x^3 + 9x^2 + 20x + 19) + C. \quad \blacktriangleright$$

It is not always possible to end with a zero appearing in the derivative column. The next example illustrates the termination of the table when the functions in the first row, to within a constant multiple, reappear in a later row.

Example 7.13 Find $\int e^{5x} \cos 3x \, dx$.

Solution Choosing to differentiate e^{5x} and integrate $\cos 3x$, and constructing the table of derivatives and integrals we have

D	I
e^{5x}	$\cos 3x$
$5e^{5x}$	$\frac{1}{3} \sin 3x$
$25e^{5x}$	$-\frac{1}{9} \cos 3x$

As the entries in the third row are a constant multiple of those found in the first row, the table is terminated after three rows. The integral is

$$\int e^{5x} \cos 3x \, dx = e^{5x} \cdot \frac{1}{3} \sin 3x - 5e^{5x} \cdot -\frac{1}{9} \cos 3x$$
$$+ \int 25e^{5x} \cdot \left(-\frac{1}{9} \cos 3x\right) dx$$
$$= \frac{1}{3} e^{5x} \sin 3x + \frac{5}{9} e^{5x} \cos 3x$$
$$- \frac{25}{9} \int e^{5x} \cos 3x \, dx,$$

giving

$$\int e^{5x} \cos 3x \, dx = \frac{e^{5x}}{34} (3 \sin 3x + 5 \cos 3x) + C. \qquad \blacktriangleright$$

The next example illustrates the termination of the table when the product between the two functions in a given row becomes an easily found integral. Typically in such cases only a single application of integration by parts is needed, so the tabular method may not give any particular advantage over normal integration by parts.

Example 7.14 Find $\int x \tan^{-1} x \, dx$.

Solution Choosing to differentiate $\tan^{-1} x$ and integrate x, and constructing the table of derivatives and integrals we have

D	I
$\tan^{-1} x$	x
$\dfrac{1}{1 + x^2}$	$\dfrac{x^2}{2}$

The table is terminated after the second row as we observe the product between the two functions in the second row leads to an integral that is relatively easy to find. Failure to spot this would result in the table going on without end. Thus

$$\int x \tan^{-1} x \, dx = \frac{x^2}{2} \tan^{-1} x - \frac{1}{2} \int \frac{x^2}{1 + x^2} \, dx$$

$$= \frac{x^2}{2} \tan^{-1} x - \frac{1}{2} \int \frac{(1 + x^2) - 1}{1 + x^2} \, dx$$

$$= \frac{x^2}{2} \tan^{-1} x - \frac{1}{2} \int dx + \frac{1}{2} \int \frac{dx}{1 + x^2}$$

$$= \frac{x^2}{2} \tan^{-1} x - \frac{x}{2} + \frac{1}{2} \tan^{-1} x + C$$

$$= \frac{1}{2}(x^2 + 1) \tan^{-1} x - \frac{x}{2} + C. \qquad \blacktriangleright$$

As our last example of the tabular method we give a situation where the rows continue without end.

Example 7.15 By applying the method of tabular integration by parts to the integral $\int_0^x e^{-t} \, dt$, show that

$$e^x = 1 + x + \frac{x^2}{2!} + \frac{x^3}{3!} + \cdots + \frac{x^n}{n!} + R_{n+1},$$

where $R_{n+1} = \dfrac{1}{n!} \displaystyle\int_0^x e^{x-t} t^n \, dt$.

Solution Ordinarily one would not use integration by parts at all to find the integral, preferring instead to directly integrate. Doing so yields

$$\int_0^x e^{-t}\, dt = -e^{-t}\Big|_0^x = 1 - e^{-x}.$$

If, however, we consider the integrand as a product with unity, differentiating e^{-t} and integrating 1, we have

D		I
e^{-t}	$+$	1
$-e^{-t}$	$-$	t
e^{-t}	$+$	$\dfrac{t^2}{2!}$
$-e^{-t}$	$-$	$\dfrac{t^3}{3!}$
e^{-t}		$\dfrac{t^4}{4!}$
$\vdots$		$\vdots$
$(-1)^{n-1}e^{-t}$	$(-1)^{n-1}$	$\dfrac{t^{n-1}}{(n-1)!}$
$(-1)^{n}e^{-t}$	$(-1)^{n}$	$\dfrac{t^{n}}{n!}$

So one has

$$\int_0^x e^{-t}\, dt = \left[te^{-t} + \frac{t^2}{2!}e^{-t} + \frac{t^3}{3!}e^{-t} + \cdots + \frac{t^n}{n!}e^{-t} \right]_0^x$$
$$+ \int_0^x e^{-t}\frac{t^n}{n!}\, dt$$
$$\Rightarrow 1 - e^{-x} = xe^{-x} + \frac{x^2}{2!}e^{-x} + \frac{x^3}{3!}e^{-x} + \cdots + \frac{x^n}{n!}e^{-x}$$
$$+ \frac{1}{n!}\int_0^x e^{-t}t^n\, dt.$$

or

$$e^x = 1 + x + \frac{x^2}{2!} + \frac{x^3}{3!} + \cdots + \frac{x^n}{n!} + \frac{1}{n!}\int_0^x e^{x-t}t^n\, dt,$$

after multiplying both sides by e^x and rearranging, as required to show. ▶

Exercises for Chapter 7

✠ **Warm-ups**

1. Let $f(x) = x$ and $g(x) = x^2$.

   (a) Find $\displaystyle\int f(x)g(x)\,dx$.

   (b) Find $\displaystyle\int f(x)\,dx \cdot \int g(x)\,dx$.

   (c) Do you expect the integrals in (a) and (b) to be equal to each other? If not, why not?

2. *A paradox resulting from integration by parts.*[2]

   Suppose the integral $\displaystyle\int \frac{dx}{x}$ is to be found using integration by parts. Setting

   $$u = \frac{1}{x} \qquad dv = dx$$

   $$du = -\frac{1}{x^2}\,dx \qquad v = x,$$

   and integrating by parts one has

   $$\int \frac{dx}{x} = 1 + \int \frac{dx}{x}.$$

   On cancellation of the two integrals, one obtains $0 = 1$! Briefly explain how this paradox is resolved.

✠ **Practice questions**

3. Use integration by parts to find

   (a) $\displaystyle\int x \sin x\,dx$ (b) $\displaystyle\int x^2 \ln x\,dx$

   (c) $\displaystyle\int x \sec x \tan x\,dx$ (d) $\displaystyle\int x \sec^2 x\,dx$

   (e) $\displaystyle\int \ln\left(\frac{1}{x}\right) dx$ (f) $\displaystyle\int \ln(x^2)\,dx$

   (g) $\displaystyle\int \tan^{-1} x\,dx$ (h) $\displaystyle\int (\ln x)^2\,dx$

   (i) $\displaystyle\int \cos 4x \sin 3x\,dx$ (j) $\displaystyle\int \cos(\ln x)\,dx$

(k) $\int e^{-3x} \cos 2x \, dx$

(l) $\int (x+1)e^x \ln x \, dx$

(m) $\int \dfrac{dx}{x + x \ln^2 x}$

(n) $\int \ln(1 + x^2) \, dx$

(o) $\int x \ln(x^2 e^{x^2}) \, dx$

(p) $\int x^3 e^{x^2} \, dx$

4. Evaluate the following integrals using integration by parts.

(a) $\int_0^{\frac{\pi}{4}} x \sin 2x \, dx$

(b) $\int_1^2 \dfrac{\ln x}{x^2} \, dx$

(c) $\int_0^{\frac{\pi}{2}} x \sin x \cos x \, dx$

(d) $\int_1^{e^2} \sqrt{x} \ln(\sqrt{x}) \, dx$

(e) $\int_1^{e^{\frac{\pi}{2}}} \sin(\ln x) \, dx$

(f) $\int_0^1 \ln(1 + x^2) \, dx$

5. (a) If $\displaystyle\int_1^2 x \ln x \, dx = a + 1$, find the value of $a \in \mathbb{R}$.

   (b) If $\displaystyle\int_0^{\frac{\pi}{2}} e^{2x} \cos x \, dx = \dfrac{1}{n}(e^\pi - 2)$, find the value of $n \in \mathbb{N}$.

6. Find $\displaystyle\int \ln(x^2 - 1) \, dx$.

7. If $n \neq 0$ show that $\displaystyle\int_0^{2\pi} x^2 \cos(nx) \, dx = \dfrac{4\pi}{n^2}$.

8. Use integration by parts to show that

$$\int \cos^2 x \, dx = \dfrac{1}{2} \sin x \cos x + \dfrac{x}{2} + C.$$

9. If $I_n = \displaystyle\int_0^{\frac{\pi}{2}} x^n \sin^n x \, dx$, evaluate I_0, I_1, and I_2.

10. Using the substitution $x = u^2$ followed by integration by parts, find

$$\int \left(e^{\sqrt{x}} + 1 \right) dx.$$

11. Suppose that f is differentiable on the interval $[1, 4]$ such that $\int_1^4 x f'(x) \, dx = 12$. If $f(1) = 3$ and $f(4) = 10$, find the value of $\int_1^4 f(x) \, dx$.

12. Let $f(x) = x$ and $g(x) = \dfrac{1}{1 + \tan^2 x}$. If $c \in \left(0, \frac{\pi}{4}\right)$ such that

$$\int_0^{\frac{\pi}{4}} f(x) g(x) \, dx = \int_0^{\frac{\pi}{4}} \frac{x}{1 + \tan^2 x} \, dx = f(c) \int_0^{\frac{\pi}{4}} \frac{dx}{1 + \tan^2 x},$$

find the value for c.

13. (a) Show that $\tan^{-1} \left(\dfrac{\sqrt{1 + x} - \sqrt{1 - x}}{\sqrt{1 + x} + \sqrt{1 - x}} \right) = \dfrac{\pi}{4} - \dfrac{1}{2} \cos^{-1} x$.

(b) Hence find $\displaystyle\int \tan^{-1} \left(\dfrac{\sqrt{1 + x} - \sqrt{1 - x}}{\sqrt{1 + x} + \sqrt{1 - x}} \right) dx$.

14. Using integration by parts, show that $\displaystyle\int_0^{\sqrt{3}} \sin^{-1} \left(\dfrac{2x}{1 + x^2} \right) dx = \dfrac{\pi}{\sqrt{3}}$.

15. If h is a continuous, twice-differentiable function such that $h(0) = 1$, $h(2) = 3$, and $h'(2) = 5$, find the value of $\displaystyle\int_0^1 x h''(x) \, dx$.

16. Suppose that $F_n(x) = \displaystyle\int_0^x \dfrac{u^n}{e^u} \, du$.

(a) Find, as a function of x, expressions for $F_1(x)$ and $F_2(x)$.

(b) Evaluate $\displaystyle\int_2^5 \left[\exp \left(\int \dfrac{F_1(x)}{F_2(x)} \right) dx \right]^2 dx$.

17. Suppose that for a certain function $f(x)$, it is known that

$$f'(x) = \frac{\cos x}{x}, \quad f\left(\frac{\pi}{2}\right) = 2 \text{ and } f\left(\frac{3\pi}{2}\right) = 4.$$

Using integration by parts, evaluate $\displaystyle\int_{\frac{\pi}{2}}^{\frac{3\pi}{2}} f(x) \, dx$.

18. In this question two alternative ways for finding $\displaystyle\int \tan^{-1} x \, dx$ will be given.

(a) Using integration by parts, show that

$$\int \tan^{-1} x \, dx = x \tan^{-1} x - \frac{1}{2} \ln(x^2 + 1) + C.$$

(b) (i) By using integration by parts, show that

$$\int \frac{2x}{1+x^2} \, dx = 2x \tan^{-1} x - 2 \int \tan^{-1} x \, dx + C.$$

(ii) By finding the indefinite integral appearing on the left side in (b)(i) directly, deduce that

$$\int \tan^{-1} x \, dx = x \tan^{-1} x - \frac{1}{2} \ln(x^2 + 1) + C.$$

19. Suppose that $I_n = \displaystyle\int_0^1 \frac{dx}{(1+x^2)^n}$ for $n \in \mathbb{N}$.

(a) Using integration by parts, show that $I_n = \dfrac{1}{2^n} + 2n \displaystyle\int_0^1 \frac{x^2}{(1+x^2)^{n+1}} \, dx.$

(b) Find values for A and B for which

$$\frac{x^2}{(1+x^2)^{n+1}} = \frac{A}{(1+x^2)^n} + \frac{B}{(1+x^2)^{n+1}}.$$

(c) Hence deduce that $I_{n+1} = \dfrac{1}{n2^{n-1}} + \left(\dfrac{2n-1}{2n}\right) I_n.$

20. Consider the integral $\displaystyle\int_0^1 \ln\left(\sqrt{1+x} + \sqrt{1-x}\right) dx.$

(a) Show that

$$\frac{d}{dx}\left[\ln\left(\sqrt{1+x} + \sqrt{1-x}\right)\right] = \frac{1}{2\sqrt{1-x^2}} \cdot \frac{\sqrt{1-x} - \sqrt{1+x}}{\sqrt{1+x} + \sqrt{1-x}}.$$

(b) By 'rationalising the numerator' of the second term appearing in (a), show that the expression for the derivative can be rewritten as

$$\frac{d}{dx}\left[\ln\left(\sqrt{1+x} + \sqrt{1-x}\right)\right] = \frac{1}{2x} - \frac{1}{2x\sqrt{1-x^2}}.$$

(c) Using integration by parts together with the result in (b), evaluate the integral.

21. (a) If $f(a) = f(b) = 0$, show that

$$\int_a^b x f(x) f'(x) \, dx = -\frac{1}{2} \int_a^b [f(x)]^2 \, dx.$$

(b) Using the result in (a), evaluate $\displaystyle\int_0^\pi x \sin 2x \, dx.$

22. If $f(0) = g(0) = 0$, f'' and g'' are continuous, and $a > 0$, use integration by parts to show that

$$\int_0^a f(x)g''(x)\,dx = f(a)g'(a) - f'(a)g(a) + \int_0^a f''(x)g(x)\,dx.$$

23. Find the value of $a > 0$ such that $\displaystyle\int_1^a x^n \ln x\,dx = \frac{1}{(n+1)^2}$, $n \in \mathbb{N}$.

24. Let f and g be twice-differentiable functions such that $f''(x) = af(x)$ and $g''(x) = bg(x)$ where a and b are constants such that $a \neq b$.

(a) Show that $\displaystyle\int f(x)g(x)\,dx = \frac{f(x)g'(x) - f'(x)g(x)}{b - a} + C$.

(b) If $m^2 \neq n^2$ use the result given in (a) to show that

$$\int \sin mx \cos nx\,dx = \frac{n \sin mx \sin nx + m \cos mx \cos nx}{n^2 - m^2} + C.$$

(c) Hence deduce that $\displaystyle\int_0^{2\pi} \sin mx \cos nx\,dx = 0$, when $m \neq n$.

25. If $\displaystyle I = \int_0^1 (1 - x^n)^{2n}\,dx$ and $\displaystyle J = \int_0^1 (1 - x^n)^{2n+1}\,dx$ where $n \in \mathbb{N}$, show that $\dfrac{I}{J} = \dfrac{n(2n+1) + 1}{n(2n+1)}$.

26. If $\displaystyle \gamma = \int_0^\pi \frac{\cos x}{(x+2)^2}\,dx$, find the value of $\displaystyle\int_0^{\frac{\pi}{2}} \frac{\sin x \cos x}{x+1}\,dx$ in terms of γ.

27. Find the following integrals by applying the method of tabular integration by parts.

(a) $\displaystyle\int x^3 \sin x\,dx$ (b) $\displaystyle\int \frac{x^3}{(1+x)^5}\,dx$ (c) $\displaystyle\int \frac{12x^2 + 36}{\sqrt[5]{3x+2}}\,dx$

(d) $\displaystyle\int x \sec^2 x\,dx$ (e) $\displaystyle\int e^{3x}(x^3 + 6x^2 + 11x + 6)\,dx$

28. By performing integration by parts, find the following integrals.

(a) $\displaystyle\int \ln(1 + \sqrt{x})\,dx$ (b) $\displaystyle\int \ln(x + \sqrt{x})\,dx$

(c) $\displaystyle\int \ln(x + \sqrt{x^2 + 1})\,dx$

29. Let $F(x) = e^x \left(\alpha \ln(x^2 + 1) + \dfrac{x}{x^2 + 1} \right)$. If $\displaystyle\int_0^1 F(x)\, dx = \dfrac{e}{2} \ln 2$, find the value of α.

30. Find

(a) $\displaystyle\int x \tan^2 x\, dx$ (b) $\displaystyle\int x \cot^2 x\, dx$

31. By making an appropriate substitution followed by integration by parts, find

(a) $\displaystyle\int x \cos(\sqrt{x})\, dx$ (b) $\displaystyle\int e^{2x} \cos(e^x)\, dx$ (c) $\displaystyle\int x \sin(\ln x)\, dx$

32. By making an appropriate substitution followed by integration by parts, evaluate $\displaystyle\int_{-\frac{\pi}{4}}^{\frac{\pi}{4}} \cos^{-1}(\tan x)\, dx$.

33. The *lower incomplete gamma function* $\gamma(x, \alpha)$ is defined by the following integral:

$$\gamma(x, \alpha) = \int_0^\alpha e^{-u} u^{x-1}\, dx, \quad x > 0,\ \alpha \geqslant 0.$$

(a) By performing tabular integration by parts, express $\gamma(4, \alpha)$ in the simplest form in terms of elementary functions.

(b) Hence show that

$$\int \gamma(4, x)\, dx = e^{-x}(x^4 + 4x^3 + 12x^2 + 24x + 24) + x\gamma(4, x) + C.$$

(c) By using integration by parts, show that the lower incomplete gamma function satisfies the following *functional relation*:

$$\gamma(x + 1, \alpha) = x\gamma(x, \alpha) - e^{-\alpha} \alpha^x.$$

34. (a) By using an appropriate substitution, find $\displaystyle\int \dfrac{dx}{(x + 1)^2}$.

(b) Using the result found in (a) together with integration by parts, find $\displaystyle\int \dfrac{xe^x}{(x + 1)^2}\, dx$.

35. Using the substitution $x = u^2$ followed by tabular integration by parts, find $\displaystyle\int x^2 \sin(\sqrt{x})\, dx$.

36. Find all values of x for which $\displaystyle\int_0^x t^2 \sin(x - t)\, dt = x^2$.

37. If f is a twice-differentiable function on $[0, 1]$ such that $f''(x) + (4x - 2)f'(x) + f(x) = 0$, find the value of $\displaystyle\int_0^1 (x^2 - x) f(x)\, dx$ in terms of $f(0)$ and $f(1)$.

38. (a) Using integration by parts, show that
$$\int_0^1 \frac{x \ln^2(1 + x^2)}{1 + x^2}\, dx = \frac{1}{6} \ln^3(2).$$

(b) Using the substitution $u = \tan x$ together with the result given in (a), deduce that
$$\int_0^{\frac{\pi}{4}} \tan x \ln^2(\cos x)\, dx = \frac{\ln^3(2)}{24}.$$

39. (a) By using an appropriate substitution, find $\displaystyle\int \frac{x}{\sqrt{4 + x^4}}\, dx$.

(b) Using the result found in (a) together with integration by parts, find
$$\int \frac{x^3}{\sqrt{4 + x^2}}\, dx.$$

40. Consider the integral $\displaystyle\int \frac{e^{x^3}}{x}\left(6x^3 \ln x + 2\right) dx$.

(a) By applying the method of integration by parts, show that
$$\int \frac{e^{x^3}}{x}\, dx = e^{x^3} \ln x - \int 3x^2 e^{x^3} \ln x\, dx.$$

(b) Hence deduce that $\displaystyle\int \frac{e^{x^3}}{x}\left(6x^3 \ln x + 2\right) dx = 2e^{x^3} \ln x + C$.

41. Find

(a) $\displaystyle\int x \ln\left(\frac{x + 1}{x}\right) dx$

(b) $\displaystyle\int \frac{x - \sin x \cos x}{\sin^2 x}\, dx$

42. Show that $\dfrac{\pi}{4} - \dfrac{1}{2} < \displaystyle\int_0^1 \frac{\sin^{-1} x}{1 + x^8}\, dx < \dfrac{\pi}{2} - 1$.

43. Given $\displaystyle\int_1^3 \frac{dx}{x^4 \sqrt{1 + x}} = k$, find the value of $\displaystyle\int_1^3 \frac{dx}{x^5 \sqrt{1 + x}}$ in terms of k.

44. Suppose $I = \int \dfrac{x^n}{(1+x)^{n+2}} \, dx$ where $n \in \mathbb{Z}^+$. Using tabular integration by parts, show that

$$I = \frac{1}{n+1} \left[\left(\frac{x}{1+x} \right)^{n+1} - 1 \right] + C.$$

✠ Extension questions and Challenge problems

45. If $I = \displaystyle\int_0^1 (1 - x^{50})^{100} \, dx$ and $J = \displaystyle\int_0^1 (1 - x^{50})^{101} \, dx$, find the value of I/J.

46. Let $u(x)$ and $v(x)$ be n-times continuously differentiable functions where n is a positive integer. Using tabular integration by parts, show that

$$\int u(x) v^{(n)}(x) \, dx = \sum_{k=0}^{n-1} (-1)^k u^{(k)}(x) v^{(n-1-k)}(x)$$

$$+ (-1)^n \int u^{(n)}(x) v(x) \, dx.$$

Here $u^{(k)}(x)$ denotes the kth order derivative of u with respect to x while $u^{(0)}(x) = u(x)$.

47. Find

(a) $\displaystyle\int \frac{x \tan^{-1} x}{(1+x^2)^3} \, dx$ (b) $\displaystyle\int \ln(1+x^2) \tan^{-1} x \, dx$

48. Let $I = \displaystyle\int_\alpha^{\alpha+\beta} \cos^{-1} \left(\frac{\tan \alpha}{\tan x} \right) \sin x \cos x \, dx$.

(a) Using integration by parts, show that

$$I = \frac{1}{2} \sin^2(\alpha + \beta) \cos^{-1} \left[\frac{\tan \alpha}{\tan(\alpha + \beta)} \right]$$

$$- \frac{1}{2} \tan \alpha \int_\alpha^{\alpha+\beta} \frac{\sin x}{\sqrt{1 - (\cos x / \cos \alpha)^2}} \, dx.$$

(b) Hence deduce that

$$I = \frac{1}{2} \sin^2(\alpha + \beta) \cos^{-1} \left[\frac{\tan \alpha}{\tan(\alpha + \beta)} \right] - \frac{1}{2} \sin \alpha \cos^{-1} \left[\frac{\cos(\alpha + \beta)}{\cos \alpha} \right].$$

49. (a) If f satisfies $x = f(x) e^{f(x)}$ for all x, evaluate $\displaystyle\int_0^e f(x) \, dx$.

(b) Suppose that $a = 1$, $b^2 = 1 + \dfrac{\ln 2}{2}$ where $b > 0$ and f is given implicitly by the equation $x^2 + f(x)e^{f(x)} = 1$ such that $f(x) > -1$. Evaluate $\displaystyle\int_a^b xf(x)\,dx$.

50. The *Bessel function* of the first kind of orders zero $J_0(x)$ and one $J_1(x)$ is known to satisfy the following relations:

$$J_0'(x) = -J_1(x) \quad \text{and} \quad J_1'(x) = J_0(x) - \frac{1}{x}J_1(x).$$

Using these relations, where needed, together with integration by parts, show the following results.

(a) $\displaystyle\int J_0(x)J_1(x)\,dx = -\frac{1}{2}J_0^2(x) + C$

(b) $\displaystyle\int J_0(x)\cos x\,dx = x\cos x J_0(x) + x\sin x J_1(x) + C$

(c) $\displaystyle\int xJ_0^2(x)\,dx = \frac{x^2}{2}\left(J_0^2(x) + J_0^2(x)\right) + C$

51. If $\displaystyle\int_0^1 \left[(1 + f(x))x + \int_0^x f(t)\,dt\right] dx = 1$, find the value of $\displaystyle\int_0^1 f(t)\,dt$.

52. Integration by parts is an integration rule corresponding to the product rule for differentiation. In this question we will find an integration rule corresponding to the quotient rule for differentiation.

(a) By the quotient rule, if $f(x)$ and $g(x)$ are differentiable functions, then

$$\frac{d}{dx}\left[\frac{f(x)}{g(x)}\right] = \frac{g(x)f'(x) - f(x)g'(x)}{[g(x)]^2}.$$

Let $u = g(x)$ and $v = f(x)$. By integrating both sides of the quotient rule, show that

$$\int \frac{dv}{u} = \frac{v}{u} + \int \frac{v}{u^2}\,du.$$

This has been called the *integration by parts quotient rule formula*.[3]

(b) Show that the integration by parts quotient rule formula can be obtained from the standard integration by parts formula by considering the integral

$$\int \frac{dv}{u}, \text{ with } U = \frac{1}{u} \text{ and } dV = dv.$$

53. Let f be a differentiable function such that

$$f(x) = -(x^2 - x + 1)e^x + \int_0^x e^{x-u} f'(u) \, du.$$

(a) Show that $f(0) = -1$.

(b) Hence show that $\int_0^x e^{-u} f(u) \, du = x^2 - x$.

(c) Hence verify that $f(x) = e^x(2x + 1)$ is a solution to the integral equation.

54. Consider the integral $\int_a^x f'(t) \, dt$. Clearly,

$$\int_a^x f'(t) \, dt = f(t) \Big|_a^x = f(x) - f(a).$$

(a) By rewriting the integral as

$$\int_a^x f'(t) \, dt = \int_a^x (-f'(t)) \cdot (-1) \, dt,$$

and choosing to differentiate the function $-f'(t)$ and integrate the function -1, after one application of tabular integration by parts we have

D	I
$-f'(t)$	-1
$-f''(t)$	$x - t$

Notice in the second row of the column I, on integrating -1 the term $x - t$ has been written. Normally one would just write $-t$ with the arbitrary constant of integration being set equal to zero. Instead, for reasons that will become apparent in a moment, for the first integration we have chosen to set the constant equal to x.

By continuing the above table, show using tabular integration by parts that

$$f(x) = f(a) + f'(a)(x-a) + \frac{f''(a)}{2!}(x-a)^2$$

$$+ \frac{f''(a)}{3!}(x-a)^3 + \cdots + \frac{f^{(n)}(a)}{n!}(x-a)^2$$

$$+ \frac{1}{n!}\int_a^x (x-t)^n f^{(n+1)}(t)\,dt$$

$$= \sum_{k=0}^n \frac{f^{(k)}(a)}{k!}(x-a)^k + \frac{1}{n!}\int_a^x (x-t)^n f^{(n+1)}(t)\,dt.$$

Here $f^{(n)}$ denotes an nth order derivative while $f^{(0)} = f$.

This result is known as *Taylor's theorem with integral remainder*. It is an important result in the calculus and has many, very important applications.

(b) (i) If $f(x) = x^3$ and $a = 2$, use the result in (a) to show that

$$x^3 = 8 + 12(x-2) + 6(x-2)^2 + (x-2)^3.$$

(ii) Hence find $\displaystyle\int \frac{x^3}{x-2}\,dx$.

(c) If $f(x) = e^x$ and $a = 0$, use the result in (a) to show that

$$e^x = 1 + x + \frac{x^2}{2!} + \frac{x^3}{3!} + \cdots + \frac{x^n}{n!} + \frac{1}{n!}\int_0^x (x-t)^n e^t\,dt.$$

Endnotes

1. The method seems to have been first explicitly stated by Karl W. Folley (1905–1991) in 1947. See: K. W. Folley, 'Integration by parts', *The American Mathematical Monthly*, **54**(9), 542–543 (1947).
2. The paradox is attributed to J. L. Walsh who first proposed it 1927. See: J. L. Walsh', 'A paradox resulting from integration by parts', *The American Mathematical Monthly*, **34**(2), 88 (1927).
3. Jennifer Switkes, 'A quotient rule integration by parts formula', *The College Mathematics Journal*, **36**(1), 58–60 (2005).

8

Trigonometric Integrals

The essence of mathematics is not to make simple things complicated,
but to make complicated things simple.

— *Stan Gudder*

We now focus our attention on integrals involving only trigonometric functions.
We structure this chapter by considering each of the main commonly encountered types. In Chapter 15 not-so-common integrals involving trigonometric
functions will be considered.

§ Powers of Sine or Cosine

Consider integrals of the form

$$\int \cos^n x \, dx \quad \text{or} \quad \int \sin^n x \, dx,$$

where n is a non-negative integer. We will do this for the case of the sine function where $n = 1, 2, 3, 4, 5$. Later, in Chapter 18, we will show how one deals
with the general case when n is a positive integer of any order when *reduction
formulae* are introduced.

Case of $n = 1$: $\int \sin x \, dx = -\cos x + C$

Case of $n = 2$: Using the well-known double-angle formula for the cosine
function, namely $\cos 2x = 1 - 2\sin^2 x$, the sine squared term appearing in the
integral can be written as

$$\int \sin^2 x \, dx = \frac{1}{2} \int (1 - \cos 2x) \, dx = \frac{1}{2} \left(x - \frac{1}{2} \sin 2x \right) + C.$$

An alternative way to integrate $\sin^2 x$ without the need for having to recall an
appropriate double-angle formula uses integration by parts (see Example 7.6
on page 85 of Chapter 7).

Case of $n = 3$: Manipulating the integral we have

$$\int \sin^3 x \, dx = \int \sin x \cdot \sin^2 x \, dx$$

$$= \int \sin x (1 - \cos^2 x) \, dx$$

$$= \int (\sin x - \sin x \cos^2 x) \, dx$$

$$= \int \sin x \, dx - \int \sin x \cos^2 x \, dx$$

$$= -\cos x - \int \sin x \cos^2 x \, dx.$$

In the last integral if we let $x = \cos x$, then $dx = -\sin x \, dx$. Thus

$$\int \sin^3 x \, dx = -\cos x + \int u^2 \, du$$

$$= -\cos x + \frac{1}{3} u^3 + C$$

$$= -\cos x + \frac{1}{3} \cos^3 x + C. \quad \text{(since } u = \cos x\text{)}.$$

Case of $n = 4$: Application of the double-angle formula for cosine involving first a sine squared term followed by a cosine squared term leads to

$$\int \sin^4 x \, dx = \int (\sin^2 x)^2 \, dx$$

$$= \int \left(\frac{1 - \cos 2x}{2} \right)^2 dx, \quad \text{since } \sin^2 x = \frac{1 - \cos 2x}{2}$$

$$= \frac{1}{4} \int (1 - 2 \cos 2x + \cos^2 2x) \, dx$$

$$= \frac{1}{4} \int \left(1 - 2 \cos 2x + \frac{1 + \cos 4x}{2} \right) dx,$$

$$\text{since } \cos^2 2x = \frac{\cos 4x - 1}{2}$$

$$= \frac{1}{4} \int \left(\frac{3}{2} - 2 \cos 2x + \frac{1}{2} \cos 4x \right) dx$$

$$= \frac{1}{4} \left(\frac{3x}{2} - \sin 2x + \frac{1}{8} \sin 4x \right) + C.$$

Case of $n = 5$: Manipulating the integral, we have

$$\int \sin^5 x \, dx = \int \sin x \cdot (\sin^2 x)^2 \, dx = \int \sin x (1 - \cos^2 x)^2 \, dx.$$

Now letting $u = \cos x, du = - \sin x \, dx$. Thus

$$\int \sin^5 x \, dx = - \int (1 - u^2)^2 \, du$$

$$= - \int (1 - 2u^2 + u^4) \, du$$

$$= - \left(u - \frac{2}{3} u^3 + \frac{1}{5} u^5 \right) + C$$

$$= - \left(\cos x - \frac{2}{3} \cos^3 x + \frac{1}{5} \cos^5 x \right) + C \quad (\text{since } u = \cos x).$$

For powers of cosine, similar strategies to those just applied to sine can be used.

§ Products of Powers of Sine and Cosine

Now consider integrals of the form

$$\int \cos^m x \cdot \sin^n x \, dx,$$

where m and n are non-negative integers. In performing such integrals two possible cases arise and are best handled separately: (i) either m or n or both are odd; or (ii) both m and n are even. As we shall see, each requires a different strategy in its solution.

Case (i): Suppose that m is odd. Then the substitution $u = \sin x$ along with the identity $\sin^2 x + \cos^2 x = 1$ can be used.

Example 8.1 Find $\int \cos^3 x \sin^4 x \, dx$.

Solution Note that

$$\int \cos^3 x \sin^4 x \, dx = \int \cos^2 x \sin^4 x \cos x \, dx$$

$$= \int (1 - \sin^2 x) \sin^4 x \cos x \, dx.$$

Now let $u = \sin x$, $du = \cos x\, dx$. Then

$$\int \cos^3 x \sin^4 x\, dx = \int (1 - u^2) u^4\, du = \int (u^4 - u^6)\, du$$

$$= \frac{u^5}{5} - \frac{u^7}{7} + C = \frac{1}{5} \sin^5 x - \frac{1}{7} \sin^7 x + C. \quad \blacktriangleright$$

If n is odd, we use the substitution $u = \cos x$ and follow the same strategy as presented above. If both m and n are odd, then one can use a substitution of either $u = \sin x$ or $u = \cos x$. If m and n are both even, then neither of these substitutions will work, meaning a different approach is needed.

Case (ii): As indicated, the case where m and n are both even requires an entirely different approach. In such cases the identities

$$\cos^2 x = \frac{1 + \cos 2x}{2} \quad \text{and} \quad \sin^2 x = \frac{1 - \cos 2x}{2}$$

are used to change the integral into a sum of integrals of the form

$$\int \cos^k (2x)\, dx, \quad \text{where } k \text{ is a positive integer.}$$

One then repeats the methods of either Case (i) or Case (ii) until each integral in the sum is easy to find.

Example 8.2 Find $\int \sin^2 x \cos^4 x\, dx$.

Solution As both indices are even, we begin by writing

$$\int \sin^2 x \cos^4 x\, dx = \int \sin^2 x (\cos^2 x)^2\, dx$$

$$= \int \left(\frac{1 - \cos 2x}{2} \right) \left(\frac{1 + \cos 2x}{2} \right)^2 dx$$

$$= \frac{1}{8} \int (1 - \cos 2x)(1 + \cos 2x)^2\, dx$$

$$= \frac{1}{8} \int (1 - \cos 2x)(1 + 2 \cos 2x + \cos^2 2x)\, dx$$

$$= \frac{1}{8} \int (1 + \cos 2x - \cos^2 2x - \cos^3 3x)\, dx$$

$$= \frac{x}{8} + \frac{\sin 2x}{16} - \frac{1}{8} \int \cos^2 2x\, dx - \frac{1}{8} \int \cos^3 2x\, dx.$$

Now the first integral is an even power of $\cos 2x$ so

$$\int \cos^2 2x \, dx = \frac{1}{2} \int (1 + \cos 4x) dx = \frac{x}{2} + \frac{\sin 4x}{8} + C_1.$$

The second integral is an odd power of $\cos 2x$. So

$$\int \cos^3 2x \, dx = \int \cos^2 2x \cos 2x \, dx = \frac{1}{2} \int (1 - \sin^2 2x) 2 \cos 2x \, dx,$$

and using the substitution $u = \sin 2x, du = 2 \cos 2x \, dx$, one has

$$\int \cos^3 2x \, dx = \frac{1}{2} \int (1 - u^2) du = \frac{u}{2} - \frac{u^3}{6} + C_2$$

$$= \frac{1}{2} \sin 2x - \frac{1}{6} \sin^3 2x + C_2 \quad (\text{since } u = \sin 2x).$$

Combining all the pieces together yields

$$\int \sin^2 x \cos^4 dx = \frac{x}{8} + \frac{\sin 2x}{16} - \frac{1}{8} \left(\frac{x}{2} + \frac{\sin 4x}{8} + C_1 \right)$$

$$- \frac{1}{8} \left(\frac{\sin 2x}{2} - \frac{\sin^3 2x}{6} + C_2 \right)$$

$$= \frac{x}{16} - \frac{1}{64} \sin 4x + \frac{1}{48} \sin^3 2x + C. \qquad \blacktriangleright$$

§ Products of Multiple Angles of Sine and Cosine

The next class of trigonometric integrals consists of integrals of the form

$$\int \cos mx \sin nx \, dx, \quad \int \cos mx \cos nx \, dx \quad \text{or} \quad \int \sin mx \sin nx \, dx,$$

where m and n are real numbers. Integrals of these forms are often found in applications. In evaluating them one of the following three prosthaphaeresis formulae (product-to-sum identities) is used:

1. $\sin A \cos B = \frac{1}{2}[\sin(A + B) + \sin(A - B)]$

2. $\cos A \cos B = \frac{1}{2}[\cos(A - B) + \cos(A + B)]$

3. $\sin A \sin B = \frac{1}{2}[\cos(A - B) - \cos(A + B)]$

After application of the prosthaphaeresis formulae the resulting integral will consist of single sine and/or cosine terms, which can be readily found.

Example 8.3 Find $\int \cos 5x \cos 3x \, dx$.

Solution Noting that

$$\cos 5x \cos 3x = \frac{1}{2}[\cos(5x - 3x) + \cos(5x + 3x)] = \frac{1}{2}(\cos 2x + \cos 8x),$$

the integral becomes

$$\int \cos 5x \cos 3x \, dx = \frac{1}{2} \int (\cos 2x + \cos 8x) dx = \frac{1}{4} \sin 2x + \frac{1}{16} \sin 8x + C.$$

▶

As an alternative to integrating multiple angles of sine and cosine using prosthaphaeresis formulae, see Exercise 24 on page 98 of Chapter 7.

§ Powers of Tangent or Cotangent

Consider integrals of the form

$$\int \tan^n x \, dx \quad \text{or} \quad \int \cot^n x \, dx,$$

where n is a non-negative integer. We will do this for the case of the tangent function where $n = 1, 2, 3, 4, 5$. The general case when n is a positive integer of any order will be dealt with in Chapter 18 after reduction formulae have been introduced.

Case of $n = 1$: $\int \tan x \, dx = \int \frac{\sin x}{\cos x} \, dx = -\ln |\cos x| + C$.

Case of $n = 2$: For the square of the tangent, the identity $1 + \tan^2 x = \sec^2 x$ is used:

$$\int \tan^2 x \, dx = \int (\sec^2 x - 1) \, dx = \tan x - x + C.$$

Case of $n = 3$: Manipulating the integral we have

$$\int \tan^3 x \, dx = \int \tan x \cdot \tan^2 x \, dx = \int \tan x (\sec^2 x - 1) \, dx$$

$$= \int \tan x \sec^2 x \, dx - \int \tan x \, dx$$

$$= \int \tan x \sec^2 x \, dx + \ln |\cos x| + C.$$

In the first of the integrals, set $u = \tan x, du = \sec^2 x \, dx$; Thus

$$\int \tan^3 x \, dx = \int u \, du + \ln|\cos x| + C = \frac{u^2}{2} + \ln|\cos x| + C$$

$$= \frac{1}{2} \tan^2 x + \ln|\cos x| + C \quad (\text{since } u = \tan x).$$

Case of $n = 4$: Again manipulating the integral we have

$$\int \tan^4 x \, dx = \int \tan^2 x \cdot \tan^2 x \, dx = \int \tan^2 x (\sec^2 x - 1) \, dx$$

$$= \int \tan^2 x \sec^2 x \, dx - \int \tan^2 x \, dx$$

$$= \int \tan^2 x \sec^2 x \, dx - (\tan x - x).$$

In the first integral, set $u = \tan x, du = \sec^2 x \, dx$, thus

$$\int \tan^4 x \, dx = \int u^2 \, du - \tan x + x = \frac{1}{3} u^3 - \tan x + x + C$$

$$= \frac{1}{3} \tan^3 x - \tan x + x + C \quad (\text{since } u = \tan x).$$

Case of $n = 5$: Once more, manipulating the integral we have

$$\int \tan^5 x \, dx = \int \tan^3 x \cdot \tan^2 x \, dx = \int \tan^3 x (\sec^2 x - 1) \, dx$$

$$= \int \tan^3 x \sec^2 x \, dx - \int \tan^3 x \, dx$$

$$= \int \tan^3 x \sec^2 x \, dx - \left(\frac{1}{2} \tan^2 x + \ln|\cos x| \right).$$

In the first integral we again set $u = \tan x, du = \sec^2 x \, dx$. Thus

$$\int \tan^5 x \, dx = \int u^3 \, dx - \frac{1}{2} \tan^2 x - \ln|\cos x|$$

$$= \frac{1}{4} u^4 - \frac{1}{2} \tan^2 x - \ln|\cos x| + C$$

$$= \frac{1}{4} \tan^4 x - \frac{1}{2} \tan^2 x - \ln|\cos x| + C \quad (\text{since } u = \tan x).$$

§ Powers of Secant or Cosecant

Consider integrals of the form

$$\int \sec^n x \, dx \quad \text{or} \quad \int \csc^n x \, dx,$$

where n is a non-negative integer.

For the integral of secant, we can write this as

$$\int \sec x \, dx = \int \sec x \cdot \frac{\sec x + \tan x}{\sec x + \tan x} \, dx = \int \frac{\sec^2 x + \sec x \tan x}{\sec x + \tan x} \, dx,$$

giving

$$\boxed{\int \sec x \, dx = \ln |\sec x + \tan x| + C}$$

For an alternative method for finding this integral, see Exercise 10 on page 60 of Chapter 5. The integral for cosecant is done in a similar way. Here

$$\int \csc x \, dx = \int \csc x \cdot \frac{\csc x + \cot x}{\csc x + \cot x} \, dx$$

$$= \int \frac{\csc^2 x + \csc x \cot x}{\csc x + \cot x} \, dx,$$

giving

$$\boxed{\int \csc x \, dx = -\ln |\csc x + \cot x| + C}$$

The integrals for the secant and cosecant squared can be readily found. Their results are

$$\int \sec^2 x \, dx = \tan x + C \quad \text{and} \quad \int \csc^2 x \, dx = -\cot x + C.$$

Integrals for secant and cosecant for integer powers greater than two can be found using integration by parts in combination with the application of the identities $1 + \tan^2 x = \sec^2 x$ or $1 + \cot^2 x = \csc^2 x$. For example, the integral of $\sec^3 x$ was found in Example 7.8 on page 86 of Chapter 7. As an example of the method, consider the case of the cosecant function raised to the power of four.

Example 8.4 Find $\int \operatorname{cosec}^4 x \, dx$.

Solution We begin by rewriting the integral as follows:

$$\int \operatorname{cosec}^4 x \, dx = \int \operatorname{cosec}^2 x \cdot \operatorname{cosec}^2 x \, dx.$$

Integrating by parts we have

$$\int \operatorname{cosec}^4 x \, dx = -\cot x \operatorname{cosec}^2 x - \int -\cot x \cdot 2 \operatorname{cosec} x \cdot (-\operatorname{cosec} x \cot x) \, dx$$

$$= -\cot x \operatorname{cosec}^2 x - 2 \int \cot^2 x \operatorname{cosec}^2 x \, dx$$

$$= -\cot x \operatorname{cosec}^2 x - 2 \int (\operatorname{cosec}^2 x - 1) \operatorname{cosec}^2 x \, dx,$$

$$(\text{since } 1 + \cot^2 x = \operatorname{cosec}^2 x)$$

$$= -\cot x \operatorname{cosec}^2 x - 2 \int \operatorname{cosec}^4 x \, dx + 2 \int \operatorname{cosec}^2 x \, dx.$$

As the original integral has reappeared we have

$$3 \int \operatorname{cosec}^4 x \, dx = -\cot x \operatorname{cosec}^2 x - 2 \cot x,$$

or finally

$$\int \operatorname{cosec}^4 x \, dx = -\frac{1}{3} \cot x \operatorname{cosec}^2 x - \frac{2}{3} \cot x + C. \qquad \blacktriangleright$$

§ Products of Powers of Secant and Tangent

Suppose now we need to find integrals of the form

$$\int \sec^m x \cdot \tan^n x \, dx,$$

where m and n are non-negative integers. As was the case for integrals consisting of products of powers of sine and cosine, it is best if the problem is divided into the following three separate cases: (i) when m is even; (ii) when m and n are both odd; and (iii) when m is odd and n is even.

Case (i): When m is even. The substitution $u = \tan x$ along with the identity $1 + \tan^2 x = \sec^2 x$ can be used.

Example 8.5 Find $\int \sec^4 x \tan^3 x \, dx$.

Solution As the power for the secant function is even ($m = 4$), we begin by rewriting the integral as

$$\int \sec^4 x \tan^3 x \, dx = \int \sec^2 x \cdot \sec^2 x \cdot \tan^3 x \, dx$$

$$= \int (1 + \tan^2 x) \tan^3 x \cdot \sec^2 x \, dx,$$

$$\text{(since } 1 + \tan^2 x = \sec^2 x)$$

$$= \int (\tan^3 x + \tan^5 x) \cdot \sec^2 x \, dx.$$

On setting $u = \tan x$, $du = \sec^2 x \, dx$ and we see that

$$\int \sec^4 x \tan^3 x \, dx = \int (u^5 + u^3) \, du = \frac{u^6}{6} + \frac{u^4}{4} + C$$

$$= \frac{1}{6} \tan^6 x + \frac{1}{4} \tan^4 x + C \quad \text{(since } u = \tan x\text{).}\quad \blacktriangleright$$

Case (ii): When m and n are both odd. The substitution $u = \sec x$ along with the identity $1 + \tan^2 x = \sec^2 x$ can be used.

Example 8.6 Find $\int \sec^5 x \tan^3 x \, dx$.

Solution As both powers are odd, we begin by rewriting the integral as

$$\int \sec^5 x \tan^3 x \, dx = \int (\sec^4 x \cdot \sec x)(\tan^2 x \cdot \tan x) \, dx$$

$$= \int \sec^4 x \tan^2 x \cdot \sec x \tan x \, dx$$

$$= \int \sec^4 x (\sec^2 x - 1) \cdot \sec x \tan x \, dx,$$

$$\text{(since } 1 + \tan^2 x = \sec^2 x)$$

$$= \int (\sec^6 x - \sec^4 x) \cdot \sec x \tan x \, dx.$$

On setting $u = \sec x, du = \sec x \tan x \, dx$, we see that

$$\int \sec^5 x \tan^3 x \, dx = \int (u^6 - u^4) \, du = \frac{u^7}{7} - \frac{u^5}{5} + C$$

$$= \frac{1}{7} \sec^7 x - \frac{1}{5} \sec^5 x + C \quad (\text{since } u = \sec x). \quad \blacktriangleright$$

Case (iii): When m is odd and n is even. Here we can write $n = 2k$ where $k \in \mathbb{N}$. Then

$$\int \sec^m x \cdot \tan^n x \, dx = \int \sec^m \cdot \tan^{2k} x \, dx = \int \sec^m x (\sec^2 x - 1)^k \, dx,$$

where the identity $1 + \tan^2 x = \sec^2 x$ has been used. Expanding the $(\sec^2 x - 1)^k$ term, after multiplying it by $\sec^m x$, gives a sum of powers of secant, which can then be found.

Example 8.7 Find $\int \sec^3 x \tan^2 x \, dx$.

Solution Rewriting the integral we have

$$\int \sec x \tan^2 x \, dx = \int \sec x (\sec^2 x - 1) \, dx, \quad (\text{since } 1 + \tan^2 x = \sec^2 x)$$

$$= \int (\sec^3 x - \sec x) \, dx = \int \sec^3 x \, dx - \int \sec x \, dx.$$

These two integrals have already been calculated earlier. The results are

$$\int \sec^3 x \, dx = \frac{1}{2} \sec x \tan x + \frac{1}{2} \ln |\sec x + \tan x| + C_1,$$

(see Example 7.8 on page 86 of Chapter 7) and

$$\int \sec x \, dx = \ln |\sec x + \tan x| + C_2.$$

Thus

$$\int \sec x \tan^2 x \, dx = \frac{1}{2} \sec x \tan x - \frac{1}{2} \ln |\sec x + \tan x| + C. \quad \blacktriangleright$$

As integrals of the form $\int \operatorname{cosec}^m x \cot^n x \, dx$ can be handled in a manner analogous to how integrals of the form $\int \sec^m x \tan^n x \, dx$ were handled, they will not be considered here.

Exercises for Chapter 8

✠ **Warm-ups**

1. The answers to the following integrals should be obvious. Write them down.

(a) $\displaystyle\int \frac{dx}{\cos^2 x}$ (b) $\displaystyle\int \frac{dx}{\sin^2 x}$

(c) $\displaystyle\int \frac{\sin x}{\cos^2 x}\, dx$ (d) $\displaystyle\int \frac{\cos x}{\sin^2 x}\, dx$

2. Of the two integrals given below, which one do you think takes the least effort to find? Give a reason for your choice.

$$\int \cos^{50} x \, dx \quad \text{or} \quad \int \cos^{100} x \sin x \, dx.$$

✠ **Practice questions**

3. (a) Let $c_n = \displaystyle\int \cos^n x \, dx$. Find c_n when $n = 1, 2, 3, 4, 5$.

(b) Let $t_n = \displaystyle\int \cot^n x \, dx$. Find t_n when $n = 1, 2, 3, 4, 5$.

4. Find the following trigonometric integrals involving powers of sine and cosine.

(a) $\displaystyle\int \cos^4 x \sin^3 x \, dx$ (b) $\displaystyle\int \sin^6 x \cos^5 x \, dx$

(c) $\displaystyle\int \cos^2 x \sin^2 x \, dx$ (d) $\displaystyle\int \sin^4 x \cos^2 x \, dx$

(e) $\displaystyle\int \cos^3 x \sin^3 x \, dx$ (f) $\displaystyle\int \cos^4 x \sin^3 x \, dx$

5. Find the following trigonometric integrals involving powers of tangent and secant.

(a) $\displaystyle\int \sec x \tan^2 x \, dx$ (b) $\displaystyle\int \sec^2 x \tan^7 x \, dx$

(c) $\displaystyle\int \sec x \tan^4 x \, dx$ (d) $\displaystyle\int \sec^3 x \tan^5 x \, dx$

6. Evaluate

(a) $\displaystyle\int_0^{\frac{\pi}{2}} \cos x \sin^5 x \, dx$

(b) $\displaystyle\int_0^{\frac{\pi}{2}} \cos^4 x \sin^2 x \, dx$

(c) $\displaystyle\int_0^{\frac{\pi}{6}} \sec^3 x \tan^2 x \, dx$

(d) $\displaystyle\int_{\frac{\pi}{6}}^{\frac{\pi}{2}} \operatorname{cosec}^4 x \cot^2 x \, dx$

7. Show that $\displaystyle\int_0^{\pi} (\cos^3 x - \sin^3 x) \cos x \, dx = \dfrac{3\pi}{8}$.

8. (a) Find $\displaystyle\int \operatorname{cosec}^4 x \cot^3 x \, dx$.

(b) Hence evaluate $\displaystyle\int_{\frac{\pi}{4}}^{\frac{\pi}{2}} \operatorname{cosec}^4 x \cot^3 x \, dx$.

9. Find

(a) $\displaystyle\int \frac{\cos^2 x}{\sin x} \, dx$

(b) $\displaystyle\int \frac{\sin^3 x}{\cos^2 x} \, dx$

(c) $\displaystyle\int \frac{\cos^4 x}{\sin^2 x} \, dx$

(d) $\displaystyle\int \frac{\sec^6 x}{\tan^2 x} \, dx$

(e) $\displaystyle\int \frac{\operatorname{cosec}^4 x}{\cot^2 x} \, dx$

(f) $\displaystyle\int \frac{\operatorname{cosec}^3 x}{\cot^2 x} \, dx$

10. Find

(a) $\displaystyle\int \sin 9x \cos 3x \, dx$

(b) $\displaystyle\int \sin 7x \sin 5x \, dx$

(c) $\displaystyle\int \cos 4x \cos 8x \, dx$

(d) $\displaystyle\int \cos x \sin 2x \, dx$

11. (a) Show that $\displaystyle\int_0^{\frac{\pi}{2}} \sin^{2n} x \cos x \, dx = \dfrac{1}{2n+1}$.

(b) By writing $\cos^{2n} x = (1 - \sin^2 x)^n$, show that
$$\int_0^{\frac{\pi}{2}} \cos^{2n+1} x \, dx = \sum_{k=0}^{n} \frac{(-1)^k}{2k+1} \binom{n}{k}.$$

(c) Hence evaluate $\displaystyle\int_0^{\frac{\pi}{2}} \cos^9 x \, dx$.

✠ **Extension Questions and Challenge problems**

12. Function *composition*, that is a function of a function, between two functions f and g can be written as $f(g(x)) = (f \circ g)(x)$.

Suppose that $\displaystyle\int (\underbrace{f \circ f \circ \cdots \circ f}_{n \text{ times}})(\cos x)\, dx$, where $f(x+1) = 8x^4 + 32x^3 + 40x^2 + 16x + 1$.

(a) Show that $f(x) = f[(x-1)+1] = 8x^4 - 8x^2 + 1$, so that $f(\cos \theta) = 8\cos^4 \theta - 8\cos^2 \theta + 1$.

(b) Show that $\cos(4\theta) = 8\cos^4 \theta - 8\cos^2 \theta + 1$.

(c) By induction on n, prove that $(f \circ f \circ \cdots \circ f)(\cos x) = \cos(4^n x)$.

(d) Hence find $\displaystyle\int (f \circ f \circ \cdots \circ f)(\cos x)\, dx$.

9

Hyperbolic Integrals

The principal advantage arising from the use of hyperbolic functions is that they bring to light some curious analogies between the integrals of certain irrational functions.

— W. E. Byerly, *Integral Calculus*

Integrals containing hyperbolic functions proceed largely in an exactly analogous matter to the integration of trigonometric functions. One important difference is that hyperbolic functions are defined in terms of exponentials; this allows for the possibility to reduce such integrals to rational functions of e^x. And as we shall see all hyperbolic functions are rational functions of the exponential function. However, it is often not desirable to reduce an integral containing hyperbolic functions to an integral consisting of a rational function in terms of the exponential function as many of the properties that exist between the hyperbolic functions can be taken advantage of in order to simplify otherwise difficult-looking integrals.

We begin by first reviewing the definition for the hyperbolic functions and some of their associated properties that will prove useful when integrating integrals containing hyperbolic functions.

§ The Hyperbolic Functions

You may recall that trigonometric functions are known as *circular functions* since sine and cosine can be defined as the coordinates of points on the unit circle $x^2 + y^2 = 1$. By analogy, the *hyperbolic functions* are constructed by replacing the unit circle with the right-hand branch of the unit hyperbola $x^2 - y^2 = 1$.

The two most common hyperbolic functions, the hyperbolic cosine and the hyperbolic sine functions are defined as follows:

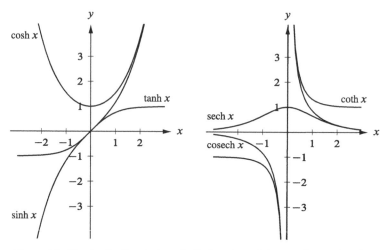

Figure 9.1. Graphs for the six hyperbolic functions sinh x, cosh x, and tanh x (left) and coth x, sech x, and cosech x (right).

Definition 9.1 The *hyperbolic cosine function* cosh : $\mathbb{R} \to \mathbb{R}$ is defined by

$$\cosh x = \frac{e^x + e^{-x}}{2}$$

Definition 9.2 The *hyperbolic sine function* sinh : $\mathbb{R} \to \mathbb{R}$, often pronounced as 'shine', is defined by

$$\sinh x = \frac{e^x - e^{-x}}{2}$$

The remaining four hyperbolic functions, namely the hyperbolic tangent function tanh, the hyperbolic cotangent function coth, the hyperbolic secant function sech, and the hyperbolic cosecant function cosech, are defined in an exactly analogous manner to their trigonometric counterparts. Thus

$$\tanh x = \frac{\sinh x}{\cosh x}, \; \coth x = \frac{1}{\tanh x}, \; \text{sech } x = \frac{1}{\cosh x}, \; \text{cosech } x = \frac{1}{\sinh x}.$$

From the definition for sinh and cosh, each of the four latter hyperbolic functions can be expressed as a rational function in terms of the exponential function. Graphs for all six of the hyperbolic functions are given in Figure 9.1.

§ Hyperbolic Identities

Just as the trigonometric functions have a rich structure in terms of the number of identities that exist between the functions, so too do the hyperbolic functions.

This should come as no surprise given the analogous nature of the two sets of functions.

The most important and famous of all the identities involving hyperbolic functions is the Pythagorean identity

$$\boxed{\cosh^2 x - \sinh^2 x = 1}$$

To prove this result, from the definition for cosh and sinh we have

$$\cosh^2 x - \sinh^2 x = \left(\frac{e^x + e^{-x}}{2}\right)^2 - \left(\frac{e^x - e^{-x}}{2}\right)^2 = 1.$$

As is the case with the trigonometric functions, two other Pythagorean identities for the hyperbolic functions can be found. Factoring out $\cosh^2 x$ and moving this term to the right-hand side gives

$$1 - \tanh^2 x = \operatorname{sech}^2 x,$$

while factoring out $\sinh^2 x$ and moving this term to the right-hand side gives

$$\coth^2 -1 = \operatorname{cosech}^2 x.$$

We now list other important results without proof analogous to those familiar for the trigonometric functions. Each can be verified by appealing to the definition for the hyperbolic functions involved. The sum and difference formulae for sinh, cosh, and tanh are

$$\sinh(x \pm y) = \sinh x \cosh y \pm \cosh x \sinh y$$

$$\cosh(x \pm y) = \cosh x \cosh y \pm \sinh x \sinh y$$

$$\tanh(x \pm y) = \frac{\tanh x \pm \tanh y}{1 \pm \tanh x \tanh y}$$

The double-angle formulae are

$$\sinh(2x) = 2 \sinh x \cosh x$$

$$\cosh(2x) = \cosh^2 x + \sinh^2 x = 2 \cosh^2 x - 1 = 1 + 2 \sinh^2 x$$

$$\tanh(2x) = \frac{2 \tanh x}{1 + \tanh^2 x}$$

while the half-angle formulae are

$$\sinh \frac{x}{2} = \pm \sqrt{\frac{\cosh x - 1}{2}}$$

$$\cosh \frac{x}{2} = \sqrt{\frac{\cosh x + 1}{2}}$$

$$\tanh \frac{x}{2} = \pm \sqrt{\frac{\cosh x - 1}{\cosh x + 1}} = \frac{\sinh x}{\cosh x + 1}.$$

In the half-angle formulae the positive sign is selected when $x \geqslant 0$ while the negative sign is chosen when $x < 0$.

You have perhaps noticed that all of these identities for the hyperbolic functions, at least in appearance, are very similar to those identities found for the trigonometric functions. Given it is more than likely that what one will actually remember are those identities relating to the trigonometric functions, the close analogy between the two sets of identities allows for a useful mnemonic to be introduced whereby the latter hyperbolic identities can be arrived at from the former trigonometric ones. Known as *Osborn's rule*[1] it states that any trigonometric identity can be converted into a corresponding hyperbolic identity by expanding it completely in terms of integral powers of sines and cosines, exchanging trigonometric functions with their hyperbolic counterparts, and then switching the sign of every term that contains a product of $2, 6, 10, 14, \ldots$ hyperbolic sines. As a simple example, from the trigonometric identity

$$\cos(x + y) = \cos x \cos y - \sin x \sin y,$$

application of Osborn's rule gives the corresponding hyperbolic identity

$$\cosh(x + y) = \cosh x \cosh y + \sinh x \sinh y,$$

the sign of the second term on the right being switched as the identity contains the product of two hyperbolic sine terms. In another example, from the trigonometric identity

$$\tan(x + y) = \frac{\tan x + \tan y}{1 - \tan x \tan y},$$

since the term $\tan x \tan y$ consists of a product of sines, namely

$$\tan x \cdot \tan y = \frac{\sin x}{\cos x} \cdot \frac{\sin y}{\cos y},$$

application of Osborn's rule gives

$$\tanh(x + y) = \frac{\tanh x + \tanh y}{1 + \tanh x \tanh y}.$$

§ Derivatives of the Hyperbolic Functions

As cosh and sinh are linear combinations of exponential functions, each is differentiable. So for their derivatives we have,

1. $\dfrac{d}{dx}(\cosh x) = \dfrac{d}{dx}\left(\dfrac{e^x + e^{-x}}{2}\right) = \dfrac{e^x - e^{-x}}{2} = \sinh x,$

2. $\dfrac{d}{dx}(\sinh x) = \dfrac{d}{dx}\left(\dfrac{e^x - e^{-x}}{2}\right) = \dfrac{e^x + e^{-x}}{2} = \cosh x.$

From the derivatives for cosh and sinh, the derivatives for the four other hyperbolic functions quickly follow. They are

3. $\dfrac{d}{dx}(\tanh x) = \dfrac{d}{dx}\left(\dfrac{\sinh x}{\cosh x}\right)$

$\qquad = \dfrac{\cosh x \cdot \cosh x - \sinh x \cdot \sinh x}{\cosh^2 x}$ (by the quotient rule)

$\qquad = \dfrac{\cosh^2 x - \sinh^2 x}{\cosh^2}$

$\qquad = \dfrac{1}{\cosh^2 x}$ (since $\cosh^2 x - \sinh^2 x = 1$)

$\qquad = \operatorname{sech}^2 x.$

4. $\dfrac{d}{dx}(\coth x) = \dfrac{d}{dx}\left(\dfrac{\cosh x}{\sinh x}\right)$

$\qquad = \dfrac{\sinh x \cdot \sinh x - \cosh x \cdot \cosh x}{\sinh^2 x}$ (by the quotient rule)

$\qquad = -\dfrac{\cosh^2 x - \sinh^2 x}{\sinh^2 x}$

$\qquad = -\dfrac{1}{\sinh^2 x}$ (since $\cosh^2 x - \sinh^2 x = 1$)

$\qquad = -\operatorname{cosech}^2 x.$

5. $\dfrac{d}{dx}(\operatorname{sech} x) = \dfrac{d}{dx}\left(\dfrac{1}{\cosh x}\right) = -(\cosh x)^{-2}\sinh x$

$\qquad = -\dfrac{1}{\cosh x} \cdot \dfrac{\sinh x}{\cosh x} = -\operatorname{sech} x \tanh x.$

6. $\dfrac{d}{dx}(\operatorname{cosech} x) = \dfrac{d}{dx}\left(\dfrac{1}{\sinh x}\right) = -(\sinh x)^{-2}\cosh x$

$\qquad = -\dfrac{1}{\sinh x} \cdot \dfrac{\cosh x}{\sinh x} = -\operatorname{cosech} x \coth x.$

§ Standard Integrals Involvng Hyperbolic Functions

From the derivatives for the hyperbolic functions this immediately gives us six standard forms for integrals involving hyperbolic functions. These are summarised in Table 9.1.

In addition to the indefinite integrals already found for cosh and sinh, indefinite integrals for the four remaining hyperbolic functions can be found. Finding these we have

(a) $\displaystyle\int \tanh x\, dx = \int \frac{\sinh x}{\cosh x}\, dx = \ln|\cosh x| + C.$

(b) $\displaystyle\int \coth x\, dx = \int \frac{\cosh x}{\sinh x}\, dx = \ln|\sinh x| + C.$

(c) $\displaystyle\int \operatorname{sech} x\, dx = \int \frac{1}{\cosh x}\, dx = \int \frac{1}{\cosh x} \cdot \frac{\cosh x}{\cosh x}\, dx$

$$= \int \frac{\cosh x}{\cosh^2 x}\, dx = \int \frac{\cosh x}{1 + \sinh^2 x}\, dx,$$

since $\cosh^2 x - \sinh^2 x = 1$. Now let $u = \sinh x,\, du = \cosh x\, dx$. Thus

$$\int \operatorname{sech} x\, dx = \int \frac{du}{1 + u^2} = \tan^{-1} u + C = \tan^{-1}(\sinh x) + C.$$

Table 9.1. *Six standard integrals involving hyperbolic functions.*

Six standard hyperbolic integrals
1. $\displaystyle\int \cosh x\, dx = \sinh x + C$
2. $\displaystyle\int \sinh x\, dx = \cosh x + C$
3. $\displaystyle\int \operatorname{sech}^2 x\, dx = \tanh x + C$
4. $\displaystyle\int \operatorname{cosech}^2 x\, dx = -\coth x + C$
5. $\displaystyle\int \operatorname{sech} x \tanh x\, dx = -\operatorname{sech} x + C$
6. $\displaystyle\int \operatorname{cosech} x \coth x\, dx = -\operatorname{cosech} x + C$

(d) $\displaystyle\int \operatorname{cosech} x \, dx = \int \frac{1}{\sinh x} \, dx = \int \frac{1}{\sinh x} \cdot \frac{\sinh x}{\sinh x} \, dx$

$\displaystyle\qquad = \int \frac{\sinh x}{\sinh^2 x} \, dx = \int \frac{\sinh x}{\cosh^2 x - 1} \, dx,$

since $\cosh^2 x - \sinh^2 x = 1$. Now let $u = \cosh x, du = \sinh x \, dx$. Thus

$$\int \operatorname{sech} x \, dx = \int \frac{du}{u^2 - 1}.$$

By writing

$$\frac{1}{u^2 - 1} = \frac{1}{(u - 1)(u + 1)} = \frac{1}{2(u - 1)} - \frac{1}{2(u + 1)},$$

we have

$$\int \operatorname{cosech} x \, dx = \int \left[\frac{1}{2(u - 1)} - \frac{1}{2(u + 1)} \right] dx$$

$$= \frac{1}{2} \ln |u - 1| - \frac{1}{2} \ln |u + 1| + C$$

$$= \ln \left| \sqrt{\frac{\cosh x - 1}{\cosh x + 1}} \right| + C.$$

And since $\tanh \frac{x}{2} = \pm \sqrt{\frac{\cosh x - 1}{\cosh x + 1}}$, the expression for the indefinite integral for cosech can be further reduced to

$$\int \operatorname{cosech} x \, dx = \ln |\tanh(x/2)| + C.$$

Note that the general method of decomposing a rational function into a partial fraction, as was done in (d), will be taken up in Chapter 11. The indefinite integrals for the six hyperbolic functions are summarised in Table 9.2.

§ Integrals Involving Hyperbolic Functions

For the remainder of the chapter we give examples of integrals containing hyperbolic functions.

Example 9.1 Find $\displaystyle\int \operatorname{sech}^2(5x + 2) \, dx$.

Solution

$$\int \operatorname{sech}^2(5x + 2) \, dx = \frac{1}{5} \tanh(5x + 2) + C. \qquad\blacktriangleright$$

Table 9.2. *Integrals for the six hyperbolic functions.*

Integrals for the six hyperbolic functions
1. $\int \cosh x \, dx = \sinh x + C$
2. $\int \sinh x \, dx = \cosh x + C$
3. $\int \tanh x \, dx = \ln \lvert \cosh x \rvert + C$
4. $\int \coth x \, dx = \ln \lvert \sinh x \rvert + C$
5. $\int \operatorname{sech} x \, dx = \tan^{-1}(\sinh x) + C$
6. $\int \operatorname{cosech} x \, dx = \ln \lvert \tanh(x/2) \rvert + C$

Example 9.2 Find $\int \sinh x \cosh^4 x \, dx$.

Solution Let $u = \cosh x$, $du = \sinh x \, dx$. Thus

$$\int \sinh x \cosh^4 x \, dx = \int u^4 \, du = \frac{1}{5} u^5 + C = \frac{1}{5} \cosh^5 x + C. \quad \blacktriangleright$$

Example 9.3 Find $\int \cosh^2 x \, dx$.

Solution Using the hyperbolic identity $\cosh 2x = 2\cosh^2 x - 1$, we have

$$\int \cosh^2 x \, dx = \frac{1}{2} \int (1 + \cosh 2x) \, dx = \frac{x}{2} + \frac{1}{4} \sinh 2x + C. \quad \blacktriangleright$$

Example 9.4 Find $\int x \sinh 3x \, dx$.

Solution Using integration by parts we have

$$\int x \sinh 3x \, dx = \frac{x}{3} \cosh 3x - \frac{1}{3} \int \cosh 3x \, dx$$

$$= \frac{x}{3} \cosh 3x - \frac{1}{9} \sinh 3x + C. \quad \blacktriangleright$$

Example 9.5 Find $\int e^x \sinh x \, dx$.

Solution As the integrand involves a product between an exponential function and a hyperbolic function, in this case the integral will be more easily found by converting sinh to its exponential form. Doing so yields

$$\int e^x \sinh x \, dx = \int e^x \left(\frac{e^x - e^{-x}}{2} \right) dx = \frac{1}{2} \int (e^{2x} - 1) \, dx$$

$$= \frac{1}{4} e^{2x} - \frac{x}{2} + C. \quad \blacktriangleright$$

Example 9.6 If $a \in \mathbb{R}$ find $\int \dfrac{\cosh(x - a)}{\cosh(x + a)} \, dx$.

Solution In this example the term appearing in the denominator can be removed by taking advantage of the difference formula for cosh. Observing that

$$\cosh(x - a) = \cosh[(x + a) - 2a] = \cosh(x + a) \cosh 2a$$
$$- \sinh(x + a) \sinh 2a,$$

we have

$$\int \frac{\cosh(x - a)}{\cosh(x + a)} \, dx = \int \frac{\cosh(x + a) \cosh 2a - \sinh(x + a) \sinh 2a}{\cosh(x + a)} \, dx$$

$$= \cosh 2a \int dx - \sinh 2a \int \frac{\sinh(x + a)}{\cosh(x + a)} \, dx$$

$$= x \cosh 2a - \sinh 2a \int \tanh(x + a) \, dx$$

$$= x \cosh 2a - \sinh 2a \ln |\cosh(x + a)| + C. \quad \blacktriangleright$$

Example 9.7 Find $\int \dfrac{\cosh x}{2 \cosh^2 x - 1} \, dx$.

Solution With a $\cosh x$ term appearing in the numerator, it being the derivative of $\sinh x$, we try to rewrite the denominator as a function of $\sinh x$ only. From the identity $\cosh^2 x - \sinh^2 x = 1$, the $\cosh^2 x$ term appearing in the denominator can be replaced with $1 + 2 \sinh^2 x$. Doing so yields

$$\int \frac{\cosh x}{2 \cosh^2 x - 1} \, dx = \int \frac{\cosh x}{2(1 + \sinh^2 x) - 1} \, dx = \int \frac{\cosh x}{1 + 2 \sinh^2 x} \, dx.$$

Now letting $u = \sinh x, du = \cosh x \, dx$. Thus

$$\int \frac{\cosh x}{2\cosh^2 x - 1} \, dx = \int \frac{du}{1 + 2u^2} = \frac{1}{\sqrt{2}} \tan^{-1}(\sqrt{2}u) + C$$

$$= \frac{1}{\sqrt{2}} \tan^{-1}(\sqrt{2} \sinh x) + C. \qquad \blacktriangleright$$

Example 9.8 Find $\displaystyle\int \frac{dx}{\tanh x - 1}$.

Solution We begin by manipulating the integrand.

$$\int \frac{dx}{\tanh x - 1} = \int \frac{1}{\tanh x - 1} \cdot \frac{1 + \tanh x}{1 + \tanh x} \, dx = \int \frac{\tanh x + 1}{\tanh^2 x - 1} \, dx.$$

But since $\tanh^2 x + \operatorname{sech}^2 x = 1$ we have

$$\int \frac{dx}{\tanh x - 1} = -\int \frac{\tanh x + 1}{\operatorname{sech}^2 x} \, dx = -\int \left(\tanh x \cdot \frac{1}{\operatorname{sech}^2 x} + \frac{1}{\operatorname{sech}^2 x} \right) dx$$

$$= -\int \left(\frac{\sinh x}{\cosh x} \cdot \cosh^2 x + \cosh^2 x \right) dx$$

$$= -\int \sinh x \cosh x \, dx - \int \cosh^2 x \, dx.$$

From the two hyperbolic identities $\sinh 2x = 2 \sinh x \cosh x$ and $\cosh 2x = 2\cosh^2 x - 1$, the integral can be rewritten as

$$\int \frac{dx}{\tanh x - 1} = -\frac{1}{2} \int \sinh 2x \, dx - \frac{1}{2} \int (\cosh 2x + 1) \, dx$$

$$= -\frac{1}{4} \cosh 2x - \frac{1}{4} \sinh 2x - \frac{x}{2} + C. \qquad \blacktriangleright$$

As an alternative to the 'hyperbolic' method presented for Example 9.8, one could have proceeded by writing tanh in terms of its exponential definition.

In our last example we will introduce as a simple trick one of the hyperbolic identities into the integrand.

Example 9.9 Find $\displaystyle\int \frac{dx}{4 + 13\sinh^2 x} \, dx$.

Solution Taking advantage of the hyperbolic identity $\cosh^2 x - \sinh^2 x = 1$, this identity can be introduced into the integrand as rewriting the 4 appearing in the denominator as $4\cosh^2 x - 4\sinh^2 x$. Thus

$$\int \frac{dx}{4 + 13\sinh^2 x} = \int \frac{dx}{(4\cosh^2 x - 4\sinh^2 x) + 13\sinh^2 x}$$

$$= \int \frac{dx}{4\cosh^2 x + 9\sinh^2 x} = \int \frac{dx}{\cosh^2 x\,(4 + 9\tanh^2 x)}$$

$$= \int \frac{\operatorname{sech}^2 x}{4 + 9\sinh^2 x}\,dx.$$

Let $u = \tanh x$, $du = \operatorname{sech}^2 x\,dx$. Thus

$$\int \frac{dx}{4 + 13\sinh^2 x} = \int \frac{du}{4 + 9u^2} = \frac{1}{9}\int \frac{du}{(\frac{2}{3})^2 + u^2}$$

$$= \frac{1}{6}\tan^{-1}\left(\frac{3u}{2}\right) + C = \frac{1}{6}\tan^{-1}\left(\frac{3\tanh x}{2}\right) + C. \quad \blacktriangleright$$

Exercises for Chapter 9

✠ Warm-ups

1. Find $\displaystyle\int x^2 \sinh x \cosh x \tanh x \operatorname{cosech} x \operatorname{sech} x \coth x\, dx$.

2. Consider the integral $\displaystyle\int e^x \cosh x\, dx$.

   Integrating by parts twice, one obtains

   $$\int e^x \cosh x\, dx = e^x \cosh x - \int e^x \sinh x\, dx$$

   $$= e^x \cosh x - e^x \sinh x + \int e^x \cosh x\, dx.$$

   Thus $0 = e^x(\cosh x - \sinh x) = e^x \cdot e^{-x} = 1$.

   Briefly explain what is wrong with the above reasoning.

✠ Practice questions

3. (a) Show that $\sinh x + \cosh x = e^x$ and $\sinh x - \cosh x = -e^{-x}$.

   (b) Find $\displaystyle\int \frac{dx}{(\sinh x + \cosh x)^2}$ and $\displaystyle\int \frac{dx}{(\sinh x - \cosh x)^2}$.

   (c) Find $\displaystyle\int \frac{\sinh x - \cosh x}{\sinh x + \cosh x}\,dx$.

4. Find

(a) $\displaystyle\int \mathrm{cosech}^2(2x + 5)\, dx$

(b) $\displaystyle\int x \sinh x\, dx$

(c) $\displaystyle\int \frac{\cosh\sqrt{1-x}}{\sqrt{1-x}}\, dx$

(d) $\displaystyle\int \frac{2^{\tanh x}}{\cosh^2 x}\, dx$

(e) $\displaystyle\int \sinh^2 x\, dx$

(f) $\displaystyle\int \cos x \sinh x\, dx$

(g) $\displaystyle\int x\, \mathrm{cosech}^2 x\, dx$

(h) $\displaystyle\int \frac{\sinh x \cosh x}{\sinh^2 x + \cosh^2 x}\, dx$

(i) $\displaystyle\int \sinh 2x \sinh 3x\, dx$

(j) $\displaystyle\int \frac{dx}{\sinh^2 x + \cosh^2 x}$

(k) $\displaystyle\int \frac{dx}{5\cosh^2 x - 1}\, dx$

(l) $\displaystyle\int \frac{1 + \cosh x}{\sinh x}\, dx$

5. Evaluate

(a) $\displaystyle\int_0^{\frac{\ln 3}{3}} \mathrm{sech}^2(3x)\, dx$

(b) $\displaystyle\int_0^{\ln 2} \tanh x\, dx$

(c) $\displaystyle\int_0^1 (1 - x)\cosh x\, dx$

(d) $\displaystyle\int_1^e \sinh(\ln x)\, dx$

(e) $\displaystyle\int_1^e \frac{\cosh(\ln x)}{x}\, dx$

(f) $\displaystyle\int_1^2 e^x \cosh x\, dx$

6. (a) For $x > 0$, show that $\tanh(\ln x) = \dfrac{x^2 - 1}{x^2 + 1}$.

(b) Hence find $\displaystyle\int \tanh(\ln x)\, dx$.

7. If $a^2 \neq b^2$ use integration by parts to show that

$$\int \cosh(ax)\cosh(bx)\, dx = \frac{1}{a^2 - b^2}\big(a\sinh(ax)\cosh(bx)$$
$$- b\sinh(bx)\cosh(ax)\big).$$

8. (a) Show that $3 \sinh x + 4 \cosh x = \frac{7}{2}e^x + \frac{1}{2}e^{-x}$.

    (b) Hence show that

    (i) $\displaystyle\int \frac{dx}{3 \sinh x + 4 \cosh x} = \frac{2}{\sqrt{7}} \tan^{-1}(\sqrt{7}e^x) + C$

    (ii) $\displaystyle\int \frac{dx}{(3 \sinh x + 4 \cosh x)^2} = -\frac{2}{7(7e^{2x} + 1)} + C$

9. If $a \in \mathbb{R}$, find

    (a) $\displaystyle\int \frac{\sinh(x - a)}{\cosh(x + a)}\, dx$ (b) $\displaystyle\int \frac{\sinh(x + a)}{\sinh(x - a)}\, dx$

10. Using tabular integration by parts, find $\displaystyle\int x^4 \cosh(2x)\, dx$.

11. Find

    (a) $\displaystyle\int \tanh^3 x\, dx$ (b) $\displaystyle\int \operatorname{sech}^4 x\, dx$ (c) $\displaystyle\int \cosh^4 x\, dx$

12. (a) Show that $\sinh 3x = 3 \sinh x + 4 \sinh^3 x$.

    (b) Hence find $\displaystyle\int \sinh^3 x\, dx$.

13. Find

    (a) $\displaystyle\int \sinh^3 x \cosh^4 x\, dx$ (b) $\displaystyle\int \sinh^3 x \cosh^5 x\, dx$

    (c) $\displaystyle\int \sinh^2 x \cosh^4 x\, dx$ (d) $\displaystyle\int \operatorname{sech} x \tanh^2 x\, dx$

14. If $n \in \mathbb{N}$, find

    (a) $\displaystyle\int \left(\frac{1 + \tanh x}{1 - \tanh x} \right)^n dx$ (b) $\displaystyle\int \left(\frac{1 + \coth x}{1 - \coth x} \right)^n dx$

15. Find

    (a) $\displaystyle\int \frac{\sinh 2x}{\sinh(\frac{x}{2})}\, dx$ (b) $\displaystyle\int \frac{1 + \sinh x + \cosh x}{1 - \sinh x - \cosh x}\, dx$

16. Consider the integral $\displaystyle\int \frac{dx}{1 - \cosh x}$.

(a) By multiplying the integrand by the factor $\dfrac{1 - \cosh x}{1 - \cosh x}$, before integrating, show that

$$\int \frac{dx}{1 - \cosh x} = \coth\left(\frac{x}{2}\right) + C_1.$$

(b) By first converting the hyperbolic cosine term into its exponential form before integrating, show that

$$\int \frac{dx}{1 - \cosh x} = \frac{2}{e^x - 1} + C_2.$$

(c) The results found in (a) and (b) appear to be different. Show that they are indeed equivalent.

✠ **Extension questions and Challenge problems**

17. Find $\displaystyle\int \cosh x \cosh 2x \cosh 4x \, dx$.

18 (The Gudermannian function). The *Gudermannian function*
$\mathrm{gd} : \mathbb{R} \to (-\frac{\pi}{2}, \frac{\pi}{2})$, named after the German mathematician Christoph Gudermann (1798–1852), relates the circular and hyperbolic functions in the real domain without the need for complex numbers. It is defined by

$$\mathrm{gd}\, x = \int_0^x \mathrm{sech}\, t \, dt.$$

(a) Find $\dfrac{d}{dx}(\mathrm{gd}\, x)$.

(b) Show that $\mathrm{gd}\, x = \tan^{-1}(\sinh x)$. Hence $\tan(\mathrm{gd}\, x) = \sinh x$ and shows how a hyperbolic function (in this case the hyperbolic sine function) is related to a trigonometric function (in this case the tangent function).

(c) Hence deduce that $\mathrm{gd}\, x = 2 \tan^{-1}\left(\tanh \frac{x}{2}\right) = \sin^{-1}(\tanh x)$.

(d) By making a substitution of $u = e^t$ in the definition for gd, show that $\mathrm{gd}\, x = 2 \tan^{-1}(e^x) - \dfrac{\pi}{2}$.

(e) Using the result in (d), find $\displaystyle\lim_{x \to -\infty} \mathrm{gd}\, x$ and $\displaystyle\lim_{x \to \infty} \mathrm{gd}\, x$.

19. In this question we will find the integral $\displaystyle\int \frac{\sinh(2n + 1)x}{\sinh x} \, dx$, where $n \in \mathbb{N}$.

(a) From the sum and difference formulae for $\sinh(x + y)$ and $\sinh(x - y)$, show that $2 \cosh x \sinh y = \sinh(x + y) - \sinh(x - y)$.

(b) Let $f_n(x) = \dfrac{1}{2}D_n(x) = \dfrac{1}{2} + \displaystyle\sum_{k=1}^{n} \cosh kx$.

By multiplying $f_n(x)$ by $2\sinh\frac{x}{2}$ and evaluating the resulting finite sum, show that

$$D_n(x) = 1 + 2\sum_{k=1}^{n} \cosh kx = \frac{\sinh(n + \frac{1}{2})x}{\sinh\left(\frac{x}{2}\right)}.$$

(c) Using the result given in (b), find $\displaystyle\int \frac{\sinh(2n + 1)x}{\sinh x}\, dx$.

(d) Hence find $\displaystyle\int \frac{\sinh 7x}{\sinh x}\, dx$.

Endnote

1. The rule is named after George Osborn who first introduced it in 1902. See: G. Osborn, 'Mnemonic for hyperbolic formulae', *The Mathematical Gazette*, **2**(34), 189 (1902).

10

Trigonometric and Hyperbolic Substitutions

Nature laughs at the difficulties of integration.

— Pierre-Simon Laplace

As we have already seen, using an appropriate substitution to find an integral is a very powerful technique of integration. Unfortunately, no systematic procedure for employing such a method to find a given integral exists. Instead, often a great deal of practice and insight is needed to recognise when to apply a substitution and what its form should be.

Integrals containing square roots of quadratics can be transformed into a sum or difference of squares, which then often yield to either a trigonometric or hyperbolic substitution. That is, integrals of the form

$$\int \mathcal{R}\left(x, \sqrt{ax^2 + bx + c}\right) dx,$$

where $\mathcal{R}$ is a rational function of x, can be reduced to integrals of the form

$$\int \mathcal{R}\left(u, \sqrt{\pm u^2 \pm m^2}\right) du,$$

where $\mathcal{R}$ is now a rational function of the variable u and can be found using one of three trigonometric or hyperbolic substitutions as listed in Table 10.1.

It should be pointed out that the use of either a trigonometric or hyperbolic substitution may not always turn out to be the best substitution to make, but it can at least be tried for integrals of the form listed in Table 10.1. A further complication is to determine which of the two types of substitutions ought to be used. If either will work and will deliver an answer, which of the two types, either trigonometric or hyperbolic, is the easiest or best to use? In general, one tends to use the substitution corresponding to the functions one is most familiar with and comfortable in using. Occasionally one type of substitution may prove

Table 10.1. *Standard types of trigonometric and hyperbolic substitutions used.*

Integral type	Trigonometric substitution	Hyperbolic substitution
1. $\int \mathcal{R}\left(u, \sqrt{m^2 - x^2}\right) dx$	$u = m \sin \theta$	$u = m \tanh t$
2. $\int \mathcal{R}\left(u, \sqrt{m^2 + x^2}\right) dx$	$u = m \tan \theta$	$u = m \sinh t$
3. $\int \mathcal{R}\left(u, \sqrt{x^2 - m^2}\right) dx$	$u = m \sec \theta$	$u = m \cosh t$

to be slightly easier to use, and knowing which is easier is something that can only be gained through experience in working with both substitution types.

We now consider a number of examples where we make use of various trigonometric and hyperbolic substitutions.

Example 10.1 Find $\int \sqrt{4 - x^2}\, dx$.

Solution As the expression in the integrand is of the form of an integral of the first type, we try the trigonometric substitution $x = 2 \sin \theta$. Since $dx = 2 \cos \theta\, d\theta$ we have

$$\int \sqrt{4 - x^2}\, dx = \int \sqrt{4 - 4\sin^2 \theta} \cdot 2 \cos \theta\, d\theta = 2 \int \sqrt{4 \cos^2 \theta} \cdot \cos \theta\, d\theta,$$

where the result $\cos^2 \theta + \sin^2 \theta = 1$ has been used. On simplifying we have

$$\int \sqrt{4 - x^2}\, dx = 4 \int \cos \theta \cdot \cos \theta\, d\theta = 4 \int \cos^2 \theta\, d\theta = 4 \int \frac{1 + \cos 2\theta}{2}\, d\theta$$

$$= 2\theta + \frac{1}{2} \sin 2\theta + C = 2\theta + \sin \theta \cos \theta + C.$$

Of course, the final answer for the indefinite integral needs to be written in terms of x. To find $\cos \theta$ in terms of x, it is easiest if a right-angled triangle containing the angle θ such that $\sin \theta = \frac{x}{2}$ is drawn.

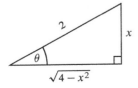

From the triangle we see that

$$\cos \theta = \frac{\sqrt{4 - x^2}}{2}.$$

And as $\theta = \sin^{-1}\left(\dfrac{x}{2}\right)$ we finally have

$$\int \sqrt{4 - x^2}\, dx = 2\sin^{-1}\left(\frac{x}{2}\right) + \frac{x}{2}\sqrt{4 - x^2} + C. \qquad \blacktriangleright$$

Example 10.2 Find $\displaystyle\int \frac{x}{\sqrt{x^2 - 4}}\, dx$.

Solution As the expression in the integrand is of the form of an integral of the third type, we try the hyperbolic substitution $x = 2\cosh t$. Since $dx = 2\sinh t\, dt$ we have

$$\int \frac{x}{\sqrt{x^2 - 4}}\, dx = \int \frac{2\cosh t}{\sqrt{4\cosh^2 t - 4}} \cdot 2\sinh t\, dt$$

$$= 2\int \frac{\cosh t}{\sinh t} \cdot \sinh t\, dt,$$

where the result $\cosh^2 t - \sinh^2 t = 1$ has been used. On simplifying we have

$$\int \frac{x}{\sqrt{x^2 - 4}}\, dx = 2\int \cosh t\, dt = 2\sinh t + C.$$

Writing the indefinite integral in terms of x, as $x = 2\cosh t$, for $\sinh t$ we have

$$\sinh t = \sqrt{\cosh^2 t - 1} = \frac{\sqrt{x^2 - 4}}{2}.$$

Thus

$$\int \frac{x}{\sqrt{x^2 - 4}}\, dx = \sqrt{x^2 - 4} + C. \qquad \blacktriangleright$$

Note that this example is a case where it is not necessary to use either a trigonometric or hyperbolic substitution. Instead, a substitution of $u = x^2 - 4$ could have been used.

Example 10.3 Find $\displaystyle\int \frac{dx}{(x^2 - 1)^{\frac{3}{2}}}$.

Solution As the expression in the integrand is of the form of an integral of the third type, we try the trigonometric substitution $x = \sec\theta$. Since $dx = \sec\theta \tan\theta\, d\theta$ we have

$$\int \frac{dx}{(x^2 - 1)^{\frac{3}{2}}}\, dx = \int \frac{\sec\theta \tan\theta}{(\sec^2\theta - 1)^{\frac{3}{2}}}\, d\theta \int \frac{\sec\theta \tan\theta}{(\tan^2\theta)^{\frac{3}{2}}},$$

where the result $\tan^2 \theta + 1 = \sec^2 \theta$ has been used. On simplifying we have

$$\int \frac{dx}{(x^2-1)^{\frac{3}{2}}}\,dx = \int \frac{\sec \theta \tan \theta}{\tan^3 \theta}\,dx = \int \frac{\sec \theta}{\tan^2 \theta}\,d\theta$$

$$= \int \frac{1}{\cos \theta} \cdot \frac{\cos^2 \theta}{\sin^2 \theta}\,d\theta = \int \frac{\cos \theta}{\sin^2 \theta}\,d\theta.$$

Now let $u = \sin \theta, du = \cos \theta\, d\theta$, which gives

$$\int \frac{dx}{(x^2-1)^{\frac{3}{2}}}\,dx = \frac{du}{u^2} = -\frac{1}{u} + C = -\frac{1}{\sin \theta} + C \quad \text{(since } u = \sin \theta\text{)}.$$

To find $\sin \theta$ in terms of x, we once again draw a right-angled triangle containing the angle θ such that $\sec \theta = x$.

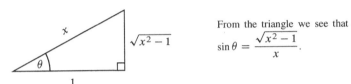

From the triangle we see that

$$\sin \theta = \frac{\sqrt{x^2-1}}{x}.$$

Thus $\displaystyle\int \frac{dx}{(x^2-1)^{\frac{3}{2}}}\,dx = -\frac{x}{\sqrt{x^2-1}} + C.$ ▶

Example 10.4 Find $\displaystyle\int \frac{\sqrt{x^2+9}}{x}.$

Solution As the expression in the integrand is of the form of an integral of the second type, we try the trigonometric substitution $x = 3 \tan \theta$. Since $dx = 3 \sec^2 \theta\, d\theta$ we have

$$\int \frac{\sqrt{x^2+9}}{x}\,dx = \int \frac{\sqrt{9 \tan^2 \theta + 9}}{3 \tan \theta} 3 \sec^2 \theta\, d\theta$$

$$= 3 \int \frac{\sec \theta}{\tan \theta} \sec^2 \theta = 3 \int \frac{\sec^3 \theta}{\tan \theta}\,dx.$$

Rewriting the secant term in terms of cosine, and the tangent term in terms of sine and cosine, one has

$$\int \frac{\sqrt{x^2+9}}{x}\,dx = 3 \int \frac{1}{\cos^3 \theta} \cdot \frac{\cos \theta}{\sin \theta}\,d\theta = 3 \int \frac{1}{\cos^2 \theta \sin \theta}\,d\theta.$$

If in the numerator of the last integral we write unity as $\cos^2 \theta + \sin^2 \theta$, the integral becomes

$$\int \frac{\sqrt{x^2 + 9}}{x} \, dx = 3 \int \frac{\cos^2 \theta + \sin^2 \theta}{\cos^2 \theta \sin \theta} \, d\theta$$

$$= 3 \int \frac{\cos^2 \theta}{\cos^2 \theta \sin \theta} \, d\theta + 3 \int \frac{\sin^2 \theta}{\cos^2 \theta \sin \theta} \, d\theta$$

$$= 3 \int \frac{d\theta}{\sin \theta} + 3 \int \frac{\sin \theta}{\cos^2 \theta} \, d\theta$$

$$= 3 \int \operatorname{cosec} \theta \, d\theta + 3 \int \frac{\sin \theta}{\cos^2 \theta} \, d\theta.$$

The first integral is well known. The result is

$$\int \operatorname{cosec} \theta \, d\theta = -\ln |\operatorname{cosec} \theta + \cot \theta| + C_1.$$

In the second integral, let $u = \cos \theta, du = -\sin \theta \, d\theta$. So

$$\int \frac{\sin \theta}{\cos^2 \theta} \, d\theta = -\int \frac{du}{u^2} = \frac{1}{u} + C_2 = \frac{1}{\cos \theta} + C_2 \quad (\text{since } u = \cos \theta).$$

Thus

$$\int \frac{\sqrt{x^2 + 9}}{x} \, dx = -3 \ln |\operatorname{cosec} \theta + \cot \theta| + \frac{3}{\cos \theta} + C.$$

As $\tan \theta = x/3, \cot \theta = 3/x$, to find $\operatorname{cosec} \theta$ in terms of x, we once again draw a right-angled triangle containing the angle θ such that $\tan \theta = \frac{x}{3}$.

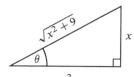

From the triangle we see that

$$\operatorname{cosec} \theta = \frac{1}{\sin \theta} = \frac{\sqrt{x^2 + 9}}{x}.$$

Thus $\int \dfrac{\sqrt{x^2 + 9}}{x} \, dx = \sqrt{x^2 + 9} - 3 \ln \left| \dfrac{3 + \sqrt{x^2 + 9}}{x} \right| + C.$ ▶

Example 10.5 Find $\displaystyle\int \frac{dx}{(x + 1)^2 \sqrt{x^2 + 2x + 5}}.$

Solution We begin by first completing the square for the quadratic term appearing under the root. Here

$$x^2 + 2x + 5 = (x + 1)^2 + 4.$$

So the integral can be written as

$$\int \frac{dx}{(x+1)^2\sqrt{x^2+2x+5}} = \int \frac{dx}{(x+1)^2\sqrt{(x+1)^2+4}}.$$

As the expression in the integrand is now of the form of an integral of the second type (in this case $u = x + 1$), we try the trigonometric substitution $x + 1 = 2\tan\theta$. Thus $dx = 2\sec^2\theta\,d\theta$ and we have

$$\int \frac{dx}{(x+1)^2\sqrt{x^2+2x+5}} = \int \frac{2\sec^2\theta}{(2\tan\theta)^2\sqrt{4\tan^2\theta+4}}\,d\theta$$

$$= \frac{1}{4}\int \frac{\sec^2\theta}{\tan^2\theta\sec\theta}\,d\theta = \frac{1}{4}\int \frac{\sec\theta}{\tan^2\theta}\,d\theta$$

$$= \frac{1}{4}\int \frac{\cos\theta}{\sin^2\theta}\,d\theta = -\frac{1}{4}\frac{1}{\sin\theta} + C.$$

To find $\sin\theta$ in terms of x, we once again draw a right-angled triangle containing the angle θ such that $\tan\theta = \dfrac{x+1}{2}$.

From the triangle we see that
$$\sin\theta = \frac{x+1}{\sqrt{x^2+2x+5}}.$$

Thus $\displaystyle\int \frac{dx}{(x+1)^2\sqrt{x^2+2x+5}} = -\frac{\sqrt{x^2+2x+5}}{4(x+1)} + C.$ ▶

Exercises for Chapter 10

✠ **Warm-ups**

1. Consider the integral $\displaystyle\int_0^1 \sqrt{1-x^2}\,dx$.

The integral can be found using either the substitution $x = \sin u$ or $x = \tanh u$. It is, however, far easier to find the value of the integral by sketching the region in the plane bounded by the curve, the x-axis, and the lines $x = 0$ and $x = 1$ and finding the corresponding area. Find the value of the integral using this latter geometric approach.

✠ Practice questions

2. By applying a suitable trigonometric or hyperbolic substitution, find the following integrals.

(a) $\displaystyle\int \sqrt{25 + x^2}\, dx$ (b) $\displaystyle\int \sqrt{x^2 - 3}\, dx$ (c) $\displaystyle\int \frac{dx}{(1 - x^2)^{\frac{3}{2}}}$

(d) $\displaystyle\int \frac{x^2}{\sqrt{9 - x^2}}\, dx$ (e) $\displaystyle\int \frac{\sqrt{1 + x^2}}{x^2}\, dx$ (f) $\displaystyle\int \frac{dx}{x\sqrt{x^2 - 1}}$

(g) $\displaystyle\int \frac{\sqrt{4 - x^2}}{x}\, dx$ (h) $\displaystyle\int \frac{x}{\sqrt{x^2 - 1}}\, dx$ (i) $\displaystyle\int \frac{dx}{x^2\sqrt{4 - x^2}}$

(j) $\displaystyle\int \frac{x^2}{(1 + x^2)^2}\, dx$ (k) $\displaystyle\int \frac{x + 8}{\sqrt{64 - x^2}}\, dx$ (l) $\displaystyle\int \frac{\sqrt{x^2 - 1}}{x^2 + x}\, dx$

(m) $\displaystyle\int \frac{x^3}{(x^2 + 1)^{\frac{3}{2}}}\, dx$ (n) $\displaystyle\int \frac{x^3}{\sqrt{x^2 - 1}}$ (o) $\displaystyle\int \frac{x^3}{\sqrt{x^2 + 4}}\, dx$

3. Evaluate

(a) $\displaystyle\int_1^{\sqrt{2}} \frac{dx}{x^2\sqrt{x^2 + 8}}\, dx$ (b) $\displaystyle\int_1^2 (x^2 - 1)\sqrt{4 - x^2}\, dx$

(c) $\displaystyle\int_0^1 \frac{dx}{(x + 1)\sqrt{1 + x^2}}$ (d) $\displaystyle\int_0^1 \frac{x^2}{(x^2 + 1)^3}\, dx$

4. By completing the square first before using an appropriate trigonometric or hyperbolic substitution, find the following integrals.

(a) $\displaystyle\int \frac{x + 3}{\sqrt{5 - 4x - x^2}}\, dx$ (b) $\displaystyle\int x\sqrt{6x - x^2 - 8}\, dx$

(c) $\displaystyle\int \frac{dx}{(2x^2 + 4x - 2)^{\frac{3}{2}}}$ (d) $\displaystyle\int \frac{x^2}{\sqrt{x^2 - 2x + 3}}\, dx$

5. Using the indicated hyperbolic substitution, find the following integrals.

(a) $\displaystyle\int \frac{x^3}{\sqrt{x^2 - 16}}\, dx$ $(x = 4\cosh t)$

(b) $\int x^3 \sqrt{16x^2 + 9}\, dx \quad (x = \frac{3}{4}\sinh t)$

(c) $\int \dfrac{dx}{x\sqrt{4 - x^2}} \quad (x = 2\tanh t)$

6. Using the indicated trigonometric substitution, find the following integrals.

(a) $\int \dfrac{dx}{(1 + x^2)\sqrt{1 + x^2}} \quad (x = \tan\theta)$

(b) $\int \dfrac{dx}{(x + 1)\sqrt{x^2 + 1}} \quad (x = \tan\theta)$

(c) $\int \dfrac{dx}{(2 - x)\sqrt{1 - x^2}} \quad (x = \sin\theta)$

(d) $\int \dfrac{x\ln x}{\sqrt{1 - x^2}}\, dx \quad (x = \sin\theta)$

7. If $-1 < a < 1$, use the substitution $x = \sin\theta$ to show that

$$\int \frac{dx}{(1 + ax)\sqrt{1 - x^2}} = \frac{2}{\sqrt{1 - a^2}}$$

$$\times \tan^{-1}\left[\frac{1}{\sqrt{1 - a^2}}\left(\frac{x}{1 + \sqrt{1 - x^2}} + a\right)\right] + C.$$

8. Consider the integral $\int \dfrac{dx}{(2 + x^2)^2}$.

(a) Find the integral using the hyperbolic substitution $x = \sqrt{2}\sinh u$.
(b) Find the integral using the trigonometric substitution $x = \sqrt{2}\tan\theta$.
(c) Of the two substitutions, which do you consider made it slightly easier to find the integral?

9. Using the substitution $x = \sin\theta$, evaluate

(a) $\int_{-1}^{1} \tan^{-1}\left(\sqrt{\dfrac{1 - x}{1 + x}}\right) \dfrac{dx}{\sqrt{1 - x^2}}$

(b) $\int_{-1}^{1} \dfrac{dx}{(1 + x^2)\sqrt{1 - x^2}}$

10. Using the substitution $x = \tan\theta$, find $\int \dfrac{\sqrt{1 + x^2}}{x^4}\ln\left(1 + \dfrac{1}{x^2}\right) dx.$

✠ **Extension questions and Challenge problems**

11 (Gunther's hyperbolic substitutions). This question develops a method for finding integrals of the form

$$\int \cos^m x \sin^n x \, dx,$$

where m and n are integers and $m + n$ is an *odd negative* integer employing one of two different hyperbolic substitutions.[1]

Depending on the form for this trigonometric integral, one of the following two hyperbolic substitutions can be made.

TYPE I: Let $\tan x = \sinh u$. Then $d(\tan x) = \sec^2 x \, dx = \cosh u \, du$, and

$$\sec x = \sqrt{1 + \tan^2 x} = \sqrt{1 + \sinh^2 u} = \cosh u,$$

for $-\pi/2 < x < \pi/2$. Also, since $e^u = \cosh u + \sinh u$ it follows that $u = \ln(\cosh u + \sinh u) = \ln|\sec x + \tan x|$. As the substitution is one-to-one on the interval $(-\pi/2, \pi/2)$, it is only valid on this interval. The substitution will, however, produce the correct antiderivative outside of this interval provided absolute values with logarithms are used.

TYPE II: Let $\cot x = \sinh u$. Then $d(\cot x) = -\operatorname{cosec}^2 x \, dx = \cosh u \, du$, and

$$\operatorname{cosec} x = \sqrt{1 + \cot^2 x} = \sqrt{1 + \sinh^2 u} = \cosh u,$$

for $0 < x < \pi/2$. Also, since $e^u = \cosh u + \sinh u$ it follows that $u = \ln(\cosh u + \sinh u) = \ln|\operatorname{cosec} x + \cot x|$. As the substitution is one-to-one on the interval $(0, \pi/2)$, it is valid only on this interval. The substitution will, however, again produce the correct antiderivative outside of this interval provided absolute values with logarithms are used.

As an example of the method, let us find the integral $\int \sec x \, dx$. We begin by noting that

$$\int \sec x \, dx = \int (\cos x)^{-1} \, dx,$$

where $m = -1$ and $n = 0$. As $m + n = -1$ is a negative odd integer, a Gunther hyperbolic substitution may be used. Write the integral as

$$\int \sec x \, dx = \int \sec x \cdot \frac{\sec x}{\sec x} \, dx = \int \frac{\sec^2 x}{\sec x} \, dx = \int \frac{d(\tan x)}{\sec x}.$$

Let $\tan x = \sinh u$, $d(\tan x) = \cosh u$, and $\sec x = \sinh u$. Thus

$$\int \sec x \, dx = \int \frac{\cosh u}{\cosh u} \, du = \int du = u + C$$

$$= \ln(\cosh u + \sinh u) + C = \ln|\tan x + \sec x| + C,$$

since $\sinh u = \tan x$ and $\cosh u = \sec x$, as expected.

Using a Gunther hyperbolic substitution, find the following integrals.

(a) $\displaystyle\int \sec^3 x \, dx$ (b) $\displaystyle\int \tan x \sec^3 x \, dx$

(c) $\displaystyle\int \tan^2 x \sec x \, dx$ (d) $\displaystyle\int \tan^2 x \sec^3 x \, dx$

(e) $\displaystyle\int \csc x \, dx$ (f) $\displaystyle\int \cot^2 x \csc x \, dx$

(g) $\displaystyle\int \frac{dx}{\sin^2 x \cos x}$ (h) $\displaystyle\int \frac{dx}{\cos^2 x \sin^3 x}$

12. A special function known as the *complete elliptic integral of the first kind* is defined by

$$\mathcal{K}(k) = \int_0^{\frac{\pi}{2}} \frac{du}{\sqrt{1 - k^2 \sin^2 u}}, \quad 0 \leqslant k < 1.$$

(a) Find the value of $\mathcal{K}(0)$.

(b) Consider the following substitution of $\tan u = \dfrac{\sin 2\theta}{k + k \cos 2\theta}$.

For this substitution it can be shown that the limits of integration change in the following manner. When $u = 0$, $\theta = 0$, and when $u = \frac{\pi}{2}$, $\theta = \frac{\pi}{2}$.

Using this substitution, show that $\mathcal{K}(k) = \dfrac{2}{1+k} \mathcal{K}\left(\dfrac{2\sqrt{k}}{1+k}\right)$.

Endnote

1. The method is so named after the American mathematician Charles O. Gunther who first popularised it in his book, *Integration by Trigonometric and Imaginary Substitutions* (D. van Nostrand, New York, 1907). See also: William K. Viertel, 'Use of hyperbolic substitution for certain trigonometric integrals', *Mathematics Magazine*, **38**(3), 141–144 (1965).

11

Integrating Rational Functions by Partial Fraction Decomposition

Unlike simple arithmetic, to solve a calculus problem – and in particular to perform integration – you have to be smart about which integration technique should be used: integration by partial fractions, integration by parts, and so on.

— *Marvin Minsky*

In this section we show how every rational function has a primitive among the elementary functions. And what is truly amazing, there is a systematic way of finding the primitive. We begin by recalling that a *rational function* f is a function that can be written of the form

$$f(x) = \frac{P(x)}{Q(x)},$$

where $P(x)$ and $Q(x)$ are polynomials in x. The function f is said to be *proper* if $\deg P < \deg Q$. Here $\deg P$ is the degree of the polynomial appearing in the numerator, while $\deg Q$ is the degree of the polynomial appearing in the denominator. If, on the other hand, $\deg P \geqslant \deg Q$ the rational function is said to be *improper*. If f is improper, synthetic division or polynomial long division can be used to write the rational function as

$$f(x) = A(x) + \frac{P(x)}{Q(x)},$$

where $A(x)$ is a polynomial and $P(x)/Q(x)$ is a rational function that is now proper.

We now consider the general problem of how to integrate any rational function that is proper. In Appendix A we show how every rational function that is proper can be decomposed into a partial fraction of the form

$$\frac{f(x)}{g(x)} = \sum_{i=1}^{m} \sum_{j=1}^{k_i} \frac{A_{ij}}{(a_i x + b_i)^j} + \sum_{i=1}^{n} \sum_{j=1}^{s_i} \frac{B_{ij} x + C_{ij}}{(\alpha_i x^2 + \beta_i x + \gamma_i)^j}.$$

Any partial fraction decomposition then consists of a sum of terms of the form

$$\frac{A}{(ax + b)^m},$$

and

$$\frac{Bx + C}{(\alpha x^2 + \beta x + \gamma)^n} = \frac{Bx}{(\alpha x^2 + \beta x + \gamma)^n} + \frac{C}{(\alpha x^2 + \beta x + \gamma)^n}.$$

So the integral of any rational function that is proper reduces to finding integrals of the following three forms:

$$\int \frac{A}{(ax + b)^m} dx, \quad \int \frac{Bx}{(\alpha x^2 + \beta x + \gamma)^n} dx, \quad \text{and} \quad \int \frac{C}{(\alpha x^2 + \beta x + \gamma)^n} dx.$$

By completing the square, using a substitution, or performing simple algebraic manipulations, each of these integrals can be put into one of the following standard forms (see Exercise 31 on page 260 of Chapter 18 for details):

$$\int x^n \, dx = \frac{x^{n+1}}{n+1} + C, \ n \neq -1, \quad \int \frac{g'(x)}{g(x)} dx = \ln |g(x)| + C,$$

or

$$\int \frac{dx}{x^2 + a^2} = \frac{1}{a} \tan^{-1} \left(\frac{x}{a} \right) + C,$$

which, as we can see, can all be readily integrated.

The integral of a rational function therefore yields rational functions, and logarithmic and inverse tangent functions. The first is what we call the *rational part* of the integral, while the second is known as the *transcendental* parts of the integral.[1]

Recall that for any rational function that is proper, only one of four separate cases need be considered to find a partial fraction decomposition. If you are not familiar with this topic it is highly recommended you read through Appendix A before proceeding any further.

Example 11.1 Find $\displaystyle\int \frac{x}{(x-2)(2x+7)}dx.$

Solution Denote the integrand by $f(x)$. As it is a proper rational function we begin by finding the partial fraction decomposition for f. Since the denominator consists of two distinct linear factors, the form of its partial fraction decomposition is

$$\frac{x}{(x-2)(2x+7)} = \frac{A}{x-2} + \frac{B}{2x+7}.$$

The two unknown coefficients can be quickly found by applying the Heaviside cover-up method (see Appendix A starting on page 321). When this is done one finds

$$\frac{x}{(x-2)(2x+7)} = \frac{\frac{2}{11}}{x-2} + \frac{\frac{7}{11}}{2x+7}.$$

To integrate f, instead of being required to integrate the rational function directly, we need only integrate each term appearing in the partial fraction decomposition separately. Thus

$$\int \frac{x}{(x-2)(2x+7)}dx = \frac{2}{11}\int \frac{dx}{x-2} + \frac{7}{22}\int \frac{2}{2x+7}dx$$

$$= \frac{2}{11}\ln|x-2| + \frac{7}{22}\ln|2x+7| + C. \quad \blacktriangleright$$

Example 11.2 Find $\displaystyle\int \frac{x^3-4x-1}{x(x-1)^3}dx.$

Solution Denote the integrand by $f(x)$. As it is a proper rational function we begin by finding the partial fraction decomposition for f. Since the denominator consists of two distinct linear factors, one of which is repeated, the form of its partial fraction decomposition is

$$\frac{x^3-4x-1}{x(x-1)^3} = \frac{A}{x} + \frac{B}{x-1} + \frac{C}{(x-1)^2} + \frac{D}{(x-1)^3}.$$

Clearing the denominator by multiplying throughout by $x(x-1)^3$ and collecting like terms in x we arrive at

$$x^3-4x-1 = (A+B)x^3 + (-3A-2B+C)x^2 + (3A+B-C+D)x - A.$$

After equating equal coefficients for x and solving, one finds

$$A = 1, B = 0, C = 3, D = -4.$$

So the partial fraction decomposition for the rational function is

$$\frac{x^3 - 4x - 1}{x(x-1)^3} = \frac{1}{x} + \frac{3}{(x-1)^2} - \frac{4}{(x-1)^3}.$$

Thus

$$\int \frac{x^3 - 4x - 1}{x(x-1)^3} dx = \int \frac{dx}{x} + 3\int \frac{dx}{(x-1)^2} - 4\int \frac{dx}{(x-1)^3}$$

$$= \ln|x| - \frac{3}{x-1} + \frac{2}{(x-1)^2} + C. \qquad \blacktriangleright$$

Example 11.3 Find $\displaystyle\int \frac{7}{(x+2)(x^2+3)} dx.$

Solution The integrand is the partial fraction considered in Example A.4 on page 318 of Appendix A where it was found that

$$\frac{7}{(x+2)(x^2+3)} = \frac{1}{x+2} - \frac{x-2}{x^2+3}.$$

On rewriting the integral we have

$$\int \frac{7}{(x+2)(x^2+3)} dx = \int \frac{dx}{x+2} - \int \frac{x-2}{x^2+3} dx$$

$$= \int \frac{dx}{x+2} - \frac{1}{2}\int \frac{2x}{x^2+3} dx - 2\int \frac{dx}{x^2+3}$$

$$= \ln|x+2| - \frac{1}{2}\ln|x^2+3| - \frac{2}{\sqrt{3}}\tan^{-1}\left(\frac{x}{\sqrt{3}}\right) + C.$$

$\qquad\qquad\qquad\qquad\qquad\qquad\qquad\qquad\qquad\qquad\qquad\qquad \blacktriangleright$

Example 11.4 Find $\displaystyle\int \frac{2x^4 - 5x^3 - 3x - 1}{(x-2)(x^2+1)} dx.$

Solution Denote the integrand by

$$f(x) = \frac{2x^4 - 5x^3 - 3x - 1}{(x-2)(x^2+1)}.$$

As the degree of the numerator is greater than the degree of the denominator, the rational function is improper. Performing polynomial long division first we have

$$
\begin{array}{r}
2x - 1 \\
x^3 - 2x^2 + x - 2\,)\;\overline{\;2x^4 - 5x^3 - 3x - 1\;} \\
\underline{-\,2x^4 + 4x^3 - 2x^2 + 4x} \\
-x^3 - 2x^2 + x - 1 \\
\underline{x^3 - 2x^2 + x - 2} \\
-4x^2 + 2x - 3.
\end{array}
$$

f can therefore be rewritten as

$$
f(x) = 2x - 1 - \frac{4x^2 - 2x + 3}{(x-2)(x^2+1)}.
$$

For the remaining rational function, which is now proper, as the denominator contains one distinct linear factor and one distinct irreducible quadratic factor, its partial fraction decomposition will be of the form

$$
\frac{4x^2 - 2x + 3}{(x-2)(x^2+1)} = \frac{A}{x-2} + \frac{Bx+C}{x^2+1},
$$

where $A, B,$ and C are three unknown constants to be determined. To find the constants, clearing the denominator by multiplying throughout by $(x-2)(x^2+1)$ yields

$$
4x^2 - 2x + 3 = A(x^2+1) + Bx(x-2) + C(x-2).
$$

Substituting obvious values for x the constants $A, B,$ and C can be readily found.

$$
x = 2:\; 4(2)^2 - 2(2) + 3 = A(2^2 + 1) \quad \Rightarrow A = 3
$$

$$
x = 0:\; \qquad\qquad\qquad 3 = A - 2C \quad \Rightarrow C = \tfrac{1}{2}A - \tfrac{3}{2} = 0
$$

$$
x = 1:\; 4(1)^2 - 2(1) + 3 = 2A - B - C \;\Rightarrow B = 2A - C - 5 = 1
$$

The complete partial fraction decomposition for f is given by

$$
\frac{2x^4 - 5x^3 - 3x - 1}{(x-2)(x^2+1)} = 2x - 1 - \frac{3}{x-2} - \frac{x}{x^2+1}.
$$

So the integral becomes

$$
\int \frac{2x^4 - 5x^3 - 3x - 1}{(x-2)(x^2+1)}\,dx = \int (2x-1)\,dx - 3\int \frac{dx}{x-2} - \frac{1}{2}\int \frac{2x}{x^2+1}\,dx
$$

$$
= x^2 - x - \ln|x-2| - \frac{1}{2}\ln|x^2+1| + C. \quad \blacktriangleright
$$

§ Ostrogradsky's Method

One of the greatest difficulties in integrating rational functions comes from having to find a factorisation for the denominator of the rational function. A method that finds the rational part of

$$\int \frac{P(x)}{Q(x)} dx,$$

without having to find a factorisation for $Q(x)$ and without having to decompose the integrand into partial fractions, is known as *Ostrogradsky's method.*[2] If $P(x)$ and $Q(x)$ are polynomials such that $\deg P < \deg Q$ then

$$\int \frac{P(x)}{Q(x)} dx = \frac{P_1(x)}{Q_1(x)} + \int \frac{P_2(x)}{Q_2(x)} dx. \qquad (*)$$

Here $Q_1(x)$ is the greatest common factor of $Q(x)$ and its derivative $Q'(x)$, while $Q_2(x) = Q(x)/Q_1(x)$. The polynomials $P_1(x)$ and $P_2(x)$ have degrees that are one less than the degrees of polynomials $Q_1(x)$ and $Q_2(x)$, respectively.

Example 11.5 Use Ostrogradsky's method to find $\int \dfrac{4x^3 - 3x^2 - 2}{(x^3 + 1)} dx.$

Solution Here $Q(x) = (x^3 + 1)^2$, $Q'(x) = 6x^2(x^3 + 1)$ so $Q_1(x) = x^3 + 1$ is the greatest common factor for $Q(x)$ and $Q'(x)$. Also

$$Q_2(x) = \frac{Q(x)}{Q_1(x)} = \frac{(x^3 + 1)^2}{x^3 + 1} = x^3 + 1.$$

In addition, the degree of $P_1(x)$ will be at most 2, while the degree of $P_2(x)$ will also be at most 2. So for this particular example we can write $(*)$ as

$$\int \frac{4x^3 - 3x^2 - 2}{(x^3 + 1)^2} dx = \frac{Ax^2 + Bx + C}{x^3 + 1} + \int \frac{Dx^2 + Ex + F}{x^3 + 1} dx.$$

Here $A, B, \ldots, F$ are six unknown constants to be determined. Finding these constants, and differentiating the above integral with respect to x, by the first fundamental theorem of calculus we have

$$\frac{4x^3 - 3x^2 - 2}{(x^3 + 1)^2} = \frac{d}{dx}\left(\frac{Ax^2 + Bx + C}{x^3 + 1}\right) + \frac{Dx^2 + Ex + F}{x^3 + 1}$$

$$= \frac{(2Ax + B)(x^3 + 1) - 3x^2(Ax^2 + Bx + C)}{(x^3 + 1)^2}$$

$$+ \frac{Dx^2 + Ex + F}{x^3 + 1}.$$

Multiplying through by $(x^3 + 1)^2$ and collecting terms with equal coefficients for x, one has

$$4x^3 - 3x^2 - 2 = Dx^5 + (E - A)x^4 + (F - 2B)x^3 + (D - 3C)x^2$$
$$+ (2A + E)x + (B + F).$$

After equating equal coefficient for x we find

$$A = 0, B = -2, C = 1, D = 0, E = 0, F = 0.$$

So the integral we seek can now be expressed as

$$\int \frac{4x^3 - 3x^2 - 2}{(x^3 + 1)^2} dx = -\frac{2x - 1}{x^3 + 1} + \int \frac{0}{x^3 + 1} dx = -\frac{2x - 1}{x^3 + 1} + C. \quad \blacktriangleright$$

Exercises for Chapter 11

✠ Warm-ups

1. Even if the integrand of an integral is a rational function, a partial fraction decomposition may not always be necessary nor desirable in order to find the integral. Integrate the following rational functions using methods other than performing a partial fraction decomposition.

(a) $\displaystyle\int \frac{2x + 1}{x^2 + x + 1} dx$ (b) $\displaystyle\int \frac{x}{x + 3} dx$ (c) $\displaystyle\int \frac{x + 7}{1 + x^2} dx$

✠ Practice questions

2. (a) Find real numbers A, B, and C such that
$$\frac{13}{(x + 2)(x^2 + 9)} = \frac{A}{x + 2} + \frac{Bx + C}{x^2 + 4}.$$
(b) Hence find $\displaystyle\int \frac{13}{(x + 2)(x^2 + 9)} dx.$

3. By employing a partial fraction decomposition first, find the following integrals.

(a) $\displaystyle\int \frac{x - 9}{(x + 5)(x - 2)} dx$ (b) $\displaystyle\int \frac{2x + 3}{(x + 1)^2} dx$

(c) $\displaystyle\int \frac{10}{(x - 1)(x^2 + 9)} dx$ (d) $\displaystyle\int \frac{x^2 - 2x - 3}{x^3 + 2x^2 + x + 2} dx$

(e) $\displaystyle\int \frac{x^4}{x^4 - 1}\, dx$

(f) $\displaystyle\int \frac{x^3 + 5}{x^2 + x}\, dx$

(g) $\displaystyle\int \frac{x^2 + x + 1}{x^2 - x + 1}\, dx$

(h) $\displaystyle\int \frac{dx}{x^4 + 1}$

(i) $\displaystyle\int \frac{5}{(x^2 + 4)(x^2 + 9)}\, dx$

(j) $\displaystyle\int \frac{x^3 + 36}{x^2 + 36}\, dx$

4. Find

(a) $\displaystyle\int \frac{x^4 - 5x^3 + 12x^2 - 21x + 35}{x^3 - 3x^2 + 4x - 12}\, dx$

(b) $\displaystyle\int \frac{dx}{(x + 1)(x + 2)(x + 3)(x + 4)}$

(c) $\displaystyle\int \frac{x^4}{(1 - x)(1 - 2x)(1 - 3x)(1 - 4x)(1 - 5x)}\, dx$

5. If $\displaystyle\int_0^2 \frac{2}{x^2 + 9x + 20}\, dx = \ln k$, find the exact value of k.

6. By first performing the substitution $u^2 = 1 + e^x$, find $\displaystyle\int \frac{dx}{\sqrt{1 + e^x}}$.

7. By first performing integration by parts followed by a substitution, find the following integrals.

(a) $\displaystyle\int \ln(1 + \sqrt[3]{x})\, dx$

(b) $\displaystyle\int \ln(x + \sqrt[3]{x})\, dx$

8. (a) By first showing that $\dfrac{1}{2} \leqslant \dfrac{1}{1 + x} \leqslant 1$ for $0 \leqslant x \leqslant 1$, deduce that

$$\frac{1}{2}\int_0^1 x^3(1 - x)^3\, dx < \int_0^1 \frac{x^3(1 - x)^3}{1 + x}\, dx < \int_0^1 x^3(1 - x)^3\, dx.$$

(b) Hence deduce that $\dfrac{1552}{2240} < \ln 2 < \dfrac{1553}{2240}$.

9. (a) Show that $\displaystyle\int_0^1 \frac{x^4(1-x)^4}{1+x^2}\,dx = \frac{22}{7} - \pi.$

(b) Hence deduce that $\dfrac{22}{7} > \pi.$

(c) By first showing that $\dfrac{1}{2} \leqslant \dfrac{1}{1+x^2} \leqslant 1$ for $0 \leqslant x \leqslant 1$, deduce that

$$\frac{x^4(1-x)^4}{2} \leqslant \frac{x^4(1-x)^4}{1+x^2} \leqslant x^4(1-x)^4.$$

(d) Hence show that $\dfrac{22}{7} - \dfrac{1}{630} < \pi < \dfrac{22}{7} - \dfrac{1}{1260}.$

10. If $\displaystyle j(x) = \int_1^x \frac{\ln t}{1+t}\,dt,\ x > 0$, show that $j(x) + j\left(\dfrac{1}{x}\right) = \dfrac{1}{2}\ln^2 x.$

11. Rather than performing synthetic division or polynomial long division, by manipulating the integrand instead, find $\displaystyle\int \frac{x^2+1}{x+1}\,dx.$

12. By adding and subtracting the term x^2 to the numerator, find $\displaystyle\int \frac{x^4+1}{x^6+1}\,dx$ without having to use partial fractions.

13. Writing x^4 as $[1 - (1-x)]^4$, by expanding this term using the binomial theorem, show that this can be used to find $\displaystyle\int \frac{x^4}{(1-x)^3}\,dx$ without the need to use partial fractions.

14. Use Ostrogradsky's method to find the following integrals.

(a) $\displaystyle\int \frac{x^2}{(x^2+2x+2)^2}\,dx$

(b) $\displaystyle\int \frac{dx}{(x^2+1)^4}$

(c) $\displaystyle\int \frac{dx}{(x^3-1)^2}$

(d) $\displaystyle\int \frac{5x^3+3x-1}{(x^3+3x+1)^3}\,dx$

15. If f is a continuous function such that $f\left(\dfrac{3x-4}{3x+4}\right) = x + 2$, find $\displaystyle\int f(x)\,dx.$

16. Using the substitution $u = \sinh x$, find $\displaystyle\int \frac{dx}{\cosh x + \coth x}$.

⚬ **Extension questions and Challenge problems**

17. For a rational function of the form

$$\frac{1}{[(x-a)(x-b)]^n},$$

where n is a positive integer greater than one, repeated application of the identity

$$\frac{1}{(x-a)(x-b)} = \frac{1}{b-a}\left(\frac{1}{x-a} - \frac{1}{x-b}\right),$$

may be preferred compared to the standard partial fraction decomposition method for repeated linear factors, particularly when the order of the index is low.

(a) As an example of this method, use it to show

$$\frac{1}{(x^2-1)^2} = \left[\frac{1}{(x-1)(x+1)}\right]^2$$

$$= \frac{1}{4}\left[\frac{1}{(x-1)^2} - \frac{1}{x-1} + \frac{1}{x+1} + \frac{1}{(x+1)^2}\right].$$

(b) Hence find $\displaystyle\int \frac{dx}{(x^2-1)^2}$.

18. In Exercise 44 of Chapter 6 an Euler substitution of the third type was used to find an integral of the form $\displaystyle\int \mathcal{R}\left(x, \sqrt{ax^2+bx+c}\right)\,dx$, where $\mathcal{R}$ is a rational function of x and $\sqrt{ax^2+bx+c}$. In this question we consider an Euler substitution of the first type.

If the polynomial ax^2+bx+c is such that $a>0$, one may choose a substitution of the form $\sqrt{ax^2+bx+c} = t \pm \sqrt{a}x$. Note here that either sign may be selected.

Using an Euler substitution of the first type, find $\displaystyle\int \frac{dx}{x+\sqrt{x^2+2x+2}}$.

19. A class of special functions known as the *polylogarithmic function* $\mathrm{Li}_s(x)$ can be defined recursively as follows:

$$\mathrm{Li}_{s+1}(x) = \int_0^x \frac{\mathrm{Li}_s(u)}{u}\,du.$$

Here s, known as the *order* of the polylogarithm, is any real number.

(a) Given that $Li_{-2}(x) = x(1 + x)/(1 - x)^3$, use the recursive definition for the polylogarithms to find expressions for $Li_{-1}(x)$ and $Li_0(x)$, and to show that

$$Li_1(x) = -\ln(1 - x).$$

As $Li_1(x)$ is related to the natural logarithm, the polylogarithms can be thought of as one of the simplest generalisations of the natural logarithmic function.

(b) Show that when $s = 2$ we have

$$Li_2(x) = -\int_0^x \frac{\ln(1 - u)}{u} du.$$

This is the so-called *dilogarithm function*. Unlike the integrals that were used to evaluate $Li_{-1}(x), Li_0(x)$, and $Li_1(x)$, this integral cannot be found in elementary terms.

20. Consider the integral $\int \frac{p(x)}{(x + 2)^6}$. Here $p(x)$ is a polynomial of degree at most 5.

(a) Since $p(x)$ is a polynomial of degree at most 5, the form for the partial fraction decomposition for the rational function appearing in the integrand will be

$$\frac{p(x)}{(x + 2)^6} = \frac{A_1}{x + 2} + \frac{A_2}{(x + 2)^2} + \frac{A_3}{(x + 2)^3} + \frac{A_4}{(x + 2)^4}$$
$$+ \frac{A_5}{(x + 5)^5} + \frac{A_6}{(x + 2)^6}.$$

Here $A_1, \ldots, A_6$ are constants. After multiplying through by $(x + 2)^6$ one obtains

$$p(x) = A_1(x + 2)^5 + A_2(x + 2)^4 + A_3(x + 2)^3$$
$$+ A_4(x + 2)^2 + A_5(x + 2) + A_6. \tag{†}$$

Show that on equating equal coefficients for x on both sides of the above identity one obtains

$$x^5 : A_1$$

$$x^4 : \binom{5}{1} 2^1 A_1 + A_2$$

$$x^3 : \binom{5}{2} 2^2 A_1 + \binom{4}{1} 2^1 A_2 + A_3$$

$$x^2 : \binom{5}{3} 2^3 A_1 + \binom{4}{2} 2^2 A_3 + \binom{3}{1} 2^1 A_3 + A_4$$

$$x^1 : \binom{5}{4} 2^4 A_1 + \binom{4}{3} 2^3 A_2 + \binom{3}{2} 2^2 A_3 + \binom{2}{1} 2^1 A_4 + A_5$$

$$x^0 : \binom{5}{5} 2^5 A_1 + \binom{4}{4} 2^2 A_2 + \binom{3}{3} 2^3 A_3 + \binom{2}{2} 2^2 A_4$$

$$+ \binom{1}{1} 2^1 A_5 + A_6.$$

(b) If the degree of $p(x)$ is small, the linear system for A_i ($1 \leqslant i \leqslant 6$) found in (a) is relatively simple to solve.

If $p(x) = 5x + 3$, use (a) to find $\int \dfrac{5x + 3}{(x + 2)^6} dx$.

(c) If, on the other hand, $p(x)$ is large (recall p is a polynomial of degree at most 5), then the work required to solve for the six unknown coefficients is considerable. As an alternative to solving the linear system of equations, if $x = -2$ is substituted into (†) one immediately finds

$$A_6 = p(-2).$$

The question now is whether the other five unknown coefficients can be found in a similar, simple way. The answer is yes. If one differentiates (†), followed by substituting $x = -2$, one finds

$$A_5 = p'(-2).$$

By repeating the process of differentiating followed by substituting $x = -2$, show that the four remaining unknown coefficients can be expressed as

$$A_4 = \frac{p^{(2)}(-2)}{2!}, \quad A_3 = \frac{p^{(3)}(-2)}{3!}, \quad A_2 = \frac{p^{(4)}(-2)}{4!},$$

$$A_1 = \frac{p^{(5)}(-2)}{5!}.$$

(d) If $p(x) = x^5 + 2x^4 + 3x + 1$ use (c) to find

$$\int \frac{x^5 + 2x^4 + 3x + 1}{(x + 2)^6} dx.$$

Endnotes

1. Recall that a transcendental function is a function that does not satisfy a polynomial equation, in contrast to an *algebraic* function. A transcendental function is therefore said to 'transcend' algebra in that it cannot be expressed in terms of a finite sequence of the algebraic operations of addition, multiplication, and root extraction. Examples of transcendental functions include the exponential and trigonometric functions and their associated inverse functions.

2. The method is named after the Ukrainian mathematician, mechanician, and physicist Mikhail Ostrogradsky (1801–1862) who first formulated the method in 1845. The method is sometimes also known as the *Ostrogradsky–Hermite* method after the French mathematician Charles Hermite (1822–1901) who independently rediscovered the method in 1872.

12

Six Useful Integrals

Why, sometimes I've believed as many as six impossible things before breakfast.

— Lewis Carroll, *Alice in Wonderland*

We now give six useful integrals with quadratics appearing in their denominators that are of the following two types:

$$\int \frac{dx}{ax^2 + bx + c} \quad \text{or} \quad \int \frac{dx}{\sqrt{ax^2 + bx + c}}.$$

Here a, b, and c take on various values from the reals with $a \neq 0$. The six useful integrals we wish to consider here in standard form are listed in Table 12.1.

It should be noted the first of these integrals may be written as

$$\int \frac{dx}{\sqrt{a^2 - x^2}} = \sin^{-1}\left(\frac{x}{a}\right) + C_1 = -\cos^{-1}\left(\frac{x}{a}\right) + C_2,$$

as can be readily verified by differentiating the expression containing the inverse cosine function. The second and third integrals can be found using trigonometric substitutions as follows. In the second integral, let $x = a \sec \theta, dx = a \sec \theta \tan \theta \, d\theta$. Thus

$$\int \frac{dx}{\sqrt{x^2 - a^2}} = \int \frac{a \sec \theta \tan \theta}{\sqrt{a^2 \sec^2 \theta - a^2}} \, d\theta = \int \frac{\sec \theta \tan \theta}{\sqrt{\sec^2 \theta - 1}} \, d\theta$$

$$= \int \sec \theta \, d\theta = \ln |\sec \theta + \tan \theta| + C_1$$

$$= \ln \left| \frac{x}{a} + \frac{\sqrt{x^2 - a^2}}{a} \right| + C_1$$

$$= \ln \left| x + \sqrt{x^2 - a^2} \right| - \ln a + C_1$$

$$= \ln \left| x + \sqrt{x^2 - a^2} \right| + C,$$

157

Table 12.1. *Six particularly useful integrals.*

Six useful integrals

1. $\displaystyle\int \frac{dx}{\sqrt{a^2 - x^2}} = \sin^{-1}\left(\frac{x}{a}\right), \quad a > 0, \; |x| < a$

2. $\displaystyle\int \frac{dx}{\sqrt{x^2 + a^2}} = \ln\left|x + \sqrt{x^2 + a^2}\right|, \quad a > 0$

3. $\displaystyle\int \frac{dx}{\sqrt{x^2 - a^2}} = \ln\left|x + \sqrt{x^2 - a^2}\right|, \quad 0 < a < x$

4. $\displaystyle\int \frac{dx}{a^2 + x^2} = \frac{1}{a}\tan^{-1}\left(\frac{x}{a}\right), \quad a \neq 0$

5. $\displaystyle\int \frac{dx}{a^2 - x^2} = \frac{1}{2a}\ln\left|\frac{a + x}{a - x}\right|, \quad a > 0, \; |x| < a$

6. $\displaystyle\int \frac{dx}{x^2 - a^2} = \frac{1}{2a}\ln\left|\frac{x - a}{x + a}\right|, \quad a > 0, \quad |x| > a$

as required. Here the constant term of $\ln a$ has been absorbed into the constant of integration. For the third integral a substitution of $x = a\tan\theta$ can be used, but as it is very similar to what we have just done for the second integral, it will not be given here. Finally, the fifth and sixth integrals can be found using a partial fraction decomposition.

We now consider some examples that make use of these six useful integrals. The general strategy is to bring the integrand into a form of one of the six useful integrals found in Table 12.1.

Example 12.1 Find $\displaystyle\int \frac{dx}{\sqrt{4 + 9x^2}}$.

Solution Writing the integrand in standard form before integrating, we have

$$\int \frac{dx}{\sqrt{4 + 9x^2}} = \int \frac{dx}{\sqrt{9\left(\frac{4}{9} + x^2\right)}} = \frac{1}{3}\int \frac{dx}{\sqrt{(2/3)^2 + x^2}}$$

$$= \frac{1}{3}\ln\left|x + \sqrt{\frac{4}{9} + x^2}\right| + C_1 = \frac{1}{3}\ln\left|3x + \sqrt{4 + 9x^2}\right| + C.$$

▶

The real utility of these six useful forms is for any integral containing a quadratic term in its denominator that is of the type

$$\int \frac{dx}{ax^2 + bx + c} \quad \text{or} \quad \int \frac{dx}{\sqrt{ax^2 + bx + c}},$$

where $a, b, c \in \mathbb{R}$ such that $a \neq 0$ can be put into standard form by first 'completing the square'. The method was already touched upon in one of the examples considered in Chapter 10 (see Example 10.5 on page 138). We now consider this method more fully in the examples that follow.

Example 12.2 By completing the square, find $\displaystyle\int \frac{dx}{\sqrt{x(2-x)}}$.

Solution First, consider the quadratic term that appears in the denominator. Completing the square we have

$$x(2 - x) = 2x - x^2 = -(x^2 - 2x) = -[(x-1)^2 - 1] = 1 - (x-1)^2.$$

So for the integral, we can write it as

$$\int \frac{dx}{\sqrt{x(2-x)}} = \int \frac{dx}{\sqrt{1 - (x-1)^2}} = \sin^{-1}(x-1) + C. \qquad \blacktriangleright$$

Example 12.3 Find $\displaystyle\int \frac{dx}{\sqrt{(x+4)(x+2)}}$.

Solution First, consider the quadratic term that appears in the denominator. Completing the square we have

$$(x + 4)(x + 2) = x^2 + 6x + 8 = (x+3)^2 - 9 + 8 = (x+3)^2 - 1.$$

So for the integral it can be written as

$$\int \frac{dx}{\sqrt{(x+4)(x+2)}} = \int \frac{dx}{\sqrt{(x+3)^2 - 1}} = \ln\left|x + 3 + \sqrt{(x+3)^2 - 1}\right| + C$$

$$= \ln\left|x + 3 + \sqrt{x^2 + 6x + 8}\right| + C. \qquad \blacktriangleright$$

Example 12.4 Find $\displaystyle\int \frac{dx}{\sqrt{4x^2 - 4x + 5}}$.

Solution For the quadratic term that appears in the denominator, on completing the square we have

$$4x^2 - 4x + 5 = 4(x^2 - x) + 5 = 4\left[\left(x - \frac{1}{2}\right)^2 - \frac{1}{4}\right] + 5 = (2x-1)^2 + 4.$$

So for the integral one has

$$\int \frac{dx}{\sqrt{4x^2 - 4x + 5}} = \int \frac{dx}{\sqrt{(2x - 1)^2 + 4}}.$$

Now let $u = 2x - 1$ so $du = 2\,dx$. Thus

$$\int \frac{dx}{\sqrt{4x^2 - 4x + 5}} = \frac{1}{2}\int \frac{dx}{\sqrt{u^2 + 2^2}} = \frac{1}{2}\ln\left|u + \sqrt{u^2 + 4}\right| + C$$

$$= \frac{1}{2}\ln\left|2x - 1 + \sqrt{(2x - 1)^2 + 4}\right| + C$$

$$= \frac{1}{2}\ln\left|2x - 1 + \sqrt{4x^2 - 4x + 5}\right| + C.$$
▶

Example 12.5 Find $\displaystyle\int \frac{dx}{9x^2 + 12x + 11}$.

Solution Again for the quadratic term that appears in the denominator, on completing the square one has

$$9x^2 + 12x + 11 = 9\left(x^2 + \frac{4}{3}x\right) + 11 = 9\left[\left(x + \frac{2}{3}\right)^2 - \frac{4}{9}\right] + 11$$

$$= (3x + 2)^2 + 7.$$

So for the integral, it can be written as

$$\int \frac{dx}{9x^2 + 12x + 11} = \int \frac{dx}{(3x + 2)^2 + 7}.$$

Now let $u = 3x + 2$ so $du = 3\,dx$. Thus

$$\int \frac{dx}{\sqrt{9x^2 + 12x + 11}} = \frac{1}{3}\int \frac{du}{\sqrt{u^2 + (\sqrt{7})^2}} = \frac{1}{3\sqrt{7}}\tan^{-1}\left(\frac{u}{\sqrt{7}}\right) + C$$

$$= \frac{1}{3\sqrt{7}}\tan^{-1}\left(\frac{3x + 2}{\sqrt{7}}\right) + C.$$
▶

The six useful integrals in standard form can readily be extended to integrals that contain a linear factor in the numerator, namely, forms of the type:

$$\int \frac{Ax + B}{ax^2 + bx + c}\,dx \quad \text{or} \quad \int \frac{Ax + B}{\sqrt{ax^2 + bx + c}}\,dx,$$

with $a, A \neq 0$. In this case the linear term $Ax + B$ is written in the form $\alpha(2ax + b) + \beta$, where α and β are two constants to be found. The idea here is to make the numerator equal to the derivative of the quadratic term that appears in the denominator. Doing so leads to two integrals. The first can be found using a substitution while the second reduces to a standard form for one of the six useful integrals. We now consider a number of examples that demonstrate this case.

Example 12.6 Find $\int \dfrac{5x - 2}{5 + 2x - x^2} \, dx$.

Solution We begin by rewriting the integral as

$$\int \frac{5x - 2}{5 + 2x - x^2} \, dx = -\frac{5}{2} \int \frac{\frac{4}{5} - 2x}{5 + 2x - x^2} \, dx = -\frac{5}{2} \int \frac{(2 - 2x) - \frac{6}{5}}{5 + 2x - x^2} \, dx$$

$$= -\frac{5}{2} \int \frac{2 - 2x}{5 + 2x - x^2} \, dx + 3 \int \frac{dx}{5 + 2x - x^2}$$

$$= -\frac{5}{2} I_1 + 3 I_2.$$

In the first integral, set $u = 5 + 2x - x^2$, $du = (2 - 2x) \, dx$, so that

$$I_1 = \int \frac{du}{u} = \ln |u| + C_1 = \ln |5 + 2x - x^2| + C_1,$$

since $u = 5 + 2x - x^2$. The second integral can be reduced to a standard form corresponding to one of the six useful integrals as follows.

$$5 + 2x - x^2 = -[x^2 - 2x - 5] = -[(x - 1)^2 - 1 - 5] = 6 - (x - 1)^2.$$

Thus

$$I_2 = \int \frac{dx}{5 + 2x - x^2} = \int \frac{dx}{6 - (x - 1)^2} = \int \frac{dx}{(\sqrt{6})^2 - (x - 1)^2}$$

$$= \frac{1}{2\sqrt{6}} \ln \left| \frac{\sqrt{6} + (x - 1)}{\sqrt{6} - (x - 1)} \right| + C_2 = \frac{1}{2\sqrt{6}} \ln \left| \frac{\sqrt{6} + x - 1}{\sqrt{6} - x + 1} \right| + C_2.$$

So for the original integral one has

$$\int \frac{5x - 2}{5 + 2x - x^2} \, dx = -\frac{5}{2} \ln |5 + 2x - x^2| + \frac{3}{2\sqrt{6}} \ln \left| \frac{\sqrt{6} + x - 1}{\sqrt{6} - x + 1} \right| + C. \blacktriangleright$$

Example 12.7 Find $\int \dfrac{3x + 4}{\sqrt{x^2 + 2x + 5}} \, dx$.

Solution We begin by rewriting the integral as

$$\int \frac{3x + 4}{\sqrt{x^2 + 2x + 5}} \, dx = \frac{3}{2} \int \frac{2x + \frac{8}{3}}{\sqrt{x^2 + 2x + 5}} \, dx = \frac{3}{2} \int \frac{(2x + 2) + \frac{2}{3}}{\sqrt{x^2 + 2x + 5}} \, dx$$

$$= \frac{3}{2} \int \frac{2x + 2}{\sqrt{x^2 + 2x + 5}} \, dx + \int \frac{dx}{\sqrt{x^2 + 2x + 5}}$$

$$= I_1 + I_2.$$

In the first integral, set $u = x^2 + 2x + 5, du = (2x + 2)\, dx$, so that

$$I_1 = \frac{3}{2} \int \frac{du}{\sqrt{u}} = 3\sqrt{u} + C_1 = 3\sqrt{x^2 + 2x + 5} + C_1.$$

In the second integral it can be reduced to a standard form for one of the six useful integrals as follows. By first completing the square in the denominator,

$$x^2 + 2x + 5 = (x + 1)^2 - 1 + 5 = (x + 1)^2 + 4.$$

Thus

$$I_2 = \int \frac{dx}{\sqrt{x^2 + 2x + 5}} = \int \frac{dx}{\sqrt{(x + 1)^2 + 2^2}}$$

$$= \ln \left| x + 1 + \sqrt{2^2 + (x + 1)^2} \right| + C_2$$

$$= \ln \left| x + 1 + \sqrt{x^2 + 2x + 5} \right| + C_2.$$

So for the integral we started with, one has

$$\int \frac{3x + 4}{\sqrt{x^2 + 2x + 5}}\, dx = 3\sqrt{x^2 + 2x + 5}$$

$$+ \ln \left| x + 1 + \sqrt{x^2 + 2x + 5} \right| + C. \qquad \blacktriangleright$$

Example 12.8 Find $\displaystyle \int \sqrt{\frac{4 - x}{2 + x}}\, dx$.

Solution Rationalising the numerator first, we have

$$\int \sqrt{\frac{4 - x}{2 + x}}\, dx = \int \sqrt{\frac{4 - x}{2 + x}} \cdot \sqrt{\frac{4 - x}{4 - x}}\, dx = \int \frac{4 - x}{\sqrt{(2 + x)(4 - x)}}\, dx$$

$$= \int \frac{4 - x}{\sqrt{8 + 2x - x^2}}.$$

Now rearranging the numerator in the integral we have

$$\int \sqrt{\frac{4 - x}{2 + x}}\, dx = \frac{1}{2} \int \frac{8 - 2x}{\sqrt{8 + 2x - x^2}}\, dx = \frac{1}{2} \int \frac{(2 - 2x) + 6}{\sqrt{8 + 2x - x^2}}\, dx$$

$$= \frac{1}{2} \int \frac{2 - 2x}{\sqrt{8 + 2x - x^2}}\, dx + 3 \int \frac{dx}{\sqrt{8 + 2x - x^2}}$$

$$= I_1 + I_2.$$

In the first integral, setting $u = 8 + 2x - x^2, du = (2 - 2x) dx$ we have

$$I_1 = \frac{1}{2} \int \frac{du}{\sqrt{u}} = \sqrt{u} + C_1 = \sqrt{8 + 2x - x^2} + C_1,$$

since $u = 8 + 2x - x^2$. In the second integral it can be reduced to a standard form for one of the six useful integrals by first completing the square in the denominator. Here

$$8 + 2x - x^2 = -[x^2 - 2x - 8] = -[(x - 1)^2 - 1 - 8] = 9 - (x - 1)^2,$$

giving

$$I_2 = 3 \int \frac{dx}{\sqrt{8 + 2x - x^2}} = 3 \int \frac{dx}{\sqrt{3^2 - (x - 1)^2}} = 3 \sin^{-1}\left(\frac{x - 1}{3}\right) + C_2.$$

So for the original integral one has

$$\int \sqrt{\frac{4 - x}{2 + x}}\, dx = \sqrt{8 + 2x - x^2} + 3 \sin^{-1}\left(\frac{x - 1}{3}\right) + C. \qquad \blacktriangleright$$

Example 12.9 If $a \neq b$, show that

$$\int \frac{dx}{a + b \cos x} = \begin{cases} \dfrac{2}{\sqrt{a^2 - b^2}} \tan^{-1}\left(\dfrac{\sqrt{a - b} \tan \frac{x}{2}}{\sqrt{a + b}}\right) + C, & a > |b| \\[4mm] \dfrac{1}{\sqrt{b^2 - a^2}} \ln\left|\dfrac{\sqrt{b - a} \tan \frac{x}{2} + \sqrt{b + a}}{\sqrt{b - a} \tan \frac{x}{2} - \sqrt{b + a}}\right| + C, & a < |b|. \end{cases}$$

Solution As $1 = \sin^2 \frac{x}{2} + \cos^2 \frac{x}{2}$ and $\cos x = \cos^2 \frac{x}{2} - \sin^2 \frac{x}{2}$ the integral can be rewritten as

$$\int \frac{dx}{a + b \cos x} = \int \frac{dx}{a\left(\sin^2 \frac{x}{2} + \cos^2 \frac{x}{2}\right) + b\left(\cos^2 \frac{x}{2} - \sin^2 \frac{x}{2}\right)}$$

$$= \int \frac{dx}{(a + b) \cos^2 \frac{x}{2} + (a - b) \sin^2 \frac{x}{2}}$$

$$= \int \frac{\sec^2 \frac{x}{2}}{(a + b) + (a - b) \tan^2 \frac{x}{2}}\, dx.$$

Letting $u = \tan \frac{x}{2}, du = \frac{1}{2} \sec^2 \frac{x}{2}\, dx$ we have

$$\int \frac{dx}{a + b \cos x} = 2 \int \frac{du}{(a + b) + (a - b)u^2}.$$

Now if $a > |b|$, then the term $(a - b)$ will be positive and the integral is

$$\int \frac{dx}{a + b \cos x} = \frac{2}{a - b} \int \frac{du}{\left(\dfrac{\sqrt{a + b}}{\sqrt{a - b}}\right)^2 + u^2} \, du$$

$$= \frac{2}{a - b} \cdot \frac{\sqrt{a - b}}{\sqrt{a + b}} \tan^{-1} \left(\frac{u \sqrt{a - b}}{\sqrt{a + b}}\right) + C$$

$$= \frac{2}{\sqrt{a^2 - b^2}} \tan^{-1} \left(\frac{\sqrt{a - b} \tan \frac{x}{2}}{\sqrt{a + b}}\right) + C.$$

And if $a < |b|$, the term $(a - b)$ will be negative so we write it as $-(b - a)$ where the term in the brackets is now positive. For the integral we now have

$$\int \frac{dx}{a + b \cos x} = \frac{2}{b - a} \int \frac{du}{\left(\dfrac{\sqrt{b + a}}{\sqrt{b - a}}\right)^2 - u^2} \, du$$

$$= \frac{2}{b - a} \cdot \frac{\sqrt{b - a}}{2\sqrt{b + a}} \ln \left|\frac{u \sqrt{b - a} + \sqrt{b + a}}{u \sqrt{b - a} - \sqrt{b + a}}\right| + C$$

$$= \frac{1}{\sqrt{b^2 - a^2}} \ln \left|\frac{\sqrt{b - a} \tan \frac{x}{2} + \sqrt{b + a}}{\sqrt{b - a} \tan \frac{x}{2} - \sqrt{b + a}}\right| + C,$$

and the result follows.   ▶

Exercises for Chapter 12

✠ Warm-up

1. By completing the square, write down all possible standard forms that the integral

$$\int \frac{dx}{\sqrt{ax^2 + bx + c}},$$

can be reduced to for various nonzero values of the coefficient a, b, and c.

✠ Practice questions

2. By completing the square where needed, find

(a) $\displaystyle\int \frac{dx}{9 - 25x^2}$

(b) $\displaystyle\int \frac{dx}{4x^2 - 36}$

(c) $\int \dfrac{dx}{\sqrt{x^2 - 7}}$

(d) $\int \dfrac{dx}{\sqrt{3 - 5x^2}}$

(e) $\int \dfrac{dx}{\sqrt{1 + 4x^2}}$

(f) $\int \dfrac{dx}{\sqrt{x(4 - 2x)}}$

(g) $\int \dfrac{dx}{\sqrt{8x + x^2}}$

(h) $\int \dfrac{dx}{\sqrt{1 - 3x - x^2}}$

(i) $\int \dfrac{dx}{1 - 3x - x^2}$

(j) $\int \dfrac{dx}{\sqrt{3x^2 - 5x + 4}}$

(k) $\int \dfrac{dx}{\sqrt{5x + 3x^2}}$

(l) $\int \dfrac{dx}{x^2 + 2x + 5}$

(m) $\int \dfrac{dx}{5 + 8x - 4x^2}$

(n) $\int \dfrac{dx}{\sqrt{(x + 3)(x + 4)}}$

(o) $\int \dfrac{dx}{\sqrt{(x + 3)(x - 4)}}$

(p) $\int \dfrac{dx}{\sqrt{(x + 3)(4 - x)}}$

(q) $\int \dfrac{dx}{x^2 + 6x + 2}$

(r) $\int \dfrac{dx}{\sqrt{x^2 - 6x + 13}}$

3. Evaluate

(a) $\int_0^1 \dfrac{dx}{\sqrt{(x + 1)(x + 3)}}$

(b) $\int_{-1}^0 \dfrac{dx}{\sqrt{1 - 2x - x^2}}$

(c) $\int_{-1}^1 \dfrac{dx}{2x^2 - 3x + 3}$

(d) $\int_{-\frac{3}{2}}^0 \dfrac{dx}{2x^2 + 6x + 15}$

4. Find the value of α if $\displaystyle\int_2^\alpha \dfrac{dx}{x^2 - 4x + 5} = \dfrac{\pi}{4}$.

5. If $a > 0$ evaluate

(a) $\int_a^{2a} \dfrac{dx}{\sqrt{x^2 + a^2}}$

(b) $\int_a^{2a} \dfrac{dx}{\sqrt{x^2 - a^2}}$

(c) $\int_a^{2a} \dfrac{dx}{x^2 + a^2}$

6. (a) Using the substitution $x = u^2$, find $\displaystyle\int \dfrac{x}{x^4 + 2x^2 + 13}\,dx$.

(b) Using the substitution $x = \dfrac{1}{u}$, find $\displaystyle\int \dfrac{dx}{x\sqrt{x^2 + x - 6}}$.

7. By rationalising the numerator first, find $\displaystyle\int \dfrac{\sqrt{1 - x^2}}{1 - x}$.

8. If $a, b, c > 0$ and $4ac - b^2 \neq 0$ find

(a) $\displaystyle\int \dfrac{dx}{\sqrt{a + bx + cx^2}}$

(b) $\displaystyle\int \dfrac{dx}{\sqrt{a + bx - cx^2}}$

(c) $\displaystyle\int \dfrac{dx}{a + bx + cx^2}$

(d) $\displaystyle\int \dfrac{dx}{a + bx - cx^2}$

9. Find

(a) $\displaystyle\int \dfrac{2 - 3x}{4 - 9x^2}\, dx$

(b) $\displaystyle\int \dfrac{4x - 1}{x^2 + 4}\, dx$

(c) $\displaystyle\int \dfrac{1 + 2x}{\sqrt{9x^2 + 16}}\, dx$

(d) $\displaystyle\int \dfrac{1 + x}{1 + x^2}\, dx$

(e) $\displaystyle\int \dfrac{2x - 5}{1 + 4x - x^2}\, dx$

(f) $\displaystyle\int \dfrac{2x + 1}{x^2 + 2x + 2}\, dx$

(g) $\displaystyle\int \dfrac{x - 1}{\sqrt{x^2 + 2x - 1}}\, dx$

(h) $\displaystyle\int \dfrac{2x + 3}{\sqrt{x^2 + 2x + 3}}\, dx$

(i) $\displaystyle\int \dfrac{x + 1}{\sqrt{2x^2 + 3x + 4}}\, dx$

(j) $\displaystyle\int \dfrac{x}{x^2 + 6x + 15}\, dx$

10. Find the value of α that satisfies the equation

$$\int_2^3 \dfrac{\alpha x + 1}{x^2 - 6x + 10} = \dfrac{7\pi}{4} - \ln 2.$$

11. Using the substitution $x = \dfrac{1}{3u}$, evaluate $\displaystyle\int_{\frac{1}{3}}^1 \dfrac{\ln x}{-3x^2 + 3x - 1}\, dx$.

12. By making use of the substitution $x = \pi - u$, find in terms of n

$$\int_0^\pi \dfrac{x \sin x}{(n - 1)(n + 1) + \sin^2 x}\, dx, \quad n \neq 0.$$

13. By making use of the substitution $u = \tanh x$, find

$$\int \frac{dx}{\sinh^2 x - 4 \sinh x \cosh x + 9 \cosh^2 x}.$$

14. (a) Find $\displaystyle\int \sqrt{\frac{x+2}{3-x}}\, dx.$

(b) Hence evaluate $\displaystyle\int_0^1 \sqrt{\frac{x+2}{3-x}}\, dx.$

15. Suppose that $\displaystyle\int_p^q \sqrt{\frac{x-a}{b-x}}\, dx$ where $b \neq a, p, q$. If $q = (a+b)/2$ and

$p = (3a+b)/4$, show that $\displaystyle\int_p^q \sqrt{\frac{x-a}{b-x}}\, dx = \frac{(b-a)(\pi - 6 + 3\sqrt{3})}{12}.$

⚙ **Extension questions and Challenge problems**

16. (a) Use a suitable substitution to show that

$$\int \frac{dx}{(1+x^2)\sqrt{1+x^2}} = \frac{x}{(2+x^2)\sqrt{1+x^2}} + C.$$

(b) Using the result in (a) together with integration by parts, evaluate

$$\int_0^1 \frac{\tan^{-1}(\sqrt{1+x^2})}{(1+x^2)\sqrt{1+x^2}}\, dx.$$

17. Let $\alpha \in (0, \pi)$.

(a) Find $\displaystyle\int \frac{1 - x\cos\alpha}{1 - 2x\cos\alpha + x^2}\, dx.$

(b) Evaluate $\displaystyle\int_{-1}^1 \frac{dx}{1 - 2x\cos\alpha + x^2}.$

18. If $a \neq b$, show that

$$\int \frac{dx}{a + b\sin x} = \begin{cases} \dfrac{2}{\sqrt{a^2 - b^2}} \tan^{-1}\left(\dfrac{a\tan\frac{x}{2} + b}{\sqrt{a^2 - b^2}}\right) + C, & |a| > |b| \\[4mm] \dfrac{1}{\sqrt{b^2 - a^2}} \ln\left|\dfrac{a\tan\frac{x}{2} + b - \sqrt{b^2 - a^2}}{a\tan\frac{x}{2} + b + \sqrt{b^2 - a^2}}\right| + C, & |a| < |b|. \end{cases}$$

19. In integral calculus texts from the nineteenth century a trigonometric function that was often found was the *versed sine function*. Denoted by vers x the function was principally used in the field of navigation. Once considered an important trigonometric function, since the advent of electronic computers and scientific calculators it has become much less significant.

The versed sine function is defined as $\text{vers}\, x = 1 - \cos x$.

(a) Show that $0 \leqslant \text{vers}\, x \leqslant 2$ for all x.

(b) Find $\dfrac{d}{dx} \text{vers}\, x$.

(c) On the interval $[0, \pi]$ an inverse to the versed sine function can be defined, denoted by $\text{vers}^{-1} x$. Show that

$$\frac{d}{dx} \text{vers}^{-1}\left(\frac{x}{a}\right) = \frac{1}{\sqrt{2ax - x^2}}, \quad \text{for } a \neq 0.$$

(d) Hence deduce that

$$\int \frac{dx}{\sqrt{2ax - x^2}} = \text{vers}^{-1}\left(\frac{x}{a}\right) + C_1, \quad a \neq 0.$$

(e) By completing the square, show that

$$\int \frac{dx}{\sqrt{2ax - x^2}} = \sin^{-1}\left(\frac{x - a}{a}\right) + C_2, \quad a \neq 0.$$

(f) From (d) and (e) deduce that $\text{vers}^{-1}\left(\dfrac{x}{a}\right) = \cos^{-1}\left(\dfrac{a - x}{a}\right)$.

13

Inverse Hyperbolic Functions and Integrals Leading to Them

Common integration is only the memory of differentiation ... the different artifices by which integration is effected, are changes, not from the known to the unknown, but from forms in which memory will not serve us to those in which it will.

— *Augustus de Morgan*

In this chapter the inverse hyperbolic functions are briefly reviewed before various standard types of integrals that lead to inverse hyperbolic functions are considered.

§ The Inverse Hyperbolic Functions

From Chapter 9 you may recall that since the functions $\sinh : \mathbb{R} \to \mathbb{R}$ and $\tanh : \mathbb{R} \to (-1, 1)$ are both increasing functions on their domain, both are one-to-one functions and accordingly will have well-defined inverses. Those inverses are denoted by $\sinh^{-1} x$ and $\tanh^{-1} x$, respectively. The function $\cosh : \mathbb{R} \to [1, \infty)$, however, is not one-to-one on its domain. To obtain an inverse its domain must be restricted. By convention the domain of the hyperbolic cosine function is restricted to the interval $[0, \infty)$.

Since the hyperbolic functions are defined in terms of the exponential function, one would expect it would be possible to write each of the inverse hyperbolic functions in terms of the logarithmic function, the logarithmic function being the inverse of the exponential function. This is indeed possible. For example, to find an expression for the inverse hyperbolic sine function in terms of the logarithmic function, set $y = \sinh^{-1} x$, then $\sinh y = x$, or

$$\sinh y = \frac{e^y - e^{-y}}{2} = x.$$

169

Table 13.1. *The six inverse hyperbolic functions.*

Function	Formula	Domain	Range
$\sinh^{-1} x$	$\ln(x + \sqrt{x^2 + 1})$	$(-\infty, \infty)$	$(-\infty, \infty)$
$\cosh^{-1} x$	$\ln(x + \sqrt{x^2 - 1})$	$[1, \infty)$	$[0, \infty)$
$\tanh^{-1} x$	$\frac{1}{2} \ln\left(\frac{1+x}{1-x}\right)$	$(-1, 1)$	$(-\infty, \infty)$
$\coth^{-1} x$	$\frac{1}{2} \ln\left(\frac{x+1}{x-1}\right)$	$(-\infty, -1) \cup (1, \infty)$	$(-\infty, 0) \cup (0, \infty)$
$\operatorname{sech}^{-1} x$	$\ln\left(\frac{1 + \sqrt{1 - x^2}}{x}\right)$	$(0, 1]$	$[0, \infty)$
$\operatorname{cosech}^{-1} x$	$\ln\left(\frac{1 \pm \sqrt{1 + x^2}}{x}\right)$	$(-\infty, 0) \cup (0, \infty)$	$(-\infty, 0) \cup (0, \infty)$

After multiplying throughout by $2e^y$ and rearranging, the following equation results:

$$e^{2y} - 2xe^y - 1 = 0,$$

which is quadratic in e^y. On applying the quadratic formula, solving for e^y gives

$$e^y = x + \sqrt{x^2 + 1}.$$

Here, since $e^y > 0$, the positive square root is taken. Taking the natural logarithm of both sides of the equation gives

$$y = \sinh^{-1} x = \ln(x + \sqrt{x^2 + 1}).$$

Inverses for the other five hyperbolic functions can be found in a similar manner. These, together with each function's associated domain and range, are summarised in Table 13.1. In the table, where a plus and minus sign appear, the positive sign is used for $x > 0$, while the negative sign is used for $x < 0$. Graphs for all six inverse hyperbolic functions are shown in Figure 13.1.

Since the inverse hyperbolic functions can be expressed in terms of logarithms, their derivatives can be readily found. For example, for the derivative

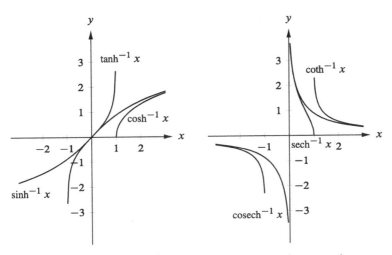

Figure 13.1. The six inverse hyperbolic functions: $\sinh^{-1} x$, $\cosh^{-1} x$, and $\tanh^{-1} x$ (left) and $\coth^{-1} x$, $\operatorname{sech}^{-1} x$, and $\operatorname{cosech}^{-1} x$ (right).

of the inverse hyperbolic sine function,

$$\frac{d}{dx}\left(\sinh^{-1} x\right) = \frac{d}{dx}\left[\ln(x + \sqrt{x^2 + 1})\right] = \frac{1 + x/\sqrt{x^2 + 1}}{1 + \sqrt{x^2 + 1}} = \frac{1}{\sqrt{x^2 + 1}}.$$

The other five derivatives for the inverse hyperbolic functions follow in a similar manner, the results of which are summarised in Table 13.2.

§ Integrals Leading to Inverse Hyperbolic Functions

From the list of derivatives for the six inverse hyperbolic functions it should now be readily apparent which integrals lead to inverse hyperbolic functions. For example, since

$$\frac{d}{dx}\left(\sinh^{-1} x\right) = \frac{1}{\sqrt{1 + x^2}},$$

then

$$\int \frac{dx}{\sqrt{1 + x^2}} = \sinh^{-1} x + C = \ln\left(x + \sqrt{1 + x^2}\right) + C.$$

More generally, for the first four derivatives for the inverse hyperbolic functions, we summarise in Table 13.3 four important integrals that lead to inverse hyperbolic functions. For brevity, the arbitrary constant of integration has not been given.

Table 13.2. *Derivatives for the six inverse hyperbolic functions.*

Derivatives for the inverse hyperbolic functions

1. $\dfrac{d}{dx}\left(\sinh^{-1} x\right) = \dfrac{1}{\sqrt{x^2 + 1}}, \quad x \in \mathbb{R}$

2. $\dfrac{d}{dx}\left(\cosh^{-1} x\right) = \dfrac{1}{\sqrt{x^2 - 1}}, \quad x > 1$

3. $\dfrac{d}{dx}\left(\tanh^{-1} x\right) = \dfrac{1}{1 - x^2}, \quad |x| < 1$

4. $\dfrac{d}{dx}\left(\coth^{-1} x\right) = \dfrac{1}{1 - x^2}, \quad |x| > 1$

5. $\dfrac{d}{dx}\left(\operatorname{sech}^{-1} x\right) = -\dfrac{1}{x\sqrt{1 - x^2}}, \quad 0 < x < 1$

6. $\dfrac{d}{dx}\left(\operatorname{cosech}^{-1} x\right) = -\dfrac{1}{|x|\sqrt{1 + x^2}}, \quad x \neq 0$

Comparing this table with Table 12.1 given on page 158 of Chapter 12, it is quickly apparent each of the four integrals given in Table 13.3 that lead to inverse hyperbolic functions provide alternative forms for the second, third, fifth, and sixth integrals given in Table 12.1. Of course the final form given is one of personal preference, as either form is correct. We now give a number of examples of integrals that lead to inverse hyperbolic functions.

Table 13.3. *Four important integrals leading to inverse hyperbolic functions.*

Integrals leading to inverse hyperbolic functions

1. $\displaystyle\int \dfrac{dx}{\sqrt{a^2 + x^2}} = \sinh^{-1}\left(\dfrac{x}{a}\right) = \ln\left(x + \sqrt{a^2 + x^2}\right), \quad a > 0$

2. $\displaystyle\int \dfrac{dx}{\sqrt{x^2 - a^2}} = \cosh^{-1}\left(\dfrac{x}{a}\right) = \ln\left(x + \sqrt{x^2 - a^2}\right), \quad 0 < a < x$

3. $\displaystyle\int \dfrac{dx}{a^2 - x^2} = \dfrac{1}{a}\tanh^{-1}\left(\dfrac{x}{a}\right) = \dfrac{1}{2a}\ln\left(\dfrac{a + x}{a - x}\right), \quad a > 0,\ |x| < a$

4. $\displaystyle\int \dfrac{dx}{x^2 - a^2} = -\dfrac{1}{a}\coth^{-1}\left(\dfrac{x}{a}\right) = \dfrac{1}{2a}\ln\left(\dfrac{x - a}{x + a}\right), \quad a > 0,\ |x| > a$

Example 13.1 Find $\displaystyle\int \frac{dx}{\sqrt{1+4x^2}}$.

Solution Writing the integrand in standard form before integrating we have

$$\int \frac{dx}{\sqrt{1+4x^2}} = \frac{1}{2}\int \frac{dx}{\sqrt{(1/2)^2 + x^2}},$$

from which it immediately follows that

$$\int \frac{dx}{\sqrt{1+4x^2}} = \frac{1}{2}\sinh^{-1}(2x) + C.$$ ▶

Example 13.2 Find $\displaystyle\int \frac{dx}{\sqrt{(x+2)(x+4)}}$.

Solution For the term $(x+2)(x+4)$, on completing the square we can write

$$(x+2)(x+4) = x^2 + 6x + 8 = (x+3)^2 - 1.$$

Setting $u = x + 3, du = dx$, we have

$$\int \frac{dx}{\sqrt{(x+2)(x+4)}} = \int \frac{du}{\sqrt{u^2 - 1}} = \cosh^{-1} u + C$$

$$= \cosh^{-1}(x+3) + C.$$ ▶

Example 13.3 (a) Find, expressed in terms of an inverse hyperbolic function, the integral $\displaystyle\int \frac{dx}{x^2 + 4x - 5}$.

(b) Hence evaluate $\displaystyle\int_{-2}^{\frac{1}{2}} \frac{dx}{x^2 + 4x - 5}$.

Solution (a) On completing the square in the denominate, we have

$$\int \frac{dx}{x^2 + 4x - 5} = \int \frac{dx}{(x+2)^2 - 3^2},$$

whose integral in terms of inverse hyperbolic functions we can write as

$$\int \frac{dx}{x^2 + 4x - 5} = \begin{cases} -\dfrac{1}{3}\tanh^{-1}\left(\dfrac{x+2}{3}\right) + C, & |x+2| < 3 \\[3mm] \dfrac{1}{3}\coth^{-1}\left(\dfrac{x+2}{3}\right) + C, & |x+2| > 3. \end{cases}$$

(b) Since $|x + 2| < 3$ corresponds to the interval $-5 < x < 1$, as this falls within the interval of integration for our definite integral, the integral in terms of the inverse hyperbolic tangent function is chosen and gives

$$\int_{-2}^{\frac{1}{2}} \frac{dx}{x^2 + 4x - 5} = \left[-\frac{1}{3} \tanh^{-1} \left(\frac{x+2}{3} \right) \right]_{-2}^{\frac{1}{2}}$$

$$= -\frac{1}{3} \tanh^{-1} \left(\frac{5}{6} \right) + \frac{1}{3} \tanh^{-1}(0) = -\frac{1}{3} \tanh^{-1} \left(\frac{5}{6} \right),$$

since $\tanh^{-1}(0) = 0$. ▶

An inverse hyperbolic function may also arise from finding an integral when a hyperbolic substitution is used. The next example gives an illustration of this.

Example 13.4 Find $\displaystyle\int \frac{x^2}{\sqrt{4 + x^2}} \, dx$.

Solution Use a hyperbolic substitution of $x = 2 \sinh u$. As $dx = 2 \cosh u \, du$, we have

$$\int \frac{x^2}{\sqrt{4 + x^2}} \, dx = \int \frac{4 \sinh^2 u \cdot 2 \cosh u}{\sqrt{4 + 4 \sinh^2 u}} \, du = 4 \int \frac{\sinh^2 u \cdot \cosh u}{\cosh u} \, du,$$

since $\cosh^2 u = 1 + \sinh^2 u$, or

$$\int \frac{x^2}{\sqrt{4 + x^2}} \, dx = 4 \int \sinh^2 u \, du.$$

Making use of the hyperbolic identity $\cosh 2u = 1 + 2 \sinh^2 u$, we have

$$\int \frac{x^2}{\sqrt{4 + x^2}} \, dx = 2 \int (\cosh 2u - 1) \, du = \sinh 2u - u + C.$$

Writing this in terms of x, as $\sinh 2u = 2 \sinh u \cosh u$ from $\sinh u = \frac{x}{2}$ we have

$$\cosh u = \sqrt{1 + \sinh^2 u} = \frac{1}{2} \sqrt{4 + x^2},$$

and $u = \sinh^{-1} \left(\frac{x}{2} \right)$. Thus

$$\int \frac{x^2}{\sqrt{4 + x^2}} \, dx = \frac{x}{2} \sqrt{4 + x^2} - \sinh^{-1} \left(\frac{x}{2} \right) + C. \qquad ▶$$

§ Integrals Involving Inverse Hyperbolic Functions

Integrals for the inverse hyperbolic functions can also be found. As was the case with inverse trigonometric functions, they can be found using integration by parts. In the next example, the integral for the inverse hyperbolic sine function is given, with all other integrals for the inverses being found in a similar fashion.

Example 13.5 Find $\int \sinh^{-1} x \, dx$.

Solution Integrating by parts, one has

$$\int \sinh^1 x \, dx = x \sinh^{-1} x - \int x \cdot \frac{1}{\sqrt{1+x^2}} \, dx.$$

The integral on the right can be found using the substitution $u = 1 + x^2$. As $du = 2x \, dx$ this gives

$$\frac{x}{\sqrt{1+x^2}} \, dx = \frac{1}{2} \int \frac{du}{\sqrt{u}} = \sqrt{u} + C = \sqrt{1+x^2} + C.$$

Thus

$$\int \sinh^{-1} x \, dx = x \sinh^{-1} x - \sqrt{1+x^2} + C. \qquad \blacktriangleright$$

Exercises for Chapter 13

✠ **Practice questions**

1. Find, in terms of inverse hyperbolic functions, the following integrals.

(a) $\displaystyle\int \frac{dx}{\sqrt{x^2 - 3}}$

(b) $\displaystyle\int \frac{dx}{\sqrt{2 + 3x^2}}$

(c) $\displaystyle\int \frac{dx}{x^2 + 6x + 5}$

(d) $\displaystyle\int \frac{dx}{\sqrt{x^2 - 2x + 5}}$

(e) $\displaystyle\int \frac{dx}{\sqrt{x^2 - 6x}}$

(f) $\displaystyle\int \frac{dx}{\sqrt{x^2 + 6x}}$

2. (a) Find $\displaystyle\int \frac{dx}{2x^2 - 1}$.

   (b) Hence evaluate $\displaystyle\int_0^{\frac{1}{2}} \frac{dx}{2x^2 - 1}$.

3. Show that

   (a) $\displaystyle\int_0^{\frac{1}{2}} \frac{dx}{1 - x^2} = \tanh^{-1}\left(\frac{1}{2}\right)$

   (b) $\displaystyle\int_2^3 \frac{dx}{1 - x^2} = \coth^{-1}(3) - \coth^{-1}(2)$

4. (a) Show that $\dfrac{d}{dx}\left[\tanh^{-1}(\ln x)\right] = \dfrac{1}{2x(1 - \ln^2 x)}$.

(b) Hence find $\displaystyle\int \dfrac{dx}{x(1 - \ln^2 x)}$.

5. By using a suitable hyperbolic substitution, find

(a) $\displaystyle\int \sqrt{x^2 + 1}\, dx$

(b) $\displaystyle\int \dfrac{x^2}{\sqrt{x^2 - 1}}\, dx$

(c) $\displaystyle\int \sqrt{(x + 2)(x + 4)}\, dx$

(d) $\displaystyle\int \dfrac{\sqrt{1 - x^2}}{x}\, dx$

(e) $\displaystyle\int x^2\sqrt{4x^2 + 1}\, dx$

(f) $\displaystyle\int \dfrac{x^4}{\sqrt{1 + x^2}}\, dx$

6. (a) Using the substitution $2x = 5\sinh t$, show that

$$\int \sqrt{4x^2 + 25}\, dx = \frac{x}{2}\sqrt{4x^2 + 25} + \frac{25}{4}\sinh^{-1}\left(\frac{2x}{5}\right) + C.$$

(b) Using the substitution $2x = 5\tan\theta$, show that

$$\int \sqrt{4x^2 + 25}\, dx = \frac{x}{2}\sqrt{4x^2 + 25} + \frac{25}{4}\ln\left|2x + \sqrt{4x^2 + 25}\right| + C.$$

(c) The results in (a) and (b) appear to be different. Show that they are indeed equivalent.

7. Find

(a) $\displaystyle\int \tanh^{-1} x\, dx$

(b) $\displaystyle\int \operatorname{sech}^{-1} x\, dx$

(c) $\displaystyle\int x\cosh^{-1} x\, dx$

(d) $\displaystyle\int x\tanh^{-1} x\, dx$, if $|x| < 1$

8. Using the substitution $x = \sinh^2 u$, find $\displaystyle\int \ln\left(\sqrt{x} + \sqrt{x + 1}\right) dx$.

9. Using a suitable substitution, show that

$$\int \frac{\sinh x}{2\cosh^2 x - 1}\, dx = -\frac{1}{\sqrt{2}}\coth^{-1}(\sqrt{2}\cosh x) + C.$$

10. By making use of the substitution $u = \tanh x$, show that

$$\int_0^1 \frac{dx}{9 + 5\sinh^2 x} = \frac{1}{6} \tanh^{-1} \left(\frac{2}{3} \tanh(1) \right).$$

11. If $a \neq 0$ show that $\displaystyle\int_0^1 \frac{dx}{x^2 + 2x \cosh a + 1} = \frac{a}{2 \sinh a}$.

12. (a) Suppose $x + a > 0, x + b > 0$, and $b > a$. Find

$$\frac{d}{dx} \cosh^{-1} \left(\frac{x+b}{x+a} \right).$$

(b) Use the result in (a) to find $\displaystyle\int \frac{dx}{(x+1)\sqrt{x+3}}$.

13. Find $\displaystyle\int \sin(\cosh^{-1} x + 1)\, dx$.

✠ **Extension questions and Challenge problems**

14. Show that $\displaystyle\int \sqrt{\tanh \left(\ln \sqrt{x} \right)}\, dx = \sqrt{x^2 - 1} - \cosh^{-1} x + C$.

15 (The inverse Gudermannian function). The *inverse Gudermannian function* $\text{gd}^{-1} : (-\frac{\pi}{2}, \frac{\pi}{2}) \to \mathbb{R}$ is defined by

$$\text{gd}^{-1} x = \int_0^x \sec x\, dx.$$

(a) By using a Gunther hyperbolic substitution (see Exercise 11 on page 142 of Chapter 10 for details), show that $\text{gd}^{-1} x = \sinh^{-1}(\tan x)$.

(b) By showing that $\sinh^{-1} u = \tanh^{-1} \left(\dfrac{u}{\sqrt{1 + u^2}} \right)$, use this to deduce that $\text{gd}^{-1} x = \tanh^{-1}(\sin x)$.

14

Tangent Half-Angle Substitution

The world's sneakiest substitution is undoubtedly $t = \tan\frac{x}{2}$.

— Michael Spivak, *Calculus*

Recall in Chapter 11 that a systematic method was given for finding any integral consisting of a rational function using the method of partial fractions. By extension, any integral that is rational in any of the six trigonometric functions can always be found by the use of a rationalising substitution that converts a rational function of sine and cosine (only these two trigonometric functions need be considered as the four remaining trigonometric functions can all be expressed in terms of sine and cosine) into a rational function in terms of the substitution variable used.

The standard rationalising substitution used to convert the integral of a rational function consisting of sine and cosine into a rational function in terms of the substitution variable is a 'tangent half-angle substitution'. Other commonly used names for the substitution are a 'Weierstrass substitution' and a 't-substitution'. The former is so named after the great German mathematician Karl Weierstrass (1815–1897), though the substitution pre-dates him and was at least known to Euler, while the latter takes its name from $t = \tan\frac{x}{2}$, the form of the substitution typically used. The idea behind the method is that it allows one to convert an integral for any rational function in terms of sine and cosine in the variable x into an integral consisting of a rational function in the substitution variable t, which, as we know from Chapter 11, can always be found.

We should, however, add a word of caution. While a tangent half-angle substitution is always guaranteed to work,[1] leading to a rational function expression in terms of t, the resulting integral often requires cumbersome partial fraction decompositions. Consequently, a t-substitution should only be used as a method of last resort in the event no other simpler alternative method or strategy is found.

Consider the substitution $t = \tan \frac{x}{2}$. To find expressions for $\cos x$ and $\sin x$ in terms of t consider the following right-angled triangle.

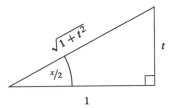

From the figure we see that

$$\cos\left(\frac{x}{2}\right) = \frac{1}{\sqrt{1+t^2}} \quad \text{and} \quad \sin\left(\frac{x}{2}\right) = \frac{t}{\sqrt{1+t^2}}.$$

From the double-angle formula $\cos 2\theta = \cos^2\theta - \sin^2\theta$ for the cosine function, if θ is replaced with $x/2$, one has

$$\cos x = \cos^2\left(\frac{x}{2}\right) - \sin^2\left(\frac{x}{2}\right) = \left(\frac{1}{\sqrt{1+t^2}}\right)^2 - \left(\frac{t}{\sqrt{1+t^2}}\right)^2 = \frac{1-t^2}{1+t^2}.$$

And from the double-angle formula $\sin 2\theta = 2\sin\theta\cos\theta$ for the sine function, if θ is replaced with $x/2$, one has

$$\sin x = 2\sin\left(\frac{x}{2}\right)\cos\left(\frac{x}{2}\right) = 2\left(\frac{t}{\sqrt{1+t^2}}\right)\left(\frac{1}{\sqrt{1+t^2}}\right) = \frac{2t}{1+t^2}.$$

Finally, for the differential dx, differentiating $t = \tan\frac{x}{2}$ with respect to x we have

$$\frac{dt}{dx} = \frac{1}{2}\sec^2\left(\frac{x}{2}\right) = \frac{1}{2}\left(1 + \tan^2\left(\frac{x}{2}\right)\right) = \frac{1}{2}(1+t^2),$$

giving

$$dx = \frac{2}{1+t^2}\,dt.$$

So, in summary, if $t = \tan\frac{x}{2}$, then

$$\boxed{\sin x = \frac{2t}{1+t^2}, \quad \cos x = \frac{1-t^2}{1+t^2}, \quad dx = \frac{2}{1+t^2}\,dt}$$

We now apply the t-substitution method to a number of integrals that are rational functions in terms of the trigonometric functions.

Example 14.1 Find $\displaystyle\int \frac{dx}{1+\sin x}$.

Solution Using a t-substitution, namely $t = \tan \frac{x}{2}$, we have $\sin x = \frac{2t}{1+t^2}$ and $dx = \frac{2}{1+t^2} \, dt$. So

$$\int \frac{dx}{1 + \sin x} = \int \frac{1}{1 + \frac{2t}{1+t^2}} \cdot \frac{2}{1+t^2} \, dt = 2 \int \frac{dt}{(t+1)^2}.$$

Evaluating the resulting integral, which is now a rational function of t, can be done either by inspection or by using the substitution $u = 1 + t$. The result is

$$\int \frac{dx}{1 + \sin x} = -\frac{2}{1+t} + C = -\frac{2}{1 + \tan \frac{x}{2}} + C.$$
▶

Example 14.2 Find $\displaystyle\int \frac{dx}{2 + \sin x + \cos x}$.

Solution Using a t-substitution, namely $t = \tan \frac{x}{2}$, we have $\sin x = \frac{2t}{1+t^2}$, $\cos x = \frac{1-t^2}{1+t^2}$, and $dx = \frac{2}{1+t^2} \, dt$. So the integral becomes

$$\int \frac{dx}{2 + \sin x + \cos x} = \int \frac{1}{2 + \frac{2t}{1+t^2} + \frac{1-t^2}{1+t^2}} \cdot \frac{2}{1+t^2} \, dt = 2 \int \frac{dt}{t^2 + 2t + 3},$$

after simplifying. Completing the square of the denominator one has

$$\int \frac{dx}{2 + \sin x + \cos x} = 2 \int \frac{dt}{(t+1)^2 + 2} = \frac{2}{\sqrt{2}} \tan^{-1} \left(\frac{t+1}{\sqrt{2}} \right) + C$$

$$= \sqrt{2} \tan^{-1} \left(\frac{\tan \frac{x}{2} + 1}{\sqrt{2}} \right) + C.$$
▶

Example 14.3 Find $\displaystyle\int \frac{dx}{3 \sin x - 4 \cos x}$.

Solution Once again when the t-substitution $t = \tan \frac{x}{2}$ is used we have $\sin x = \frac{2t}{1+t^2}$, $\cos x = \frac{1-t^2}{1+t^2}$, and $dx = \frac{2}{1+t^2} dt$. So the integral becomes

$$\int \frac{dx}{3 \sin x - 4 \cos x} = \int \frac{1}{3 \left(\dfrac{2t}{1+t^2} \right) - 4 \left(\dfrac{1-t^2}{1+t^2} \right)} \cdot \frac{2}{1+t^2} dt$$

$$= 2 \int \frac{dt}{3(2t) - 4(1 - t^2)} = \int \frac{dt}{2t^2 + 3t - 2}$$

$$= \int \frac{dt}{(2t - 1)(t + 2)}.$$

Using the Heaviside cover-up method it can be readily shown that

$$\frac{1}{(2t-1)(t+2)} = \frac{\frac{2}{5}}{2t-1} + \frac{-\frac{1}{5}}{t+2}.$$

Thus

$$\int \frac{dx}{3\sin x - 4\cos x} = \frac{2}{5}\int \frac{1}{2t-1}\,dt - \frac{1}{5}\int \frac{1}{t+2}\,dt$$

$$= \frac{1}{5}\ln|2t-1| - \frac{1}{5}\ln|t+2| + C$$

$$= \frac{1}{5}\ln\left|\frac{2t-1}{t+2}\right| + C = \frac{1}{5}\ln\left|\frac{2\tan(\frac{x}{2})-1}{\tan(\frac{x}{2})+2}\right| + C. \quad \blacktriangleright$$

Example 14.4 Find $\displaystyle\int \frac{\sin x}{1+\sin x}\,dx$.

Solution When the t-substitution $t = \tan\frac{x}{2}$ is used, we have $\sin x = \dfrac{2t}{1+t^2}$ and $dx = \dfrac{2}{1+t^2}\,dt$ so that the integral becomes

$$\int \frac{\sin x}{1+\sin x}\,dx = \int \frac{\frac{2t}{1+t^2}}{1+\frac{2t}{1+t^2}} \cdot \frac{2}{1+t^2}\,dt = 4\int \frac{t}{(1+t)^2 + 2t(1+t^2)}\,dt$$

$$= 4\int \frac{t}{(1+t^2)(t^2+2t+1)}\,dt = 4\int \frac{t}{(t^2+1)(t+1)^2}\,dt.$$

The partial fraction decomposition for the rational function appearing in the integrand in terms of t can be found. The result is

$$\frac{t}{(t^2+1)(t+1)^2} = \frac{1}{2(t^2+1)} - \frac{1}{2(t+1)^2}.$$

Thus

$$\int \frac{\sin x}{1+\sin x}\,dx = 2\int \frac{dt}{t^2+1} - 2\int \frac{dt}{(t+1)^2} = 2\tan^{-1}t + \frac{2}{t+1} + C$$

$$= 2\tan^{-1}\left(\tan\frac{x}{2}\right) + \frac{2}{1+\tan\frac{x}{2}} + C = x + \frac{2}{1+\tan\frac{x}{2}} + C.$$

$$\blacktriangleright$$

Exercises for Chapter 14

✠ **Warm-ups**

1. Express the remaining four trigonometric functions $\tan x$, $\cot x$, $\sec x$, and $\csc x$ in terms of t if the substitution $t = \tan \frac{x}{2}$ is used.

2. Using a purely algebraic method, as opposed to the partially geometric method that was used in the chapter, find expressions for $\sin x$ and $\cos x$ in terms of t when the substitution $t = \tan \frac{x}{2}$ is used.

✠ **Practice questions**

3. By using a t-substitution find the following integrals.

   (a) $\displaystyle\int \csc x \, dx$

   (b) $\displaystyle\int \frac{dx}{\sin x + \tan x}$

   (c) $\displaystyle\int \frac{dx}{1 + \sin x + \cos x}$

   (d) $\displaystyle\int \sec x \, dx$

4. (a) By using a t-substitution, show that
$$\int_0^{\frac{\pi}{2}} \frac{dx}{5 - 4\cos x} = \frac{2}{3}\tan^{-1}(3).$$

   (b) Hence find the value of $\displaystyle\int_0^{\frac{\pi}{2}} \frac{\cos x}{5 - 4\cos x}\,dx$.

5. Use a t-substitution to find the following integrals.

   (a) $\displaystyle\int \frac{dx}{3 + 4\cos x}$

   (b) $\displaystyle\int \frac{dx}{4 + 3\cos x}$

   (c) $\displaystyle\int \frac{\cos x}{3 + 4\cos x}\,dx$

   (d) $\displaystyle\int \frac{dx}{4\sin x + 3\cos x}$

   (e) $\displaystyle\int \frac{1 + \cos x}{1 + \sin x}\,dx$

   (f) $\displaystyle\int \frac{\sec x + \tan x}{2\sec x + 3\tan x}\,dx$

6. Evaluate

   (a) $\displaystyle\int_0^{\frac{\pi}{4}} \frac{x + \sin x}{1 + \cos x}\,dx$ (b) $\displaystyle\int_0^{\frac{\pi}{2}} \frac{dx}{\sqrt{1 + \cos x}\sqrt{\sin x + \cos x}}$

7. Using a t-substitution, if $|a| > |b|$ show that

$$\int_0^{\frac{\pi}{2}} \frac{d\theta}{a + b \sin \theta} = \frac{2}{\sqrt{a^2 - b^2}} \tan^{-1} \sqrt{\frac{a - b}{a + b}}.$$

8. In this question we will evaluate the integral $\displaystyle\int \frac{\sin x}{3 + 4 \cos x} dx$ in two different ways.

(a) By using a simple substitution, show that

$$\int \frac{\sin x}{3 + 4 \cos x} dx = -\frac{1}{4} \ln(3 + 4 \cos x) + C.$$

(b) Using a t-substitution, show that

$$\int \frac{\sin x}{3 + 4 \cos x} dx = \frac{1}{4} \ln \left| \frac{1 + \tan^2(\frac{x}{2})}{7 - \tan^2(\frac{x}{2})} \right| + C.$$

(c) The results found in parts (a) and (b) appear to be different. Show that they are in fact equivalent.

9. Use a t-substitution to find the following integrals.

(a) $\displaystyle\int \frac{\sec x + \tan x}{\csc x + \cot x} dx$

(b) $\displaystyle\int \frac{x + \sin x}{1 + \cos x} dx$

(c) $\displaystyle\int \frac{dx}{1 + \tan x + \sec x}$

(d) $\displaystyle\int \frac{x + \cos x}{1 + \sin x} dx$

10. Suppose that $f : \mathbb{R} \to \mathbb{R}$ is given by $f(x) = \dfrac{1}{3 + 2 \cos x}$.

(a) Show that f is a monotonically increasing function on the interval $[\frac{\pi}{3}, \frac{\pi}{2}]$.

(b) Using the result of (a), deduce that

$$\frac{\pi}{24} \leqslant \int_{\frac{\pi}{3}}^{\frac{\pi}{2}} \frac{dx}{3 + 2 \cos x} \leqslant \frac{\pi}{18}.$$

(c) By using a t-substitution, find the exact value for $\displaystyle\int_{\frac{\pi}{3}}^{\frac{\pi}{2}} \frac{dx}{3 + 2 \cos x}$.

11 (The inverse Gudermannian function). The *inverse Gudermannian function* $\text{gd}^{-1} : (-\frac{\pi}{2}, \frac{\pi}{2}) \to \mathbb{R}$ is defined as

$$\text{gd}^{-1} x = \int_0^x \sec t \, dt.$$

By making use of a t-substitution, show that

$$\mathrm{gd}^{-1} x = 2 \tanh^{-1}\left(\tan \frac{x}{2}\right).$$

12. Consider an integral of the following form involving rational functions of sine and cosine:

$$\int \frac{P(\sin\theta, \cos\theta)}{Q(\sin\theta, \cos\theta)} d\theta.$$

When the t-substitution $t = \tan\frac{\theta}{2}$ is made the above integral can always be turned into an integral of a rational function of t as

$$\int \frac{P(\sin\theta, \cos\theta)}{Q(\sin\theta, \cos\theta)} d\theta = \int \frac{P\left(\frac{2t}{1+t^2}, \frac{1-t^2}{1+t^2}\right)}{Q\left(\frac{2t}{1+t^2}, \frac{1-t^2}{1+t^2}\right)} \cdot \frac{2}{1+t^2} dt.$$

As already noted in the chapter, the integral can always be solved, at least in principle, using the method of partial fractions. Often the effort required to do so may be considerable, depending on the work needed to perform the partial fraction decomposition.

As an example of the considerable work that can arise as a result of directly applying a t-substitution, consider the following integral

$$I = \int \frac{d\theta}{\cos^3\theta + 2\sin 2\theta - 5\cos\theta},$$

which is a rational function of sine and cosine.

(a) By applying a t-substitution, show that the above integral can be written as

$$I = \frac{1}{2}\int \frac{(1+t^2)^2}{(t^2-1)(t^2-t+1)^2} dt.$$

Performing the partial fraction decomposition here would take considerable effort.

(b) Instead of applying a t-substitution, show that the integral can be rewritten as

$$I = \int \frac{\cos\theta}{(-\sin^2\theta + 4\sin\theta - 4)(1 - \sin^2\theta)} d\theta,$$

by manipulating terms in the integrand.

(c) By using the substitution $u = \sin \theta$, show that the integral appearing in (b) reduces to

$$I = -\int \frac{1}{(2-u)^2(1-u^2)}\, du,$$

and is a far simpler partial fraction to decompose compared to the one that would have needed to be done in (a) if a t-substitution was applied directly to the integral.

(d) Find the integral.

✠ Extension questions and Challenge problems

13. Given the analogous nature of the trigonometric functions and the hyperbolic functions it should come as no surprise to learn that an analogous *hyperbolic tangent half-argument substitution* can be applied to integrals consisting of rational functions of hyperbolic functions. In this instance the 'hyperbolic t-substitution' of $t = \tanh \frac{x}{2}$ is used.

(a) Show that for the hyperbolic t-substitution, we have

$$\sinh x = \frac{2t}{1-t^2}, \quad \cosh x = \frac{1+t^2}{1-t^2}, \quad dx = \frac{2}{1-t^2}\, dt.$$

(b) By using a hyperbolic t-substitution, show that

(i) $\displaystyle\int \frac{dx}{(1+\cosh x)^2} = \frac{\sinh x (\cosh x + 2)}{3(\cosh x + 1)^2} + C$

(ii) $\displaystyle\int \frac{1+\sinh x}{2+\cosh x}\, dx = \frac{2}{\sqrt{3}} \tanh^{-1}\left(\frac{\tanh \frac{x}{2}}{\sqrt{3}}\right) + \ln(\cosh x + 2) + C$

14. Consider the following integral $I = \displaystyle\int \frac{\alpha + \beta \cos x + \gamma \sin x}{a + b \cos x + c \sin x}\, dx.$

When evaluated directly using a t-substitution, integrals of this form can be extremely tedious, as one can easily end up with a rational function of degree four in the denominator.

As an alternative to a direct evaluation using a t-substitution, begin by writing the numerator as

$$\alpha + \beta \cos x + \gamma \sin x = p(a + b \cos x + c \sin x)$$

$$+ q\frac{d}{dx}(a + b \cos x + c \sin x) + r.$$

Here p, q, and r are constants to be determined and are found by equating equal coefficients for $\cos x$, $\sin x$, and the constant term.

With the numerator in this form the integral can be rewritten as

$$I = \int \left[p + \frac{q \dfrac{d}{dx}(a + b\cos x + c\sin x)}{a + b\cos x + c\sin x} \right] dx$$

$$+ \int \frac{r}{a + b\cos x + c\sin x}\,dx,$$

and can be evaluated far more easily compared to a direct approach using a t-substitution. The last of the terms appearing in the above integral is still typically evaluated using a standard t-substitution.

As an example let us apply the method to the evaluation of the integral

$$\int \frac{\sin x + 2}{\cos x + \sin x}\,dx.$$

(a) Show that the numerator of the integral can be written as

$$\sin x + 2 = (p + q)\cos x + (p - q)\sin x + r,$$

where $p = 1/2, q = -1/2$, and $r = 2$.

(b) Hence find the integral.

15. Suppose that $f_{k-1} = \displaystyle\int_{-1}^{1} \frac{\sqrt{1 - x^2}}{\sqrt{k} - x}\,dx, \, k \in \mathbb{N}$.

(a) Show that $f_0 = \pi$.

(b) By multiplying the integrand of f_{k-1} by $\dfrac{\sqrt{1 - x^2}}{\sqrt{1 - x^2}}$ show that

$$f_{k-1} = J_1 - (k - 1)J_2, \, k \geqslant 2,$$

where $J_1 = \displaystyle\int_{-1}^{1} \frac{k - x^2}{(\sqrt{k} - x)\sqrt{1 - x^2}}\,dx$ and

$$J_2 = \int_{-1}^{1} \frac{dx}{(\sqrt{k} - x)\sqrt{1 - x^2}}.$$

(c) Hence show that $J_1 = \pi\sqrt{k}$.

(d) By using the substitution $x = \sin\theta$ followed by a t-substitution, show that

$$J_2 = \frac{2}{\sqrt{k} - 1}\left[\tan^{-1}\left(\frac{\sqrt{k} - 1}{\sqrt{k} - 1}\right) + \tan^{-1}\left(\frac{\sqrt{k} + 1}{\sqrt{k} - 1}\right) \right].$$

(e) Let $u = \dfrac{\sqrt{k}-1}{\sqrt{k-1}}$.

   (i) Show that $\dfrac{1}{u} = \dfrac{\sqrt{k}+1}{\sqrt{k-1}}$, and

   (ii) $\tan^{-1}\left(\dfrac{\sqrt{k}-1}{\sqrt{k-1}}\right) + \tan^{-1}\left(\dfrac{\sqrt{k}+1}{\sqrt{k-1}}\right) = \dfrac{\pi}{2}$.

(f) Hence show that $f_{k-1} = \pi\sqrt{k} - \pi\sqrt{k-1}$ for $k \geqslant 2$.

(g) Hence find $\displaystyle\sum_{k=1}^{100} f_{k-1}$.

Endnote

1. A proof as to why this is the case can be found on pages 56–58 of G. H. Hardy's *The Integration of Functions of a Single Variable* (Cambridge University Press, Cambridge, 1916).

15

Further Trigonometric Integrals

Integration is hard. Very hard.

— *Words of warning given by the author to students*
on meeting integration for the first time

The wealth of identities that exist for the trigonometric functions provide one with plenty of scope in which to integrate expressions containing such functions. Integrals containing trigonometric functions have already been considered briefly as particular examples arising from other integration techniques and more systematically in Chapter 8. In this chapter we consider integrals of this type in greater detail. Their evaluation depends on making use of known trigonometric identities, coupled to, where needed, any of the methods of integration we have considered so far. We also introduce a set of rules, known as *Bioche's rules*, that can be used to help decide if an integral consisting of a rational function of sine and cosine can be integrated using one of the trigonometric substitutions of $t = \cos x$, $t = \sin x$, or $t = \tan x$.

We begin by considering three examples whose integrals consist of trigonometric functions, which show some of the ideas and techniques that can be used to find such integrals.

Example 15.1 Find $\displaystyle\int \frac{dx}{1 + \sin x}\, dx$.

Solution In Example 14.1 on page 179 of Chapter 14 a t-substitution was used to find the integral. Here we will instead show how it is possible to find the integral by a combination of manipulating the integrand and using a number of trigonometric identities.

$$\int \frac{dx}{1 + \sin x}\, dx = \int \frac{1}{1 + \sin x} \cdot \frac{1 - \sin x}{1 - \sin x}\, dx = \int \frac{1 - \sin x}{1 - \sin^2 x}\, dx$$

$$= \int \frac{1 - \sin x}{\cos^2 x}\, dx = \int \left(\frac{1}{\cos^2 x} - \frac{1}{\cos x} \cdot \frac{\sin x}{\cos x} \right) dx$$

$$= \int \left(\sec^2 x - \sec x \tan x \right) dx = \tan x - \sec x + C. \quad \blacktriangleright$$

Example 15.2 Evaluate $\displaystyle\int_0^{\frac{\pi}{4}} \frac{dx}{9 - 5\sin^2 x}\, dx.$

Solution Taking advantage of the trigonometric identity $\sin^2 x + \cos^2 x = 1$, the first term appearing in the denominator can be rewritten as $9\sin^2 x + 9\cos^2 x$. Thus

$$\int_0^{\frac{\pi}{4}} \frac{dx}{9 - 5\sin^2 x}\, dx = \int_0^{\frac{\pi}{4}} \frac{dx}{(9\sin^2 x + 9\cos^2 x) - 5\sin^5 x}$$

$$= \int_0^{\frac{\pi}{4}} \frac{dx}{4\sin^2 x + 9\cos^2 x} = \int_0^{\frac{\pi}{4}} \frac{dx}{\cos^2 x (4\tan^2 x + 9)}$$

$$= \int_0^{\frac{\pi}{4}} \frac{\sec^2 x}{9 + 4\tan^2 x}\, dx.$$

Let $u = \tan x, du = \sec^2 x\, dx$ while for the limits of integration we have when $x = 0, u = 0$ and when $x = \frac{\pi}{4}, u = 1$. Thus

$$\int_0^{\frac{\pi}{4}} \frac{dx}{9 - \sin^2 x}\, dx = \int_0^1 \frac{du}{9 + 4u^2} = \frac{1}{4} \int_0^1 \frac{du}{(\frac{3}{2})^2 + u^2}$$

$$= \frac{1}{6} \left[\tan^{-1} \left(\frac{2u}{3} \right) \right]_0^1 = \frac{1}{6} \tan^{-1} \left(\frac{2}{3} \right). \quad \blacktriangleright$$

Example 15.3 Find $\displaystyle\int \frac{\sin^3 x + \sin^2 x - 2\sin x - 2}{\sin^2 x + 2\sin x + 1}\, dx.$

Solution We will initially proceed by factorising as far as possible before manipulating the integrand. Doing so we have

$$\int \frac{\sin^3 x + \sin^2 x - 2\sin x - 2}{\sin^2 x + 2\sin x + 1}\, dx = \int \frac{\sin^2 x (\sin x + 1) - 2(\sin x + 1)}{(\sin x + 1)^2}\, dx$$

$$= \int \frac{(\sin x + 1)(\sin^2 x - 2)}{(\sin x + 1)^2}\, dx$$

$$= \int \frac{\sin^2 x - 2}{\sin x + 1}\, dx = \int \frac{(\sin^2 x - 1) - 1}{\sin x + 1}\, dx$$

$$= \int \frac{(\sin x - 1)(\sin x + 1) - 1}{\sin x + 1}\, dx$$

$$= \int (\sin x - 1)\, dx - \int \frac{dx}{1 + \sin x}$$

$$= I_1 - I_2.$$

The first integral is readily found. The result is

$$I_1 = -\cos x - x + C_1.$$

The second integral was considered previously in Example 15.1 where it was found:

$$I_2 = -\sec x + \tan x + C_2.$$

Thus

$$\int \frac{\sin^3 x + \sin^2 x - 2\sin x - 2}{\sin^2 x + 2\sin x + 1}\, dx = -x - \cos x + \sec x - \tan x + C. \ \blacktriangleright$$

§ Règles de Bioche (The Bioche Rules)

For integrals consisting of rational functions of sine and cosine (one only need consider rational functions of sine and cosine as the four other trigonometric functions can be written in terms of sine and cosine), it may be possible to find such integrals using a substitution $t = \varphi(x)$ where $\varphi(x) = \{\sin x, \cos x, \tan x\}$. However knowing which substitution to use may not always be immediately obvious.

In cases where the substitution to be used is not forthcoming, a set of rules known as *Bioche's rules* can be used to guide one towards the most effective substitution to use.[1]

Writing the integral as

$$\int f(\sin x, \cos x)\, dx,$$

and referring to the term $f(\sin x, \cos x)\, dx$ as a *differential form*, we test to see whether or not the differential form remains invariant (unchanged) under one of the three substitutions: $x \mapsto -x, \pi - x, \pi + x$. For the substitution where the differential form is invariant one sets $t = \varphi(x)$, where $\varphi(x)$ is the function $\cos x$, $\sin x$, or $\tan x$ that also remains invariant under the same substitution. That is, set $t = \cos x$ when $x \mapsto -x$ since $\cos(-x) = \cos x$; set $t = \sin x$ when $x \mapsto \pi - x$ since $\sin(\pi - x) = \sin x$; and set $t = \tan x$ when $x \mapsto \pi + x$ since $\tan(\pi + x) = \tan x$. The rules of Bioche are summarised in Table 15.1.

Table 15.1. *The Bioche rules.*

Change of variable	Resulting invariant differential form	Substitution to be used
$x \mapsto -x$	$f\big(\sin(-x), \cos(-x)\big) \cdot (-dx)$	$t = \cos x$
$x \mapsto \pi - x$	$f\big(\sin(\pi - x), \cos(\pi - x)\big) \cdot (-dx)$	$t = \sin x$
$x \mapsto \pi + x$	$f\big(\sin(\pi + x), \cos(\pi + x)\big) \cdot (dx)$	$t = \tan x$

Note it is the differential form, and not just the function corresponding to the integrand, which must be invariant under the change of variable. One then writes the integrand as a product between a term consisting of a rational function of only terms containing the substitution $\varphi(x)$ and its derivative $\varphi'(x)$. To check these conditions the following trigonometric identities will need to be recalled:

$$\sin(-x) = -\sin x, \quad \sin(\pi - x) = \sin x, \quad \sin(\pi + x) = -\sin x$$

$$\cos(-x) = \cos x, \quad \cos(\pi - x) = -\cos x, \quad \cos(\pi + x) = -\cos x.$$

In the event that more than one of the initial substitutions leaves the differential form unchanged, the differential form will be unchanged under all three substitutions. In this case while any one of the substitutions $t = \cos x, t = \sin x$, or $t = \tan x$ can be used, it is usually more efficient to use the substitution $t = \cos 2x$ since in all cases all three of the initial substitutions of $x \mapsto -x, \pi - x$, and $\pi + x$ leave $\cos 2x$ unchanged. On the other hand, if none of the substitutions work, and in the event that no other alternative method or strategy can be found, as a last resort a t-substitution can always be used. We now give several examples that make use of Bioche's rules.

Example 15.4 Find $\displaystyle\int \frac{1 - \cos x}{(1 + \cos x)\sin x}\, dx.$

Solution We first observe that the integrand is a rational function of sine and cosine. In this example the differential form is

$$f(\sin x, \cos x)\, dx = \frac{1 - \cos x}{(1 + \cos x)\sin x}\, dx,$$

so on applying the Bioche rules we have the following:

Change of variable	Resulting differential form	Invariant?
$x \mapsto -x$	$\dfrac{1 - \cos(-x)}{(1 + \cos(-x))\sin(-x)} \cdot (-dx)$ $= \dfrac{1 - \cos x}{(1 + \cos x)\sin x} dx$	Yes
$x \mapsto \pi - x$	$\dfrac{1 - \cos(\pi - x)}{(1 + \cos(\pi - x))\sin(\pi - x)} \cdot (-dx)$ $= -\dfrac{1 + \cos x}{(1 - \cos x)\sin x} dx$	No
$x \mapsto \pi + x$	$\dfrac{1 - \cos(\pi + x)}{(1 + \cos(\pi + x))\sin(\pi + x)} \cdot (dx)$ $= \dfrac{1 + \cos x}{(1 - \cos x)\sin x} dx$	No

We are therefore led to try a substitution of the form $t = \cos x$. As $dt = -\sin x\, dx$ we rewrite the integrand as the product between a rational function consisting of $\cos x$ terms and a single sine term. Thus

$$\int \frac{1 - \cos x}{(1 + \cos x)\sin x}\, dx = \int \frac{1 - \cos x}{(1 + \cos x)\sin x} \cdot \frac{\sin x}{\sin x}\, dx$$

$$= \int \frac{(1 - \cos x)\sin x}{(1 + \cos x)\sin^2 x}\, dx$$

$$= \int \frac{(1 - \cos x)\sin x}{(1 + \cos x)(1 - \cos^2 x)}\, dx$$

$$= \int \frac{(1 - \cos x)\sin x}{(1 + \cos x)^2(1 - \cos x)}\, dx$$

$$= \int \frac{\sin x}{(1 + \cos x)^2}\, dx.$$

Now let $t = \cos x, dt = -\sin x\, dx$ so that

$$\int \frac{1 - \cos x}{(1 + \cos x)\sin x}\, dx = -\int \frac{dt}{t^2} = \frac{1}{t} + C = \frac{1}{1 + \cos x} + C. \quad \blacktriangleright$$

Example 15.5 Find $\displaystyle\int \frac{dx}{1 + \sin^2 x}$.

Solution In this example the integrand is a rational function of sine only, with a differential form given by

$$f(\sin x, \cos x)\, dx = \frac{dx}{1 + \sin^2 x}.$$

Applying the Bioche rules to this differential form we have the following:

Change of variable	Resulting differential form	Invariant?
$x \mapsto -x$	$\dfrac{1}{1 + \sin^2(-x)} \cdot (-dx) = -\dfrac{1}{1 + \sin^2 x}\, dx$	No
$x \mapsto \pi - x$	$\dfrac{1}{1 + \sin^2(\pi - x)} \cdot (-dx) = -\dfrac{1}{1 + \sin^2 x}\, dx$	No
$x \mapsto \pi + x$	$\dfrac{1}{1 + \sin^2(\pi + x)} \cdot (dx) = \dfrac{1}{1 + \sin^2 x}\, dx$	Yes

We are therefore led to try a substitution of the form $t = \tan x$. As $dt = \sec^2 x\, dx$ we rewrite the integrand as the product between a rational function consisting of $\tan x$ terms and a $\sec^2 x$ term. So

$$\int \frac{dx}{1 + \sin^2 x} = \int \frac{1}{1 + \sin^2 x} \cdot \frac{\sec^2 x}{\sec^2 x}\, dx = \int \frac{\sec^2 x}{\sec^2 x + \tan^2 x}\, dx$$

$$= \int \frac{\sec^2 x}{(1 + \tan^2 x) + \tan^2 x}\, dx = \int \frac{\sec^2 x}{1 + 2\tan^2 x}\, dx.$$

Now let $t = \tan x$, $dt = \sec^2 t\, dt$ so that

$$\int \frac{dx}{1 + \sin^2 x} = \int \frac{dt}{1 + 2t^2}\, dx = \frac{1}{2} \int \frac{dt}{(1/\sqrt{2})^2 + t^2}$$

$$= \frac{1}{\sqrt{2}} \tan^{-1}(t\sqrt{2}) + C = \frac{1}{\sqrt{2}} \tan^{-1}(\sqrt{2}\tan x) + C. \;\blacktriangleright$$

Example 15.6 Find $\displaystyle\int \frac{\sec^2 x}{\sec x + \tan x}\, dx.$

Solution The integrand can be converted into a function consisting of sines and cosines only. Noting that

$$\frac{\sec^2 x}{\sec x + \tan x} = \frac{1}{\cos^2 x} \cdot \frac{1}{1/\cos x + \sin x/\cos x} = \frac{1}{\cos x(1 + \sin x)},$$

the integrand is therefore a rational function of sine and cosine. In this case the differential form is given by

$$f(\sin x, \cos x)\, dx = \frac{dx}{\cos x(1 + \sin x)}.$$

Applying the Bioche rules to the differential form we have the following:

Change of variable	Resulting differential form	Invariant?
$x \mapsto -x$	$\dfrac{1}{\cos(-x)(1 + \sin(-x))} \cdot (-dx)$	No
	$= -\dfrac{dx}{\cos x(1 - \sin x)}\, dx$	
$x \mapsto \pi - x$	$\dfrac{1}{\cos(\pi - x)(1 + \sin(\pi - x))} \cdot (-dx)$	Yes
	$= \dfrac{dx}{\cos x(1 + \sin x)}\, dx$	
$x \mapsto \pi + x$	$\dfrac{1}{\cos(\pi + x)(1 + \sin(\pi + x))} \cdot (dx)$	No
	$= -\dfrac{dx}{\cos x(1 - \sin x)}\, dx$	

We are therefore led to try a substitution of the form $t = \sin x$. As $dt = \cos x\, dx$ we rewrite the integrand as the product between a rational function consisting of $\sin x$ terms and a single cosine term. So

$$\int \frac{\sec^2 x}{\sec x + \tan x}\, dx = \int \frac{dx}{\cos x(1 + \sin x)} = \int \frac{1}{\cos x(1 + \sin x)} \cdot \frac{\cos x}{\cos x}\, dx$$

$$= \int \frac{\cos x}{\cos^2 x(1 + \sin x)}\, dx = \int \frac{\cos x}{(1 - \sin^2 x)(1 + \sin x)}\, dx$$

$$= \int \frac{\cos x}{(1 - \sin x)(1 + \sin x)^2}\, dx.$$

Now let $t = \sin x, dt = \cos t\, dt$ so that

$$\int \frac{\sec^2 x}{\sec x + \tan x}\, dx = \int \frac{dt}{(1 - t)(1 + t)^2}.$$

The resultant integral can be found by employing a partial fraction decomposition. For the integrand, using partial fractions, it can be readily shown that

$$\frac{1}{(1-t)(1+t)^2} = \frac{1}{4(1-t)} + \frac{1}{4(1+t)} + \frac{1}{2(1+t)^2}.$$

Thus

$$\int \frac{\sec^2 x}{\sec x + \tan x}\,dx = \frac{1}{4}\int \frac{dt}{1-t} + \frac{1}{4}\int \frac{dt}{1+t} + \frac{1}{2}\int \frac{dt}{(1+t)^2}$$

$$= -\frac{1}{4}\ln|1-t| + \frac{1}{4}\ln|1+t| - \frac{1}{2(1+t)} + C$$

$$= \frac{1}{4}\ln\left|\frac{1+\sin x}{1-\sin x}\right| - \frac{1}{2(1+\sin x)} + C. \qquad \blacktriangleright$$

The final two examples give situations where the differential form is not invariant under any of the transformations and then where it is invariant under all three transformations.

Example 15.7 Find $\displaystyle\int \frac{\sin x}{1+\sin x}\,dx.$

Solution The integrand is a rational function of sine only with differential form

$$f(\sin x, \cos x)\,dx = \frac{\sin x}{1+\sin x}\,dx.$$

Applying Bioche's rules to the differential form we have the following:

Change of variable	Resulting differential form		Invariant?
$x \mapsto -x$	$\dfrac{\sin(-x)}{1+\sin(-x)} \cdot (-dx) =$	$\dfrac{\sin x}{1-\sin x}\,dx$	No
$x \mapsto \pi - x$	$\dfrac{\sin(\pi-x)}{1+\sin(\pi-x)} \cdot (-dx) =$	$-\dfrac{\sin x}{1+\sin x}\,dx$	No
$x \mapsto \pi + x$	$\dfrac{\sin(\pi+x)}{1+\sin(\pi+x)} \cdot (dx) =$	$-\dfrac{\sin x}{1-\sin x}\,dx$	No

As the differential form is not invariant under any of the transformations, Bioche's rules tell us the integral cannot be found using any of the substitutions $t = \cos x, t = \sin x$, or $t = \tan x$. In the event no other alternative method is seen, as the integral is a rational function of sine, it can always be solved using a tangent half-angle substitution (as was done in Example 14.4 on page 181 of Chapter 14). ▶

Example 15.8 Find $\displaystyle\int \frac{\cos x \sin^3 x}{1 - \sin^2 x \cos^2 x}\, dx$.

Solution As the integrand consists of a rational function of sine and cosine, Bioche's rules can be applied. On doing so we have

Change of variable	Resulting differential form	Invariant?
$x \mapsto -x$	$\dfrac{\cos x \sin^3 x}{1 - \sin^2 x \cos^2 x}\, dx$	Yes
$x \mapsto \pi - x$	$\dfrac{\cos x \sin^3 x}{1 - \sin^2 x \cos^2 x}\, dx$	Yes
$x \mapsto \pi + x$	$\dfrac{\cos x \sin^3 x}{1 - \sin^2 x \cos^2 x}\, dx$	Yes

As the differential form is invariant under all three change of variables, the substitution $t = \cos 2x$ can be used. As $dt = -2 \sin 2x\, dx$ we rewrite the integrand as the product between a rational function consisting of $\cos 2x$ terms and a single $\sin 2x$ term. In doing so we start by writing the integral as

$$\int \frac{\cos x \sin^3 x}{1 - \sin^2 x \cos^2 x}\, dx = \int \frac{\sin^2 x \cdot \sin x \cos x}{1 - \sin^2 x \cos^2 x}\, dx.$$

Recalling

$$\sin x \cos x = \frac{\sin 2x}{2}, \quad \sin^2 x = \frac{1 - \cos 2x}{2}, \quad \text{and}\quad \cos^2 x = \frac{1 + \cos 2x}{2},$$

the integral can be rewritten as

$$\int \frac{\cos x \sin^3 x}{1 - \sin^2 x \cos^2 x}\, dx = -\frac{1}{4}\int \frac{\left(\frac{1 - \cos 2x}{2}\right)}{1 - \left(\frac{1 - \cos 2x}{2}\right)\left(\frac{1 + \cos 2x}{2}\right)}\, d(\cos 2x).$$

Setting $t = \cos 2x$ gives

$$\int \frac{\cos x \sin^3 x}{1 - \sin^2 x \cos^2 x}\, dx = -\frac{1}{2}\int \frac{1 - t}{4 - (1 - t)(1 + t)}\, dt = -\frac{1}{2}\int \frac{1 - t}{3 + t^2}\, dt$$

$$= -\frac{1}{2}\int \frac{dt}{3 + t^2} + \frac{1}{2}\int \frac{t}{3 + t^2}\, dt.$$

Each of these integrals is now in standard form and can be readily found. The result is

$$\int \frac{\cos x \sin^3 x}{1 - \sin^2 x \cos^2 x} \, dx = -\frac{1}{2\sqrt{3}} \tan^{-1}\left(\frac{\cos 2x}{\sqrt{3}}\right) + \frac{1}{4} \ln\left(3 + \cos^2 2x\right) + C.$$

▶

To finish this section we give a few more miscellaneous examples of integrals containing only trigonometric functions that illustrate other possible ideas for tackling such integrals not considered so far.

Example 15.9 Find $\int \dfrac{\sin 4x + \sin 5x}{2\cos 3x + 1} \, dx.$

Solution In finding the integral we will make use of the following sum-to-product identity:

$$\sin \theta - \sin \varphi = 2 \sin\left(\frac{\theta - \varphi}{2}\right) \cos\left(\frac{\theta + \varphi}{2}\right).$$

To find this integral our general strategy is to try and remove the term in the denominator. We will do this through a suitable factorisation. As the denominator contains a $\cos 3x$ term, we need to add such a term into the numerator. Observing that $(5 + 1)/2 = 3$ and $(4 + 2)/2 = 3$, as each of the numerical factors of $3, 4$, and 5 appear in the arguments of the trigonometric functions, it suggests that the following two sum-to-product identities be used:

$$\sin 5x - \sin x = 2\sin 2x \cos 3x \quad \text{and} \quad \sin 4x - \sin 2x = 2 \sin x \cos 3x.$$

Now, on rewriting the integral I we have

$$I = \int \frac{(2\sin x \cos 3x + \sin 2x) + (2\sin 2x \cos 3x + \sin x)}{2\cos 3x + 1} \, dx$$

$$= \int \frac{2\cos 3x (\sin x + \sin 2x) + (\sin x + \sin 2x)}{2\cos 3x + 1} \, dx$$

$$= \int \frac{(2\cos 3x + 1)(\sin x + \sin 2x)}{2\cos 3x + 1} \, dx = \int (\sin x + \sin 2x) \, dx,$$

giving

$$\int \frac{\sin 4x + \sin 5x}{2\cos 3x + 1} \, dx = -\cos x - \frac{1}{2}\cos 2x + C.$$

▶

Example 15.10

(a) Let $f_n(x) = \frac{1}{2} D_n(x) = \frac{1}{2} + \sum_{k=1}^{n} \cos kx$. By multiplying $f_n(x)$ by $2 \sin \frac{x}{2}$ and evaluating the resulting finite sum, show that

$$D_n(x) = 1 + 2 \sum_{k=1}^{n} \cos kx = \frac{\sin(n + \frac{1}{2})x}{\sin(\frac{x}{2})}.$$

The collection of functions $D_n(x)$ is known as the *Dirichlet kernel*[2] and has many, very important applications in the field of mathematical analysis.

(b) Using the result given in (a), find $\int \dfrac{\sin(2n + 1)x}{\sin x} \, dx$.

Solution (a) On multiplying $f_n(x)$ by the term $2 \sin \frac{x}{2}$ one has

$$2 \sin \frac{x}{2} f_n(x) = \sin \frac{x}{2} + 2 \sin \frac{x}{2} \sum_{k=1}^{n} \cos kx$$

$$= \sin \frac{x}{2} + \sum_{k=1}^{n} 2 \cos kx \sin \frac{x}{2}$$

$$= \sin \frac{x}{2} + \left[2 \cos x \sin \frac{x}{2} + 2 \cos 2x \sin \frac{x}{2} + \cdots \right.$$

$$\left. \cdots + 2 \cos nx \sin \frac{x}{2} \right].$$

On applying the following prosthaphaeresis formula

$$2 \cos A \sin B = \sin(A + B) - \sin(A - B),$$

to each term appearing in the square brackets, one has

$$2 \sin \frac{x}{2} f_n(x) = \sin \frac{x}{2} + \left[\left(\sin \frac{3x}{2} - \sin \frac{x}{2} \right) + \left(\sin \frac{5x}{2} - \sin \frac{3x}{2} \right) \right.$$

$$+ \left(\sin \frac{7x}{2} - \sin \frac{5x}{2} \right) + \cdots$$

$$\cdots + \left(\sin \left(nx - \frac{x}{2} \right) - \sin \left(nx - \frac{3x}{2} \right) \right)$$

$$\left. + \left(\sin \left(nx + \frac{x}{2} \right) - \sin \left(nx - \frac{x}{2} \right) \right) \right].$$

The type of sum we have here is an example of a *telescoping series*. Here we can see that all the terms in the square brackets will cancel out, except for the

$-\sin \frac{x}{2}$ term in the first bracket and the $\sin(nx + \frac{x}{2})$ term in the last bracket. Thus

$$2 \sin \frac{x}{2} f_n(x) = \sin \frac{x}{2} + \left[-\sin \frac{x}{2} + \sin \left(n + \frac{1}{2} \right) x \right] = \sin \left(n + \frac{1}{2} \right) x$$

$$\Rightarrow f_n(x) = \frac{\sin \left(n + \frac{1}{2} \right) x}{2 \sin \frac{x}{2}},$$

or

$$D_n(x) = 1 + 2 \sum_{k=1}^{n} \cos kx = \frac{\sin(n + \frac{1}{2})x}{\sin \frac{x}{2}},$$

as required to show.

(b) Letting $u = 2x$ in the integral, as $du = 2\,dx$ we have

$$\int \frac{\sin(2n + 1)x}{\sin x} dx = \frac{1}{2} \int \frac{\sin(n + \frac{1}{2})u}{\sin \frac{u}{2}} du$$

$$= \frac{1}{2} \int \left[1 + 2 \sum_{k=1}^{n} \cos ku \right] du$$

(using the result from (a))

$$= \frac{1}{2} \int du + \sum_{k=1}^{n} \int \cos ku\, du$$

$$= \frac{u}{2} + \sum_{k=1}^{n} \frac{1}{k} \sin ku + C$$

$$= x + \sum_{k=1}^{n} \frac{1}{k} \sin 2kx + C. \qquad \blacktriangleright$$

Exercises for Chapter 15

✠ Warm-ups

1. Using the result given in Example 15.10, evaluate $\displaystyle\int_{0}^{\frac{\pi}{2}} \frac{\sin(2017x)}{\sin x} dx$.

2. By exploiting double-angle formulae for sine and cosine, find

$$\int \frac{4\cos^2 x + \cos x - 2}{\sin^2 x (1 + 2\cos x)^2} dx.$$

✠ Practice questions

3. Find the following trigonometric integrals.

(a) $\displaystyle\int \frac{dx}{1 + \cos x}$

(b) $\displaystyle\int \frac{\csc^2 x}{\cot^2 x}\, dx$

(c) $\displaystyle\int \frac{dx}{\sec x \tan^2 x}$

(d) $\displaystyle\int \sin^2 x \tan x\, dx$

(e) $\displaystyle\int \frac{\cos 2x}{\cos^2 x}\, dx$

(f) $\displaystyle\int \frac{\cos^2 x}{1 + \sin x}\, dx$

(g) $\displaystyle\int \tan x \sin 2x\, dx$

(h) $\displaystyle\int \frac{\cos 3x}{\cos^3 x}\, dx$

(i) $\displaystyle\int \frac{\sec x}{\sec x + 1}\, dx$

(j) $\displaystyle\int \frac{dx}{\sin x + \tan x}$

(k) $\displaystyle\int \sec x \sec 2x\, dx$

(l) $\displaystyle\int \frac{\sec^3 x}{\tan^2 x}\, dx$

(m) $\displaystyle\int \frac{\sin 8x}{\sin x}\, dx$

(n) $\displaystyle\int \frac{\sin^2 x}{\sin x + 2\cos x}\, dx$

(o) $\displaystyle\int \frac{\cos^2 x}{1 + \sin^2 x}\, dx$

(p) $\displaystyle\int \frac{1 - \sin x}{(1 + \sin x)\cos x}\, dx$

(q) $\displaystyle\int \frac{\tan^2 x}{1 - \tan^2 x}\, dx$

(r) $\displaystyle\int \frac{\cos x}{(\sin x + \cos x)^2}\, dx$

(s) $\displaystyle\int \frac{\cos 5x + \cos 4x}{1 - 2\cos 3x}\, dx$

(t) $\displaystyle\int \sec 2x \csc 4x\, dx$

(u) $\displaystyle\int \left(\tan^{n-1} x + \tan^{n+1} x\right) dx, n \neq 0$

(v) $\displaystyle\int \frac{\sin^2 x - 4\sin x \cos x + 3\cos^2 x}{\sin x + \cos x}\, dx$

4. (a) Find $\displaystyle\int \frac{\cos x}{2\sin^2 x - 1}\, dx$.

(b) Using the substitution $x = \sec\theta$, together with the result found in (a), show that

$$\int \frac{dx}{(x^2 - 2)\sqrt{x^2 - 1}} = \frac{1}{2\sqrt{2}} \ln\left|\frac{\sqrt{2}\sqrt{x^2 - 1} - x}{\sqrt{2}\sqrt{x^2 - 1} + x}\right| + C.$$

5. Consider the integral $\int \dfrac{\sin^2 x \cos x}{\sin x + \cos x} dx$.

By multiplying the integrand by the term $\dfrac{\cos x - \sin x}{\cos x - \sin x}$, use this to show that

$$\int \frac{\sin^2 x \cos x}{\sin x + \cos x} dx = \frac{1}{8}\ln|\sec 2x + \tan 2x| - \frac{1}{8}\sin 2x$$

$$+ \frac{1}{8}\ln|\cos 2x| - \frac{1}{8}\cos 2x + C.$$

6. (a) If $a \neq b$, find real numbers A and B such that

$$\frac{\sin x}{\sin(x - a)\sin(x - b)} = \frac{A}{\sin(x - a)} + \frac{B}{\sin(x - b)}.$$

(b) Use the result found in (a) to find $\displaystyle\int \frac{\sin x}{\sin(x - a)\sin(x - b)} dx$.

7. Find the following trigonometric integrals using the indicated trigonometric substitution. Note it may be necessary to manipulate the integrand first before the substitution is made.

(a) $\displaystyle\int \frac{\cos x}{6 + 2\sin x - \cos^2 x} dx \qquad (u = \sin x)$

(b) $\displaystyle\int \frac{\sec^2 x}{(\sec x + \tan x)^{\frac{5}{2}}} \qquad (u = \sec x + \tan x)$

(c) $\displaystyle\int \frac{\sin^2 x}{\cos^4 x + \cos x \sin^3 x} dx \qquad (u = \tan x)$

(d) $\displaystyle\int \frac{\sin 2x}{1 - \sin^3 x} dx \qquad (u = \sin x)$

(e) $\displaystyle\int \frac{\cos^2 x}{\sin^2 x + 4\sin x \cos x} dx \qquad (u = \cot x)$

(f) $\displaystyle\int \sin x \sec^{\frac{3}{2}}(2x) dx \qquad (u = \cos x)$

8. Evaluate $\displaystyle\int_0^{100\pi} \sqrt{1 - \cos 2x}\, dx$.

9. Consider the integral $\displaystyle\int_0^{\frac{\pi}{4}} \frac{dx}{\cos^2 x + \sin 2x + 5\sin^2 x}$.

(a) Using the substitution $u = \tan x$, show that

$$\int_0^{\frac{\pi}{4}} \frac{dx}{\cos^2 x + \sin 2x + 5 \sin^2 x} = \frac{1}{2} \left[\tan^{-1}(3) - \tan^{-1}\left(\frac{1}{2}\right) \right].$$

(b) Let $z = (1 + 2i) \cdot (1 + 3i)$ where $z \in \mathbb{C}$.
   (i) Express z in the form $a + ib$ where $a, b \in \mathbb{R}$.
   (ii) By finding $\arg(z)$, show that $\tan^{-1}(2) + \tan^{-1}(3) = \dfrac{3\pi}{4}$.

(c) Hence deduce that $\displaystyle\int_0^{\frac{\pi}{4}} \frac{dx}{\cos^2 x + \sin 2x + 5 \sin^2 x} = \frac{\pi}{8}$.

10. Let $I = \displaystyle\int \sqrt{\tan x}\, dx$ and $J = \displaystyle\int \sqrt{\cot x}\, dx$.

(a) Show that

$$I + J = \sqrt{2} \sin^{-1}(\sin x - \cos x) + C,$$

and

$$I - J = -\sqrt{2} \ln \left| \sqrt{\sin 2x} + \sin x + \cos x \right| + C.$$

(b) Hence deduce that

$$\int \sqrt{\tan x}\, dx = \frac{1}{\sqrt{2}} \sin^{-1}(\sin x - \cos x)$$

$$- \frac{1}{\sqrt{2}} \ln \left| \sqrt{\sin 2x} + \sin x + \cos x \right| + C,$$

and

$$\int \sqrt{\cot x}\, dx = \frac{1}{\sqrt{2}} \sin^{-1}(\sin x - \cos x)$$

$$+ \frac{1}{\sqrt{2}} \ln \left| \sqrt{\sin 2x} + \sin x + \cos x \right| + C.$$

11. (a) Find $\displaystyle\int \tan(x - a) \tan(x - b)\, dx$, $a \neq b$.

(b) Hence find $\displaystyle\int \tan^2(x - a) \tan^2(x - b)\, dx$, $a \neq b$.

12. In Example 15.8 the integral was found using Bioche's rules. In this example the differential form was invariant under all three of the initial substitutions of $x \mapsto x$, $\pi - x$, and $\pi + x$. In such a case it is usually more efficient to find such an integral using a substitution of $t = \cos 2x$ as the rational function in t that the rationalising substitution produces tends to lead to integrals that are less tedious to find. However, as was pointed

out, integrals where the three initial substitutions remain unchanged can be found using any of the three substitutions $t = \cos x$, $t = \sin x$, or $t = \tan x$. Find the integral in Example 15.8 using each of the three substitutions of $t = \cos x$, $t = \sin x$, and $t = \tan x$.

13. If $a \in \mathbb{R}$ find

(a) $\displaystyle\int \frac{\sin(x - a)}{\sin(x + a)}\, dx$

(b) $\displaystyle\int \frac{\cos 2x - \cos 2a}{\cos x - \cos a}\, dx$

(c) $\displaystyle\int \frac{\sin x - \sin 2a}{\sin x - \sin a}\, dx$

(d) $\displaystyle\int \frac{\tan a - \tan x}{\tan a + \tan x}\, dx$

(e) $\displaystyle\int \frac{\cos(x + a)}{\cos(x - a)}\, dx$

(f) $\displaystyle\int \sqrt{\frac{\sin(x - a)}{\sin(x + a)}}\, dx$

14. Using a substitution of $u = \cot x$, find the value of

$$\int_{\frac{\pi}{4}}^{\frac{\pi}{2}} \frac{dx}{\sin^2 x (\sin^2 x + 1)(\sin^2 x + 2)}.$$

15. Let $I_n = \displaystyle\int_0^{\frac{\pi}{2}} \frac{\sin(2n + 1)x}{\sin x}\, dx$ where $n \in \mathbb{N}$.

(a) Show that $I_n - I_{n-1} = 0$.

(b) Hence show that $\displaystyle\int_0^{\frac{\pi}{2}} \frac{\sin(2n + 1)x}{\sin x}\, dx = \frac{\pi}{2}$.

16. Let $I_n = \displaystyle\int_0^{\pi} \frac{\sin(nx)\cos\left(\frac{x}{2}\right)}{\sin\left(\frac{x}{2}\right)}\, dx$ where $n \in \mathbb{N}$.

(a) Show that $I_{n+1} - I_n = 0$.

(b) Hence deduce that $\displaystyle\int_0^{\pi} \frac{\sin(nx)\cos\left(\frac{x}{2}\right)}{\sin\left(\frac{x}{2}\right)}\, dx = \pi$.

17. If $a - b \ne \pi k$, $k \in \mathbb{Z}$, show that

$$\int \frac{dx}{\sin(x + a)\sin(x + b)} = \frac{1}{\sin(a - b)} \ln \left| \frac{\sin(x + a)}{\sin(x + b)} \right| + C.$$

18. (a) Show that $\sin x \sin(2k - 1)x = \dfrac{\cos(2k - 2)x - \cos 2kx}{2}$, $k \in \mathbb{N}$.

(b) Use the result in (a) to show that $\displaystyle\sum_{k=1}^{n} \sin(2k - 1)x = \frac{\sin^2 nx}{\sin x}$.

(c) Using the result in (b), find $\displaystyle\int \frac{\sin^2 nx}{\sin x}\,dx$.

(d) Hence deduce that $\displaystyle\int_0^\pi \frac{\sin^2 4x}{\sin x}\,dx = \frac{352}{105}$.

19. (a) If $\alpha \neq 0$, find $\displaystyle\int \frac{1-\alpha^2}{1-2\alpha\cos x + \alpha^2}\,dx$.

(b) Hence find $\displaystyle\int \frac{\alpha - \cos x}{1 - 2\alpha\cos x + \alpha^2}\,dx$.

20. Consider the integral $\displaystyle\int_0^{\frac{\pi}{2}} \frac{dx}{1 + \cos\alpha\sin x}$.

(a) Evaluate the integral if $\alpha = 2n\pi$, where $n \in \mathbb{Z}$.

(b) Evaluate the integral if $\alpha \neq n\pi$, where $n \in \mathbb{Z}$.

21. If $\displaystyle J_n = \int \frac{dx}{\sin^n x + \cos^n x}$ find J_n when $n = 1, 2, 3, 4$, and 6.

22. Given that $\displaystyle\int_0^{\frac{\pi}{2}} \tan^{-1}(\sin^2 x)\,dx = \gamma$, find the value of

$$\int_0^{\frac{\pi}{2}} \tan^{-1}(1 - \sin^2 x \cos^2 x)\,dx, \quad \text{in terms of } \gamma.$$

23. In this question we will find the value for the definite integral $\displaystyle\int_0^{\frac{\pi}{2}} x\cot x\,dx$ and consider one consequence stemming from this result.

(a) Begin by writing the integral as

$$\int_0^{\frac{\pi}{2}} x\cot x\,dx = \int_0^{\frac{\pi}{4}} x\cot x\,dx + \int_{\frac{\pi}{4}}^{\frac{\pi}{2}} x\cot x\,dx.$$

By making a substitution of $x = \dfrac{\pi}{2} - u$ in the second integral on the right, show that

$$\int_0^{\frac{\pi}{2}} x\cot x\,dx = \frac{\pi}{4}\ln 2 + \int_0^{\frac{\pi}{4}} x\cot x\,dx - \int_0^{\frac{\pi}{4}} x\tan x\,dx.$$

(b) Hence deduce that $\displaystyle\int_0^{\frac{\pi}{2}} x\cot x\,dx = \frac{\pi}{2}\ln 2$.

(c) Starting with the result given in (b), using a suitable substitution, evaluate $\displaystyle\int_0^1 \frac{\sin^{-1} x}{x}\,dx$.

24. (a) By using integration by parts, show that

$$\int \cot^{-1} x \, dx = x \cot^{-1} x + \frac{1}{2} \ln(1 + x^2) + C.$$

(b) By writing down the angle sum identity formula for $\tan(\alpha + \beta)$, show that

$$\cot(\alpha + \beta) = \frac{\cot \alpha \cot \beta - 1}{\cot \alpha + \cot \beta}.$$

(c) Use the result in (b) to show that $\cot^{-1} \alpha - \cot^{-1} \beta = \cot^{-1} \left(\frac{\alpha\beta + 1}{\alpha - \beta} \right).$

(d) Hence find $\int \cot^{-1}(x^2 + x + 1) \, dx.$

⌖ **Extension Questions and Challenge problems**

25. Find $\displaystyle\int \frac{2 \cos^2 x + 2}{\sin^3 x} \, dx.$

26. For all values of $a \in \mathbb{R}$, find $\displaystyle\int \frac{dx}{1 + a^2 \tan^2 x}.$

27. Using sum-to-product identities for the trigonometric functions, find

(a) $\displaystyle\int \frac{\cos 5x - \cos 4x}{1 - \cos 3x} \, dx$ 　　　(b) $\displaystyle\int \frac{\sin 5x - \sin 4x}{1 + \cos 2x} \, dx$

28. Use an appropriate substitution to evaluate

$$\int_{\frac{\pi}{4}}^{\frac{\pi}{3}} \frac{(\sin^3 x - \cos^3 x - \cos^2 x)(\sin x + \cos x + \cos^2 x)^5}{\sin^7 x \cos^7 x} \, dx.$$

29. Let $I = \displaystyle\int \frac{\sin^3 x}{\sin^3 x - \cos^3 x} \, dx$ and $J = \displaystyle\int \frac{\cos^3 x}{\sin^3 x - \cos^3 x} \, dx.$

(a) Show that $I - J = x + C_1.$

(c) By making use of the substitution $u = \sin x - \cos x$, show that

$$I + J = \frac{1}{3} \ln \left| \frac{\sin x - \cos x}{[3 - (\sin x - \cos x)^2]^2} \right| + C_2.$$

(d) Hence deduce that

$$\int \frac{\sin^3 x}{\sin^3 x - \cos^3 x} \, dx = \frac{x}{2} + \frac{1}{3} \ln \left| \frac{\sin x - \cos x}{[3 - (\sin x - \cos x)^2]^2} \right| + C,$$

and

$$\int \frac{\cos^3 x}{\sin^3 x - \cos^3 x} \, dx = -\frac{x}{2} + \frac{1}{3} \ln \left| \frac{\sin x - \cos x}{[3 - (\sin x - \cos x)^2]^2} \right| + C.$$

30. (a) Let $S_n(x) = \sin x + \sin 2x + \sin 3x + \cdots + \sin(n-1)x$.

By multiplying $S_n(x)$ by the term $2 \sin \frac{x}{2}$ and summing the resulting telescoping series, show that

$$S_n(x) = \sum_{k=1}^{n-1} \sin kx = 2 \sin \frac{nx}{2} \sin \left(\frac{n-1}{2} \right) x.$$

(b) Let $C_n(x) = \cos x + \cos 2x + \cos 3x + \cdots + \cos(n-1)x$.

By multiplying $C_n(x)$ by the term $2 \sin \frac{x}{2}$ and summing the resulting telescoping series, show that

$$C_n(x) = \sum_{k=1}^{n-1} \cos kx = 2 \cos \frac{nx}{2} \sin \left(\frac{n-1}{2} \right) x.$$

(c) Hence deduce that

$$\int \frac{\sin x + \sin 2x + \cdots + \sin(n-1)x}{\cos x + \cos 2x + \cdots + \cos(n-1)x} \, dx = \frac{2}{n} \ln \left| \sec \frac{nx}{2} \right| + C,$$

where $n \in \mathbb{N}$.

(d) Using (c), evaluate $\displaystyle\int_0^{\frac{\pi}{16}} \frac{\sin x + \sin 2x + \sin 3x + \sin 4x}{\cos x + \cos 2x + \cos 3x + \cos 4x} \, dx$.

31. Let $P_1(x) = x^2 - 2$ and $P_k(x) = P_1(P_{k-1}(x))$ for $k = 2, 3, \ldots, n$.

(a) Find $\displaystyle\int P_n(x) \, dx$. Here $n \in \mathbb{N}$.

(b) Hence deduce that $\displaystyle\int_0^2 P_n(x) \, dx = \frac{4}{1 - 2^{2n}}$.

32. Suppose that $I : \mathbb{R} \to \mathbb{R}$ is the function defined by

$$I(x) = \int_0^{\pi} \ln(1 - 2x \cos \theta + x^2) \, d\theta.$$

(a) Show that $I(0) = 0$.

(b) Show that I is an even function, namely that $I(x) = I(-x)$.

(c) If $x \neq 0$ show that $I(x) = 2\pi \ln |x| + I \left(\frac{1}{x} \right)$.

(d) Show that $I(x) = \frac{1}{2} I(x^2)$.

(e) Hence deduce that $I(x) = \frac{1}{2^n} I \left(x^{2^n} \right)$ for $n \in \mathbb{N}$.

(f) Use the result in (d) to show that $I(1) = I(-1) = 0$.

(g) For $|x| < 1$, show $I(x) = 0$. Show this by showing that from (e), as $n \to \infty$, $I(x) \to 0$ for all $|x| < 1$.

(h) Hence deduce that $I(x) = \begin{cases} 0, & |x| \leqslant 1 \\ 2\pi \ln |x|, & |x| > 1. \end{cases}$

(i) Use the result in (h) to show that

$$\int_0^\pi \ln(\sin \theta) \, d\theta = \int_0^\pi \ln(\cos \theta) \, d\theta = -\pi \ln 2.$$

33. (a) By using the substitution $x = 2 \cos t$, find

$$\int \underbrace{\sqrt{2 + \sqrt{2 + \cdots + \sqrt{2 + x}}}}_{n \text{ terms}} \, dx.$$

(b) By using the substitution $x = \cos t$, find

$$\int \underbrace{\sqrt{\frac{1}{2} + \frac{1}{2}\sqrt{\frac{1}{2} + \cdots + \frac{1}{2}\sqrt{\frac{1+x}{2}}}}}_{n \text{ terms}} \, dx.$$

34. (a) Show that $\tan x = \cot x - 2 \cot 2x$.

(b) Use the result in (a) to show that $2 \tan 2x = 2 \cot 2x - 4 \cot 4x$.

(c) Hence deduce that

$$\tan(x) = \cot(x) - 2 \cot(2x)$$
$$2 \tan(2x) = 2 \cot(2x) - 2^2 \cot(2^2 x)$$
$$2^2 \tan(2^2 x) = 2^2 \cot(2^2 x) - 2^3 \cot(2^3 x)$$
$$2^3 \tan(2^3 x) = 2^3 \cot(2^3 x) - 2^4 \cot(2^4 x)$$

$$\vdots$$

$$2^{n-1} \tan(2^{n-1} x) = 2^{n-1} \cot(2^{n-1} x) - 2^n \cot(2^n x).$$

(d) By summing the first n terms given in (c) together, show that

$$\tan x + 2 \tan 2x + \cdots + 2^{n-1} \tan(2^{n-1} x) = \cot x - 2^n \cot(2^n x).$$

(e) Using the result in (d), find

$$\int (\tan x + 2 \tan 2x + 4 \tan 4x + 8 \tan 8x) \, dx.$$

Endnotes

1. The rules are named after the French mathematician Charles Bioche (1859–1949) who first developed them in 1902 in connection with solving a certain class of trigonometric equations. See Charles Bioche, 'Sur les équations trigonométriques', *Journal de Mathématiques Élémentaires* **26**(13), 105 (1902).
2. It is named after the German mathematician Johann Peter Gustav Lejeune Dirichlet (1805–1859) who did important work in the areas of mathematical analysis and analytic number theory.

16

Further Properties for Definite Integrals

So what's the point of calculating definite integrals since you can't possibly do them all?...Well, what's next – you can't possibly add together all possible pairs of the real numbers, so why bother learning to add?

— Paul J. Nahin, *Inside Interesting Integrals*

In Chapters 2 and 4 a number of basic properties for the definite integral were given. In this chapter we consider a number of other important properties for the definite integral. As we will see, in certain situations these properties may prove to be very useful since they may allow one to find the value for a definite integral, even though it may not be possible to express the primitive for the corresponding indefinite integral in elementary terms.

Property 1 – Invariance Under Translation
If the set of ordinates of a function f are shifted or translated by an amount k, the resulting set of ordinates is another function g related to f by the equation $g(x) = f(x - k)$. If f is defined on the interval $[a, b]$, then g will be defined on the interval $[a + k, b + k]$, and the fact that their sets of ordinates are the same means their definite integrals will also be the same, namely

$$\int_a^b f(x)\,dx = \int_{a+k}^{b+k} f(x - k)\,dx$$

Proof Let $x = u - k, dx = du$ while for the limits of integration, when $x = b, u = b + k$ while when $x = a, u = a + k$. On substituting into the

209

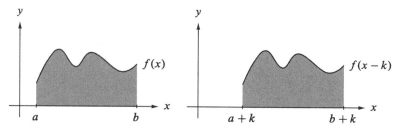

Figure 16.1. Illustration of the translational property of the definite integral.

integral on the left, one has

$$\int_a^b f(x)\,dx = \int_{a+k}^{b+k} f(u-k)\,du = \int_{a+k}^{b+k} f(x-k)\,dx,$$

on changing the dummy variable u back to x, as required to prove. ■

The translational property for the definite integral is illustrated in Figure 16.1.

Property 2 – Change in the Interval of Integration

The homogeneous property given in Chapter 2 explains what happens to a definite integral resulting from a change of scale on the y-axis. A change of scale on the x-axis (either an 'expansion' or a 'contraction') corresponds to a change in the interval of integration. Here one has

$$\boxed{\int_a^b f(x)\,dx = \frac{1}{k}\int_{ka}^{kb} f\left(\frac{x}{k}\right) dx, \quad k \neq 0}$$

Proof Let $x = \dfrac{u}{k}, dx = \dfrac{du}{k}$ where $k \neq 0$. For the limits of integration, when $x = b, u = kb$ and when $x = a, u = ka$. On substituting into the integral on the left, one has

$$\int_a^b f(x)\,dx = \frac{1}{k}\int_{ka}^{kb} f\left(\frac{u}{k}\right) du = \frac{1}{k}\int_{ka}^{kb} f\left(\frac{x}{k}\right) dx,$$

on changing the dummy variable u back to x, as required to prove. ■

Property 3 – Reflectivity Property

A special case of Property 2 occurs when $k = -1$. It is referred to as the *reflectivity property* for the definite integral since the graph of the function g given by $g(x) = f(-x)$ is obtained from that of f by a reflection about the y-axis. It states that

$$\boxed{\int_a^b f(x)\,dx = \int_{-b}^{-a} f(-x)\,dx}$$

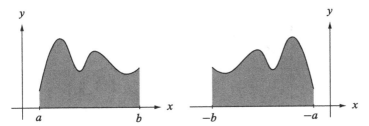

Figure 16.2. Illustration of the reflective property of the definite integral.

Proof On setting $k = -1$ in Property 2 we have

$$\int_a^b f(x)\,dx = -\int_{-a}^{-b} f(-x)\,dx = \int_{-b}^{-a} f(-x)\,dx,$$

on interchanging the limits of integration, as required to prove. ∎

The reflectivity property for the definite integral is illustrated in Figure 16.2.

Property 4 – Odd and Even Functions

The next two properties exploit symmetry in the function to be integrated. Knowing if a function is *odd* or *even* can often lead to a relatively simple definite integral to be evaluated. Let f be a function whose domain contains $-x$ whenever it contains x. Recall that f is said to be *even* if

$$f(-x) = f(x),$$

and *odd* if

$$f(-x) = -f(x),$$

for all x in the domain of f. If f is a continuous function on $[0, a]$ where $a > 0$, then

$$\boxed{\int_{-a}^{a} f(x)\,dx = 2\int_0^a f(x)\,dx, \quad \text{if } f \text{ is even}}$$

and

$$\boxed{\int_{-a}^{a} f(x)\,dx = 0, \quad \text{if } f \text{ is odd}}$$

Proof Suppose that f is an even function on the interval $[-a, a]$, namely $f(-x) = f(x)$ for $x \in [-a, a]$. Write the integral as

$$\int_{-a}^{a} f(x)\,dx = \int_{-a}^{0} f(x)\,dx + \int_0^a f(x)\,dx.$$

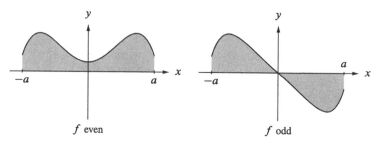

f even f odd

Figure 16.3. Graphical illustration of the even (left) and odd (right) function properties for the definite integral between symmetric limits.

Now substituting $x = -u$ into the first integral, we have

$$\int_{-a}^{a} f(x)\,dx = -\int_{a}^{0} f(-u)\,du + \int_{0}^{a} f(x)\,dx$$

$$= -\int_{a}^{0} f(u)\,du + \int_{0}^{a} f(x)\,dx, \quad \text{since } f \text{ is even}$$

$$= \int_{0}^{a} f(u)\,du + \int_{0}^{a} f(x)\,dx = 2\int_{0}^{a} f(x)\,dx,$$

where the dummy variable u in the first integral has been changed back to x, as required to prove. As the proof for the case of an odd function is done in a similar manner, it will not be given here. ∎

Geometrically these two results can be easily seen to be true from a consideration of the signed area that each definite integral represents. For example, in the case of a function f that is odd, integrating f over the interval $[-a, a]$, by symmetry one sees the positive area is exactly equal to the negative area, leading to a cancellation. Consequently the integral will be zero. For an even function f, integrating f over the interval $[-a, a]$, by symmetry, the area of the region on the interval $[-a, 0]$ is exactly equal to the area of the region on the interval $[0, a]$. Consequently the integral is equal to twice the value of the integral on the interval $[0, a]$. Both these properties for the definite integral of odd and even functions between symmetric limits are illustrated in Figure 16.3.

Example 16.1 Evaluate $\displaystyle\int_{-\pi}^{\pi} x^2 \sin^3 x\,dx$.

Solution Let $f(x) = x^2 \sin^3 x$. Since

$$f(-x) = (-x)^2 \sin^3(-x) = -x^2 \sin^3 x = -f(x),$$

the function is odd. Thus between symmetric limits we have for the integral

$$\int_{-\pi}^{\pi} x^2 \sin^3 x \, dx = 0. \qquad \blacktriangleright$$

Example 16.2 Evaluate $\int_{-\frac{\pi}{4}}^{\frac{\pi}{4}} \left(\cos x + \sqrt{1 + x^2} \sin^3 x \cos^3 x \right) dx.$

Solution Rewrite the integral as

$$\int_{-\frac{\pi}{4}}^{\frac{\pi}{4}} \left(\cos x + \sqrt{1 + x^2} \sin^3 x \cos^3 x \right) dx$$

$$= \int_{-\frac{\pi}{4}}^{\frac{\pi}{4}} \cos x \, dx + \int_{-\frac{\pi}{4}}^{\frac{\pi}{4}} \sqrt{1 + x^2} \sin^3 x \cos^3 x \, dx.$$

Observe the integrand of the first integral is even while the second is odd. As one is integrating between symmetric limits, the integral therefore reduces to

$$\int_{-\frac{\pi}{4}}^{\frac{\pi}{4}} \left(\cos x + \sqrt{1 + x^2} \sin^3 x \cos^3 x \right) dx = 2 \int_{0}^{\frac{\pi}{4}} \cos x \, dx$$

$$= 2 \sin x \Big|_{0}^{\frac{\pi}{4}} = \sqrt{2}. \qquad \blacktriangleright$$

Property 5 – Periodic Functions
A function f is said to be *periodic* with *period* a if

$$f(x + a) = f(x),$$

for all x in the domain of f. Two well-known examples of periodic functions are the sine and cosine functions. Here each function has a period of 2π since $\sin(x + 2\pi) = \sin x$ and $\cos(x + 2\pi) = \cos x$. If f is a continuous bounded function that is periodic with period a for all x in the domain of f, then

$$\boxed{\int_{b}^{b+a} f(x) \, dx = \int_{0}^{a} f(x) \, dx}$$

Proof Suppose that f is a periodic function with period a for all x in its domain, namely $f(x + a) = f(x)$. Using the additivity property with respect to the integral of integration (see Property 10 on page 41 of Chapter 4) the integral $\int_{0}^{b+a} f(x) \, dx$ can be rewritten as

$$\int_{0}^{a} f(x) \, dx + \int_{a}^{b+a} f(x) \, dx = \int_{0}^{b} f(x) \, dx + \int_{b}^{b+a} f(x) \, dx.$$

In the second integral on the left, on making a substitution of $x = u + \mathfrak{a}$ we see that $dx = du$ and the limits of integration become $x = \mathfrak{a}, u = 0$ and $x = b + \mathfrak{a}, u = b$. So one has

$$\int_0^\mathfrak{a} f(x)\, dx + \int_0^b f(u + \mathfrak{a})\, du = \int_0^b f(x)\, dx + \int_b^{b+\mathfrak{a}} f(x)\, dx$$

$$\int_0^\mathfrak{a} f(x)\, dx + \int_0^b f(u)\, du = \int_0^b f(x)\, dx + \int_b^{b+\mathfrak{a}} f(x)\, dx,$$

$$\text{since } f(u + \mathfrak{a}) = f(u)$$

$$\Rightarrow \int_b^{b+\mathfrak{a}} f(x)\, dx = \int_0^\mathfrak{a} f(x)\, dx,$$

as required to prove.   ■

Example 16.3 Evaluate $\displaystyle\int_{\frac{\pi}{3}}^{\frac{\pi}{3}+2\pi} \sin^3 x\, dx$.

Solution If $f(x) = \sin^3 x$ we begin by noting that f is both odd and periodic with a period equal to 2π. Thus

$$\int_{\frac{\pi}{3}}^{\frac{\pi}{3}+2\pi} \sin^3 x\, dx = \int_0^{2\pi} \sin^3 x\, dx, \quad \text{(Property 5 with } b = \frac{\pi}{3}\text{)}$$

$$= \int_{-\pi}^{-\pi+2\pi} \sin^3 x\, dx, \quad \text{(Property 5 with } b = -\pi\text{)}$$

$$= \int_{-\pi}^{\pi} \sin^3 x\, dx = 0. \quad \blacktriangleright$$

Property 6 – Symmetric Border Flip

The *symmetric border flip* property is an extremely useful result that can often be used to evaluate a definite integral by exploiting a reflective property. By preserving the interval of integration, a border flip changes the form of the integrand, which then may be either more amenable to being integrated directly or can be added to the original integral, resulting in an integral that is more amenable to being integrated.

Suppose f is a continuous function on the interval $[a, b]$ $(b > a)$, and consider the graph of the function $f(a + b - x)$. The graph of the function $f(a + b - x)$ is a reflection of $f(x)$ about the vertical line passing through the mid-point of the interval. So in a sense the graph of $f(a + b - x)$ has been 'flipped' (reflected) about the line $y = (a + b)/2$ compared to the graph of $f(x)$. On integrating, the area of the region under the curve $f(x)$ and

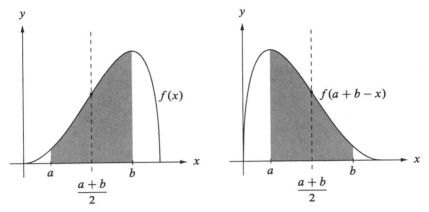

Figure 16.4. Graphical illustration of the symmetric border flip property for the definite integral.

the x-axis between $x = a$ and $x = b$ is the same as the area of the region under the curve $f(a + b - x)$ and the x-axis from $x = a$ to $x = b$ with its 'border' (its interval of integration) reversed. The property is therefore area preserving.

An illustration depicting the border flip property is shown in Figure 16.4. As the 'centre of symmetry' for each function is about the mid-point of the interval of integration, geometrically each region bounded by its corresponding curve and the x-axis between $x = a$ and $x = b$ has been reversed but the area remains unchanged. Thus it is readily seen that one must have

$$\int_a^b f(x)\, dx = \int_a^b f(a + b - x)\, dx$$

We now give a simple proof of the symmetric border flip property.

Proof Let $x = a + b - u, dx = -du$ while for the limits of integration, when $x = b, u = a$ and when $x = a, u = b$. Thus

$$\int_a^b f(x)\, dx = -\int_b^a f(a + b - u)\, du = \int_a^b f(a + b - x)\, dx,$$

where the dummy variable of integration has been changed back to x after interchanging the limits of integration, as required to prove. ∎

Example 16.4 Evaluate $\displaystyle\int_1^4 x^2 \sqrt{5-x}\, dx$.

Solution On applying the result $\displaystyle\int_a^b f(x)\, dx = \int_a^b f(a+b-x)\, dx$, to the integral we have

$$\int_1^4 x^2 \sqrt{5-x}\, dx = \int_1^4 (1+4-x)^2 \sqrt{5-(1+4-x)}\, dx$$

$$= \int_1^4 (5-x)^2 \sqrt{x}\, dx = \int_1^4 \sqrt{x}\left(25 - 10x + x^2\right) dx$$

$$= \int_1^4 \left(25x^{\frac{1}{2}} - 10x^{\frac{3}{2}} + x^{\frac{5}{2}}\right) dx$$

$$= \left[25 \cdot \frac{2}{3}x^{\frac{3}{2}} - 10 \cdot \frac{2}{5}x^{\frac{5}{2}} + \frac{2}{7}x^{\frac{7}{2}}\right]_1^4$$

$$= \frac{50}{3}\cdot 2^3 - 4\cdot 2^5 + \frac{2}{7}\cdot 2^7 - \left(\frac{50}{3} - 4 + \frac{2}{7}\right) = \frac{608}{21}. \blacktriangleright$$

Example 16.5 Evaluate $\displaystyle\int_{\frac{\pi}{6}}^{\frac{\pi}{3}} \frac{\sin^3 x}{\sin^3 x + \cos^3 x}\, dx$.

Solution Let $I = \displaystyle\int_{\frac{\pi}{6}}^{\frac{\pi}{3}} \frac{\sin^3 x}{\sin^3 x + \cos^3 x}\, dx$. On applying the symmetric border flip property to the integral I we have

$$I = \int_{\frac{\pi}{6}}^{\frac{\pi}{3}} \frac{\sin^3(\frac{\pi}{2} - x)}{\sin^3(\frac{\pi}{2} - x) + \cos^3(\frac{\pi}{2} - x)}\, dx.$$

Recalling that $\sin\left(\frac{\pi}{2} - x\right) = \cos x$ and $\cos\left(\frac{\pi}{2} - x\right) = \sin x$, the above integral reduces to

$$I = \int_{\frac{\pi}{6}}^{\frac{\pi}{3}} \frac{\cos^3 x}{\cos^3 x + \sin^3 x}\, dx.$$

Adding the above result for I to the original integral we obtain

$$I + I = \int_{\frac{\pi}{6}}^{\frac{\pi}{3}} \frac{\sin^3 x}{\sin^3 x + \cos^3 x}\, dx + \int_{\frac{\pi}{6}}^{\frac{\pi}{3}} \frac{\cos^3 x}{\cos^3 x + \sin^3 x}\, dx$$

$$2I = \int_{\frac{\pi}{6}}^{\frac{\pi}{3}} \frac{\sin^3 x + \cos^3 x}{\sin^3 x + \cos^3 x}\, dx,$$

giving

$$I = \frac{1}{2} \int_{\frac{\pi}{6}}^{\frac{\pi}{3}} dx = \frac{1}{2} \cdot x \Big|_{\frac{\pi}{6}}^{\frac{\pi}{3}} = \frac{1}{2} \left(\frac{\pi}{3} - \frac{\pi}{6} \right) = \frac{\pi}{12}.$$ ▶

As a special case, when the lower limit of integration is zero, namely $a = 0$, the symmetric border flip property reduces to

$$\boxed{\int_0^b f(x)\, dx = \int_0^b f(b - x)\, dx}$$

We refer to this special form as the *reduced symmetric border flip property*.

Example 16.6 Evaluate $\displaystyle\int_0^5 \frac{\sqrt{x}}{\sqrt{5 - x} + \sqrt{x}}\, dx$.

Solution Let $I = \displaystyle\int_0^5 \frac{\sqrt{x}}{\sqrt{5 - x} + \sqrt{x}}\, dx$. On applying the reduced symmetric border flip property to the integral, we have

$$I = \int_0^5 \frac{\sqrt{5 - x}}{\sqrt{x} + \sqrt{5 - x}}\, dx.$$

Adding the above result for I to the original integral one obtains

$$I + I = \int_0^5 \frac{\sqrt{x}}{\sqrt{5 - x} + \sqrt{x}}\, dx + \int_0^5 \frac{\sqrt{5 - x}}{\sqrt{x} + \sqrt{5 - x}}\, dx$$

$$2I = \int_0^5 \frac{\sqrt{x} + \sqrt{5 - x}}{\sqrt{x} + \sqrt{5 - x}}\, dx,$$

which reduces to

$$I = \frac{1}{2} \int_0^5 dx = \frac{1}{2} \cdot x \Big|_0^5 = \frac{5}{2}.$$ ▶

Example 16.7 Evaluate $\displaystyle\int_0^{\frac{\pi}{2}} \frac{\sin^2 x + \sin x}{\sin x + \cos x + 1}\, dx$.

Solution Let $I = \displaystyle\int_0^{\frac{\pi}{2}} \frac{\sin^2 x + \sin x}{\sin x + \cos x + 1}\, dx$. On applying the reduced symmetric border flip property to the integral, we have

$$I = \int_0^{\frac{\pi}{2}} \frac{\sin^2(\frac{\pi}{2} - x) + \sin(\frac{\pi}{2} - x)}{\sin(\frac{\pi}{2} - x) + \cos(\frac{\pi}{2} - x) + 1}\, dx = \int_0^{\frac{\pi}{2}} \frac{\cos^2 x + \cos x}{\cos x + \sin x + 1}\, dx,$$

upon recalling that $\sin\left(\frac{\pi}{2} - x\right) = \cos x$ and $\cos\left(\frac{\pi}{2} - x\right) = \sin x$. Adding the above result for I to the original integral, one has

$$I + I = \int_0^{\frac{\pi}{2}} \frac{\sin^2 x + \sin x}{\sin x + \cos x + 1}\, dx + \int_0^{\frac{\pi}{2}} \frac{\cos^2 x + \cos x}{\cos x + \sin x + 1}\, dx$$

$$2I = \int_0^{\frac{\pi}{2}} \frac{\sin^2 x + \sin x + \cos^2 x + \cos x}{\sin x + \cos x + 1}\, dx$$

$$2I = \int_0^{\frac{\pi}{2}} \frac{\sin x + \cos x + 1}{\sin x + \cos x + 1}\, dx,$$

giving

$$I = \frac{1}{2} \int_0^{\frac{\pi}{2}} dx = \frac{1}{2} \cdot x \Big|_0^{\frac{\pi}{2}} = \frac{\pi}{4}. \qquad \blacktriangleright$$

A second special case occurs when one sets $a = -b$. With the limits of integration written in terms of a (instead of b), one has

$$\boxed{\int_{-a}^{a} f(x)\, dx = \int_{-a}^{a} f(-x)\, dx}$$

We refer to this special form as the *symmetric interval border flip property*.

Example 16.8 Evaluate $\int_{-1}^{1} \frac{dx}{(e^x + 1)(x^2 + 1)}\, dx.$

Solution Let $I = \int_{-1}^{1} \frac{dx}{(e^x + 1)(x^2 + 1)}\, dx.$ On applying the symmetric interval border flip property to the integral, we have

$$I = \int_{-1}^{1} \frac{dx}{(e^{-x} + 1)(x^2 + 1)}.$$

Adding the above result for I to the original integral gives

$$I + I = \int_{-1}^{1} \frac{dx}{(e^x + 1)(x^2 + 1)}\, dx + \int_{-1}^{1} \frac{dx}{(e^{-x} + 1)(x^2 + 1)}$$

$$2I = \int_{-1}^{1} \left[\frac{1}{e^x + 1} + \frac{1}{e^{-x} + 1} \right] \frac{dx}{1 + x^2}$$

$$= \int_{-1}^{1} \frac{e^x + e^{-x} + 2}{e^x + e^{-x} + 2} \frac{dx}{1 + x^2} = \int_{-1}^{1} \frac{dx}{1 + x^2} = 2 \int_0^{1} \frac{dx}{1 + x^2}$$

$$= 2 \tan^{-1} x \Big|_0^{1} = 2 \cdot \frac{\pi}{4},$$

giving $I = \pi/4.$ $\qquad \blacktriangleright$

§ Definite Integrals Involving Absolute Values

Evaluating definite integrals that involve the absolute value of a function in the integrand requires special care. Extra attention needs to be paid to the interval of integration, which depends on whether the value for the function is either positive or negative. The integral is broken up into a number of parts, and for each there is a separate part where the value of the function within the absolute value sign is either positive or negative. As an aid to evaluating such integrals, sketching a diagram is often very useful.

Before proceeding with a number of examples illustrating the techniques involved, recall that the absolute value function is defined by

$$|x| = \begin{cases} x, & x \geqslant 0 \\ -x, & x < 0. \end{cases}$$

Example 16.9 Evaluate $\displaystyle\int_{-1}^{9} |x - 3| \, dx$.

Solution Replacing x with $x - 3$ in the definition for the absolute value function gives

$$|x - 3| = \begin{cases} x - 3, & x \geqslant 3 \\ -x + 3, & x < 3 \end{cases}.$$

Since the interval of integration is from -1 to 9 we need to consider two separate intervals. Thus

$$\int_{-1}^{9} |x - 3| \, dx = \int_{-1}^{3} |x - 3| \, dx + \int_{3}^{9} |x - 3| \, dx$$

$$= \int_{-1}^{3} (-x + 3) \, dx + \int_{3}^{9} (x - 3) \, dx$$

$$= \left[-\frac{x^2}{2} + 3x \right]_{-1}^{3} + \left[\frac{x^2}{2} - 3x \right]_{3}^{9}$$

$$= -\frac{9}{2} + 9 - \left(-\frac{1}{2} - 3 \right) + \frac{81}{2} - 27 - \left(\frac{9}{2} - 9 \right) = 26. \quad \blacktriangleright$$

Example 16.10 Evaluate $\displaystyle\int_{-4}^{1} x|2x + 4| \, dx$.

Solution Replacing x with $2x + 4$ in the definition for the absolute value function gives

$$|2x + 4| = \begin{cases} 2x + 4, & 2x + 4 \geqslant 0 \\ -2x - 4, & 2x + 4 < 0 \end{cases}$$

$$= \begin{cases} 2x + 4, & x \geqslant -2 \\ -2x - 4, & x < -2. \end{cases}$$

Since the integral of integration is from -4 to 1 we again need to consider two separate intervals. Thus

$$\int_{-4}^{1} x|2x + 4|\, dx = \int_{-4}^{-2} x|2x + 4|\, dx + \int_{-2}^{1} x|2x + 4|\, dx$$

$$= \int_{-4}^{-2} x(-2x - 4)\, dx + \int_{-2}^{1} x(2x + 4)\, dx$$

$$= \int_{-4}^{-2} (-2x^2 - 4x)\, dx + \int_{-2}^{1} (2x^2 + 4x)\, dx$$

$$= \left[-\frac{2x^3}{3} - 2x^2 \right]_{-4}^{-2} + \left[\frac{2x^3}{3} + 2x^2 \right]_{-2}^{1}$$

$$= -\frac{2}{3} \cdot (-2)^3 - 2 \cdot (-2)^2 - \left(-\frac{2}{3} \cdot (-4)^3 - 2 \cdot (-4)^2 \right)$$

$$+ \frac{2}{3} + 2 - \left(\frac{2}{3} \cdot (-2)^3 + 2 \cdot (-2)^2 \right) = -\frac{40}{3}. \quad \blacktriangleright$$

Example 16.11 Evaluate $\int_{-2}^{5} \left| -x^2 + 6x - 8 \right| dx$.

Solution Let $g(x) = -x^2 + 6x - 8 = -(x - 2)(x - 4)$ for $x \in [-2, 5]$. The roots of g are 2 and 4. Now determine the sign of g on the interval $[-2, 5]$.

   I. For $[-2, 2]$, since $x - 2 < 0, x - 4 < 0$, g will be negative.
  II. For $[2, 4]$, since $x - 2 > 0, x - 4 < 0$, g will be positive.
 III. For $[4, 5]$, since $x - 2 > 0, x - 4 > 0$, g will be negative.

$$\text{Thus } |g(x)| = \begin{cases} -g(x), & -2 \leqslant x \leqslant 2 \\ g(x), & 2 \leqslant x \leqslant 4 \\ -g(x), & 4 \leqslant x \leqslant 5 \end{cases}.$$

So for the integral we have

$$\int_{-2}^{5} |g(x)| \, dx = \int_{-2}^{2} -g(x) \, dx + \int_{2}^{4} g(x) \, dx + \int_{4}^{5} -g(x) \, dx$$

$$= \int_{-2}^{2} (x^2 - 6x + 8) \, dx + \int_{2}^{4} (-x^2 + 6x - 8) \, dx$$

$$+ \int_{4}^{5} (x^2 - 6x + 8) \, dx$$

$$= \left[\frac{x^3}{3} - 3x^2 - 8x \right]_{-2}^{2} + \left[-\frac{x^3}{3} + 3x^2 - 8x \right]_{2}^{4}$$

$$+ \left[-\frac{x^3}{3} + 3x^2 - 8x \right]_{4}^{5} = 40.$$

▶

Exercises for Chapter 16

✠ **Warm-ups**

1. Each of the following statements is either true or false. Answer each statement with either TRUE or FALSE and give a brief reason for your answer.

(a) $\displaystyle\int_{-1}^{1} x\sqrt{4 - x^2} \, dx > 0$

(b) $\displaystyle\int_{-2}^{2} \tan^{-1} x \, dx = 0$

(c) $\displaystyle\int_{-\pi}^{\pi} (|2x| + 3) \, dx > 0$

(d) $\displaystyle\int_{-3}^{3} \frac{|x|}{e^x + 1} \, dx < 0$

2. If f is a continuous periodic function with period a, show that the function $f(x + na)$ with $n \in \mathbb{N}$ is also a periodic function with period a.

3. Evaluate

(a) $\displaystyle\int_{-\pi}^{\pi} x^{100} \sin x \, dx$

(b) $\displaystyle\int_{-1}^{1} \frac{x^2 \sin x}{x^4 - 2x^2 - 1} \, dx$

(c) $\displaystyle\int_{-3}^{3} \frac{x}{1 + |x|} \, dx$

(d) $\displaystyle\int_{-2}^{2} x \sin(|x|) \, dx$

✠ Practice questions

4. If f is a continuous function on the interval where it is to be integrated, prove the following results.

(a) $\displaystyle\int_{a+k}^{b+k} f(x)\,dx = \int_a^b f(x+k)\,dx$

(b) $\displaystyle\int_a^b f(c-x)\,dx = \int_{c-b}^{c-a} f(x)\,dx$

(c) $\displaystyle\int_{ka}^{kb} f(x)\,dx = k\int_a^b f(kx)\,dx$

5. If f is a continuous function on the interval $[a,b]$, prove the following results.

(a) Conversion to a unit interval:

$$\int_a^b f(x)\,dx = (b-a)\int_0^1 f[a+(b-a)x]\,dx.$$

(b) Conversion to a symmetric unit interval about the origin:

$$\int_a^b f(x)\,dx = \frac{b-a}{2}\int_{-1}^1 f\left[\frac{b+a+(b-a)x}{2}\right]dx.$$

(c) Centring about the origin:

$$\int_a^b f(x)\,dx = \int_{-\frac{(b-a)}{2}}^{\frac{b-a}{2}} f\left(\frac{a+b}{2}-x\right)dx.$$

(d) Centring about the mid-point of the interval:

$$\int_a^b f(x)\,dx = \int_0^{\frac{b-a}{2}}\left[f\left(\frac{a+b}{2}-x\right)+f\left(\frac{a+b}{2}+x\right)\right]dx.$$

6. Evaluate the following integrals using the reduced symmetric border flip property for the definite integral.

(a) $\displaystyle\int_0^{\frac{\pi}{2}} \frac{\sin^2 x}{\sin x + \cos x}\,dx$

(b) $\displaystyle\int_0^{\frac{\pi}{2}} \frac{\sqrt{\sin x}}{\sqrt{\sin x} + \sqrt{\cos x}}\,dx$

(c) $\displaystyle\int_0^{\pi} \frac{x\sin x}{1+\sin^2 x}\,dx$

(d) $\displaystyle\int_0^{\pi} \frac{(2x+3)\sin x}{1+\cos^2 x}\,dx$

(e) $\displaystyle\int_0^{\frac{\pi}{2}} \frac{dx}{1+\tan^{\sqrt{2}} x}$

(f) $\displaystyle\int_0^1 \frac{\ln(1+x)}{\ln(2+x-x^2)}\,dx$

(g) $\displaystyle\int_0^{\frac{\pi}{2}} \frac{x}{\sin x + \cos x}\, dx$

(h) $\displaystyle\int_0^{\pi} \frac{x \tan x}{\sec x + \cos x}\, dx$

(i) $\displaystyle\int_0^{\frac{\pi}{2}} \left(\sqrt{\sin x} - \sqrt{\cos x}\right) dx$

(j) $\displaystyle\int_0^2 \frac{dx}{x + \sqrt{x^2 - 2x + 2}}$

(k) $\displaystyle\int_0^{\frac{\pi}{2}} x \left(\frac{\sin x}{1 + \cos^2 x} + \frac{\cos x}{1 + \sin^2 x}\right) dx$

(l) $\displaystyle\int_0^{\frac{\pi}{4}} (\cos^4 2x + \sin^4 2x) \ln(1 + \tan x)\, dx$

7. Evaluate the following integrals using the symmetric border flip property for the definite integral.

(a) $\displaystyle\int_1^{2017} \frac{\sqrt{x}}{\sqrt{2018 - x} + \sqrt{x}}\, dx$

(b) $\displaystyle\int_1^3 x^2(x - 4)^{20}\, dx$

(c) $\displaystyle\int_2^6 \frac{\sqrt[3]{6 - x}}{\sqrt[3]{6 - x} + \sqrt[3]{x - 2}}\, dx$

(d) $\displaystyle\int_{\frac{\pi}{6}}^{\frac{\pi}{3}} \frac{dx}{1 + \tan^8 x}$

(e) $\displaystyle\int_{\frac{\pi}{6}}^{\frac{\pi}{2}} \ln\left(\sqrt{3}\tan x - 1\right) dx$

(f) $\displaystyle\int_{\frac{\pi}{8}}^{\frac{3\pi}{8}} \ln(\tan x)\, dx$

(g) $\displaystyle\int_{\frac{\pi}{6}}^{\frac{\pi}{3}} \frac{(\tan x)^{\cot x}}{(\tan x)^{\cot x} + (\cot x)^{\tan x}}\, dx$

(h) $\displaystyle\int_0^2 (3x^2 - 3x + 1) \cos(x^3 - 3x^2 + 4x - 2)\, dx$

8. Evaluate the following integrals using the symmetric interval border flip property for definite integrals.

(a) $\displaystyle\int_{-\frac{\pi}{2}}^{\frac{\pi}{2}} \frac{\cos^7 x}{e^x + 1}\, dx$

(b) $\displaystyle\int_{-\frac{1}{2}}^{\frac{1}{2}} \cos x \ln\left(\frac{1 + x}{1 - x}\right) dx$

(c) $\displaystyle\int_{-\frac{\pi}{2}}^{\frac{\pi}{2}} \frac{dx}{1 + e^{\sin x}}$

(d) $\displaystyle\int_{-\pi}^{\pi} \tan^{-1}(\pi^x)\, dx$

(e) $\displaystyle\int_{-1}^1 \cos^{-1}(x^5)\, dx$

(f) $\displaystyle\int_{-\pi}^{\pi} \frac{x^2}{1 + \sin x + \sqrt{1 + \sin^2 x}}\, dx$

9. Evaluate

(a) $\displaystyle\int_{-\pi}^{\pi} \frac{\tan^{-1} x}{x^4 - x^2 + 4}\, dx$
(b) $\displaystyle\int_{-1}^{1} \ln\left(\frac{2 - x}{2 + x}\right) dx$

10. Using the substitution $u = x - 1$, evaluate $\displaystyle\int_0^2 x\sqrt{2x - x^2}\, dx$.

11. Evaluate $\displaystyle\int_0^{\frac{\pi}{2}} \frac{a\cos x + b\sin x}{\cos x + \sin x}\, dx$. Here $a, b \in \mathbb{R}$.

12. Using only properties for the definite integral, show that

$$\int_0^{2\pi} \sin x(1 - \cos x)\sqrt{1 - \cos x}\, dx = 0.$$

13. Evaluate $\displaystyle\int_{-\pi}^{\pi} \frac{\sin(nx)}{(1 + 2^x)\sin x}\, dx$ for all $n \in \mathbb{Z}^+$.

14. Let f be a continuous function on the interval $[0, a]$. Here $a > 0$.

(a) Show that $\displaystyle\int_0^a \frac{f(x)}{f(x) + f(a - x)}\, dx = \frac{a}{2}$.

(b) Use the result in (a) to evaluate

(i) $\displaystyle\int_0^2 \frac{x^2}{x^2 - 2x + 2}\, dx$
(ii) $\displaystyle\int_0^{\frac{\pi}{2}} \frac{\sin x}{\sin x + \cos x}\, dx$

(iii) $\displaystyle\int_0^1 \frac{x^4 + 1}{2x^4 - 4x^3 + 6x^2 - 4x + 3}\, dx$

15. Let f be a continuous function on the interval $[0, a]$. Here $a > 0$.

(a) Show that $\displaystyle\int_0^a x f(x)\, dx = \frac{a}{2}\int_0^a f(x)\, dx$.

(b) Using the result in (a), evaluate $\displaystyle\int_0^{\pi} \frac{x\sin x}{1 + \cos^2 x}\, dx$.

16. Let f be a continuous function on the interval $[-a, a]$. Here $a > 0$.

(a) Show that $\displaystyle\int_{-a}^a f(x)\, dx = \int_0^a \left(f(x) + f(-x)\right) dx$.

(b) Using the result in (a), evaluate $\displaystyle\int_{-\frac{\pi}{2}}^{\frac{\pi}{2}} \frac{e^x \cos^2 x}{1 + e^x}\, dx$.

17. Suppose $f : \mathbb{R} \to \mathbb{R}$ is a continuous function satisfying

$$\int_{-2}^{3} f(x)\, dx = 6 \quad \text{and} \quad \int_{-2}^{0} f(x)\, dx = 14.$$

Find $\int_{2}^{3} f(x)\, dx$ if:

(a) $f(x) = -3$ for all $x \in [0, 2]$.
(b) f is an odd function.
(c) f is a periodic function such that $f(x) = f(x + 2)$ for all x.

18. If $f(x) = \int_{0}^{\frac{\pi}{4}} \ln(1 + x \tan t)\, dt$, $x > -1$, find the value of

$$f\left(\frac{1}{2}\right) + f\left(\frac{1}{3}\right).$$

19. Evaluate $\int_{0}^{1} \sqrt[3]{2x^3 - 3x^2 - x + 1}\, dx$ by first centring the integral about the origin. Centring the integral about the origin is done by using the result already proved in Exercise 5 (c).

20. (a) If f is a continuous function on the interval $[0, 1]$, by centring the integral about the origin, evaluate $\int_{0}^{\pi} f(\sin x) \cos x\, dx$.

(b) Hence deduce that $\int_{0}^{\pi} (\sin x)^{\ln(1+\sin x)} \cos x\, dx = 0$.

21. Evaluate $\int_{-1}^{1} \frac{x^2}{(1 + x^4)} \frac{dx}{\sqrt{1 - x^4}}$.

22. Let f be a continuous function on the interval $[a, b]$.

(a) Show that $\int_{a}^{b} \frac{f(x)}{f(a + b - x) + f(x)}\, dx = \frac{b - a}{2}$.

(b) Hence evaluate

$$\int_{2}^{4} \frac{\sqrt{\ln(9 - x)}}{\sqrt{\ln(9 - x)} + \sqrt{\ln(3 + x)}}\, dx.$$

23. For a continuous function f on the interval $[1, a]$ where $a > 1$, the *Wolstenholme transformation*[1] states that

$$\int_{1}^{a} f\left(x + \frac{a^2}{x}\right) \frac{dx}{x} = \int_{1}^{a} f\left(x^2 + \frac{a^2}{x^2}\right) \frac{dx}{x}.$$

Prove this result.

24. *Serret's integral*[2] is given by $\displaystyle\int_0^1 \frac{\ln(1+x)}{1+x^2}\,dx.$

(a) By employing the substitution $x = \tan u$, show that Serret's integral becomes

$$\int_0^1 \frac{\ln(1+x)}{1+x^2}\,dx = \int_0^{\frac{\pi}{4}} \ln(1+\tan u)\,du.$$

(b) By applying a symmetric border flip to the integral found in (a), use this to show that

$$\int_0^1 \frac{\ln(1+x)}{1+x^2}\,dx = \frac{\pi}{8}\ln 2.$$

(c) By integrating the integral appearing in (b) by parts, show that

(i) $\displaystyle\int_0^1 \frac{\tan^{-1} x}{1+x}\,dx = \frac{\pi}{8}\ln 2$　　　(ii) $\displaystyle\int_0^1 \frac{\cot^{-1} x}{1+x}\,dx = \frac{3\pi}{8}\ln 2$

(d) If $a > 0$, in a similar manner as to what was done in (a) and (b), find the value for the generalised Serret integral $\displaystyle\int_0^a \frac{\ln(a+x)}{a^2+x^2}\,dx.$

25. The *Bessel function* of the first kind of order zero, denoted by $J_0(x)$, is defined by

$$J_0(x) = \frac{1}{2\pi}\int_0^{2\pi} \cos(x\sin\theta)\,d\theta.$$

(a) By exploiting the periodicity of $\sin\theta$, show that

$$J_0(x) = \frac{1}{\pi}\int_0^{\pi} \cos(x\sin\theta)\,d\theta.$$

(b) If f is a continuous function on the interval $[0, 2a]$ where $a > 0$, show that

$$\int_0^{2a} f(x)\,dx = \int_0^a \big(f(x) + f(2a - x)\big)\,dx.$$

(c) By applying the result given in (b) to the integral given in (a), show that

$$J_0(x) = \frac{2}{\pi}\int_0^{\frac{\pi}{2}} \cos(x\sin\theta)d\theta.$$

26. Let f be a continuous function on the interval $[0, a]$. Here $a > 0$.

(a) Show that $\displaystyle\int_0^a f(x)\,dx = \int_0^{\frac{a}{2}} \big(f(x) + f(a - x)\big)\,dx.$

(b) If $f(a - x) = f(x)$, use the result in (a) to show that

$$\int_0^a f(x)\,dx = 2\int_0^{\frac{a}{2}} f(x)\,dx.$$

(c) Use the result in (b) to evaluate $\displaystyle\int_0^\pi \frac{a\sin^2 x + b\cos^2 x}{a^2\sin^2 x + b^2\cos^2 x}\,dx.$ Here $a, b \in \mathbb{R}$.

27. Let p and q be two continuous functions on the interval $[-a, a]$.

(a) If p is even and $q(x) \cdot q(-x) = 1$, show that

$$\int_{-a}^a \frac{p(x)}{1 + q(x)}\,dx = \int_0^a p(x)\,dx.$$

(b) Using the result in (a), evaluate $\displaystyle\int_{-1}^1 \frac{dx}{(1 + x^2)(1 + \tan x + \sec x)}.$

28. (a) Show that if f is continuous on $[a, b]$ such that $f(x) + f(a + b - x)$ is constant for all $x \in [a, b]$, then

$$\int_a^b f(x)\,dx = (b - a)f\left(\frac{a + b}{2}\right) = \frac{1}{2}(b - a)[f(a) + f(b)].$$

(b) Using the result in (a), evaluate the following integrals.

(i) $\displaystyle\int_{-1}^1 \cos^{-1}(x^3)\,dx$ (ii) $\displaystyle\int_0^{2\pi} \frac{dx}{1 + e^{\sin x}}$ (iii) $\displaystyle\int_0^4 \frac{dx}{4 + 2^x}$

29. If $u = \displaystyle\int_0^{\frac{\pi}{4}} \left(\frac{\cos x}{\sin x + \cos x}\right)^2 dx$ and $v = \displaystyle\int_0^{\frac{\pi}{4}} \left(\frac{\sin x + \cos x}{\cos x}\right)^2 dx$, find the value of u/v.

30. Let f be a continuous even function on the interval $[-a, a]$. Here $a > 0$.

(a) Show that $\displaystyle\int_{-a}^a \frac{f(x)}{1 + e^x}\,dx = \int_0^a f(x)\,dx.$

(b) Using the result in (a), evaluate the following integrals.

(i) $\displaystyle\int_{-2}^2 \frac{x^2}{1 + e^x}\,dx$ (ii) $\displaystyle\int_{-\frac{\pi}{2}}^{\frac{\pi}{2}} \frac{\cos 2x}{1 + e^x}\,dx$

(c) The result given in (a) can be generalised. If f is a continuous even function on the interval $[-a, a]$ while g is a continuous function on the same interval such that $g(x) + g(-x) = 1$, show that

$$\int_{-a}^{a} g(x) f(x) \, dx = \int_{0}^{a} f(x) \, dx.$$

(d) Using the result given in (c), evaluate the following:

(i) $\displaystyle\int_{-1}^{1} \tan^{-1}(e^x) \, dx$

(ii) $\displaystyle\int_{-1}^{1} x^2 \cos^{-1}(x^3) \, dx$

(iii) $\displaystyle\int_{-\frac{\pi}{2}}^{\frac{\pi}{2}} \frac{e^{-x} \sin^2 x}{1 + e^{-x}} \, dx$

(iv) $\displaystyle\int_{-\ln 2}^{\ln 2} \frac{\cosh x}{2 \sinh x + 1 + e^{-x}} \, dx$

31. Let g be a continuous odd function on the interval $[-a, a]$. Here $a > 0$.

(a) Show that $\displaystyle\int_{-a}^{a} \tan^{-1}(e^{g(x)}) \, dx = \frac{a\pi}{2}$.

(b) Using the result in (a), evaluate the following integrals.

(i) $\displaystyle\int_{-2}^{2} \tan^{-1}(e^{x^3}) \, dx$

(ii) $\displaystyle\int_{-3}^{3} \tan^{-1}\left(\frac{3-x}{3+x}\right) \, dx$

32. Evaluate

(a) $\displaystyle\int_{0}^{1} \frac{dx}{(x^2 - x + 1)(e^{2x-1} + 1)}$

(b) $\displaystyle\int_{\sqrt{\ln 2}}^{\sqrt{\ln 3}} \frac{x \sin(x^2)}{\sin(x^2) + \sin(\ln 6 - x^2)} \, dx$

33. Let f be a continuous function on the interval $[-\frac{\pi}{4}, \frac{\pi}{4}]$ such that $f(x) f(-x) = 1$ for all $x \in [-\frac{\pi}{4}, \frac{\pi}{4}]$ and $f(\pm x) + 1 \neq 0$. Evaluate

$$\int_{-\frac{\pi}{4}}^{\frac{\pi}{4}} \frac{dx}{(1 + 2\sin^2 x)(1 + f(x))}.$$

34. Evaluate

(a) $\displaystyle\int_{0}^{4} \left(|2 - 2x| + 1\right) dx$

(b) $\displaystyle\int_{-2}^{3} 2x|x + 1| \, dx$

(c) $\displaystyle\int_{-\frac{1}{2}}^{\frac{1}{2}} \left(x - 3|x|\right) x^2 \, dx$

(d) $\displaystyle\int_{-1}^{1} \sqrt{|x| + x} \, dx$

(e) $\displaystyle\int_0^4 \left| \sqrt{x} - 1 \right| dx$ (f) $\displaystyle\int_0^{\frac{\pi}{2}} \left| \sin x - \cos 2x \right| dx$

(g) $\displaystyle\int_0^3 \left| x^4 - 4 \right| dx$ (h) $\displaystyle\int_0^5 \frac{|x - 1|}{|x - 2| + |x - 4|}\, dx$

(i) $\displaystyle\int_{-1}^3 \frac{x}{1 + |x|}\, dx$ (j) $\displaystyle\int_{-5}^2 \left| |2x - 2| + 4x \right| dx$

35. Evaluate $\displaystyle\int_0^{n\pi} \left| x \sin x \right| dx$ if $n \in \mathbb{N}$.

36. In this question a value for the definite integral $\displaystyle\int_0^{\frac{\pi}{2}} \ln(\sin x)\, dx$ will be found and some consequences stemming from this result explored.

(a) Show that $\displaystyle\int_0^{\frac{\pi}{2}} \ln(\sin x)\, dx = \int_0^{\frac{\pi}{2}} \ln(\cos x)\, dx$.

(b) On adding the two integrals in (a) together before making an appropriate substitution and simplifying, show that

$$\int_0^{\frac{\pi}{2}} \ln(\sin x)\, dx = -\frac{\pi}{4} \ln 2 + \frac{1}{4} \int_0^{\pi} \ln(\sin x)\, dx.$$

(c) By writing the integral appearing in (b) as

$$\int_0^{\pi} \ln(\sin x)\, dx = \int_0^{\frac{\pi}{2}} \ln(\sin x)\, dx + \int_{\frac{\pi}{2}}^{\pi} \ln(\sin x)\, dx,$$

and after using the substitution $x = \dfrac{\pi}{2} + u$ in the integral farthest to the right, deduce that

$$\int_0^{\frac{\pi}{2}} \ln(\sin x)\, dx = \int_0^{\frac{\pi}{2}} \ln(\cos x)\, dx = -\frac{\pi}{2} \ln 2.$$

(d) Hence deduce that $\displaystyle\int_0^{\frac{\pi}{2}} \ln(\tan x)\, dx = 0$.

(e) Show that $\displaystyle\int_0^{\pi} \ln(\sin x)\, dx = \int_0^{\pi} \ln(1 + \cos x)\, dx = -\pi \ln 2$.

(f) Hence deduce that $\displaystyle\int_0^{\pi} x \ln(\sin x)\, dx = -\frac{\pi^2}{2} \ln 2$.

37. Let f be a continuous function on the interval $[0, 1]$. Find the value of k if

$$\int_0^\pi x f(\sin x)\, dx = k \int_0^{\frac{\pi}{2}} f(\sin x)\, dx.$$

38. (a) Let f be a continuous function on $[0, 1]$ and $f(x) + f(1 - x) \neq 0$ for all $x \in [0, 1]$. Show that

$$\int_0^1 \frac{f(x)}{f(1 - x) + f(x)}\, dx = \frac{1}{2}.$$

   (b) Let g be a continuous function on $[\sqrt{5}, \sqrt{7}]$ and $g(\sqrt{9 - x}) + g(\sqrt{x + 3}) \neq 0$ for all $x \in [2, 4]$. Show that

$$\int_2^4 \frac{g(\sqrt{9 - x})}{g(\sqrt{9 - x}) + g(\sqrt{x + 3})}\, dx = 1.$$

39. By using a substitution of $x = \tan \theta$ before applying the symmetric border flip property to the definite integral, show that

$$\int_2^3 \frac{\ln(x - 1)}{x^2 + 1}\, dx = \frac{\ln 2}{2} \tan^{-1}\left(\frac{1}{7}\right).$$

40. If $n \in \mathbb{N}, \alpha > 0$, evaluate $\displaystyle\int_{-\frac{\pi}{2}}^{\frac{\pi}{2}} \frac{\sin^{2n} x}{\sin^{2n} x + \cos^{2n} x} \frac{dx}{1 + \alpha^x}$.

41. If $a \neq 0$, evaluate $\displaystyle\int_{-a}^a \frac{dx}{1 + x^5 + \sqrt{1 + x^{10}}}$.

42. (a) If $0 < a < b$, show that $\displaystyle\int_a^b \frac{dx}{\sqrt{(x - a)(b - x)}} = \pi$.

   (b) Hence deduce that

$$\int_\alpha^\beta \frac{dx}{\sqrt{(e^x - e^\alpha)(e^\beta - e^x)}} = \pi \exp\left(-\frac{\alpha + \beta}{2}\right), \quad \alpha < \beta.$$

✠ Extension Questions and Challenge problems

43. Evaluate $\displaystyle\int_0^2 \frac{x(16 - x^2)}{16 - x^2 + \sqrt{(16 - x^2)(12 + x^2)}}\, dx$.

44. Suppose f is a continuous function such that $f(2 + x) = f(2 - x)$ and $f(4 + x) = f(4 - x)$. If $\displaystyle\int_0^2 f(x)\, dx = 5$, find the value of $\displaystyle\int_{10}^{50} f(x)\, dx$.

45. The *floor function*, denoted by $\lfloor x \rfloor$, gives the greatest integer less than or equal to the real number x. So, for example, $\lfloor 14 \rfloor = 14$, $\lfloor e \rfloor = 2$, $\lfloor -5.1 \rfloor = -6$. When contained in the integrand, as we shall see, the floor function turns continuous integration problems into discrete problems.

As an example, consider the problem of $\int_0^3 \lfloor x \rfloor \, dx$. On applying the definition for the floor function on the interval $[0, 3]$, we have

$$\lfloor x \rfloor = \begin{cases} 0, & 0 \leqslant x < 1 \\ 1, & 1 \leqslant x < 2 \\ 2, & 2 \leqslant x < 3 \end{cases}.$$

So for the integral one has

$$\int_0^3 \lfloor x \rfloor \, dx = \int_0^1 0 \, dx + \int_1^2 1 \, dx + \int_2^3 2 \, dx = 3.$$

Evaluate the following integrals that contain floor functions in their integrands.

(a) $\displaystyle\int_0^5 x \lfloor x \rfloor \, dx$

(b) $\displaystyle\int_0^5 (x - \lfloor x \rfloor)^2 \, dx$

(c) $\displaystyle\int_0^5 (1 + \lfloor x \rfloor)^3 \, dx$

(d) $\displaystyle\int_{\frac{1}{5}}^5 \left\lfloor \frac{1}{x} \right\rfloor \, dx$

(e) $\displaystyle\int_0^a 2^{\lfloor x \rfloor} \, dx, \quad a \in \mathbb{N}$

46. Evaluate

(a) $\displaystyle\int_{-\frac{\pi}{3}}^{\frac{\pi}{3}} \frac{\pi + 2x^3}{2 - \cos\left(|x| + \frac{\pi}{3}\right)}.$

(b) $\displaystyle\int_{\frac{\pi}{2}}^{\frac{5\pi}{2}} \frac{e^{\tan^{-1}(\sin x)}}{e^{\tan^{-1}(\sin x)} + e^{\tan^{-1}(\cos x)}} \, dx$

(c) $\displaystyle\int_{-\pi}^{\pi} \frac{x \sin x \tan^{-1}(e^x)}{1 + \cos^2 x} \, dx$

(d) $\displaystyle\int_0^1 \left[(e - 1)\sqrt{\ln(1 + ex - x)} + e^{x^2} \right] dx$

47. For $-\pi/2 < a < \pi/2$, evaluate $\displaystyle\int_0^a \frac{x}{\cos x \cos(a - x)} \, dx.$

48. Define $u_1(x) = |x|$ and for integers $n \geqslant 2$, $u_n(x) = |x + u_{n-1}(x)|$. Let $V_n = \int_{-1}^{1} u_n(x) \, dx$ for $n \in \mathbb{N}$. To evaluate V_n we first find an expression for $u_n(x)$.

(a) For $x \geqslant 0$, using induction, show that $u_n(x) = nx$.

(b) For $x < 0$ and n even, namely $n = 2m$ where $m \in \mathbb{N}$, using induction show that $u_n(x) = 0$.

(c) For $x < 0$ and n odd, namely $n = 2m + 1$ where $m \in \mathbb{N}$, using induction show that $u_n(x) = -x$.

(d) Hence deduce that $V_n = \begin{cases} \dfrac{n}{2}, & n \text{ even} \\ \dfrac{n+1}{2}, & n \text{ odd}. \end{cases}$

49. Let f be a bounded continuous function on the interval $[0, 1]$.

(a) Show that $\displaystyle\int_0^{\pi} x f(\sin x) \, dx = \frac{\pi}{2} \int_0^{\pi} f(\sin x) \, dx$.

(b) Hence evaluate $\displaystyle\int_0^{\pi} \frac{x}{1 + \sin x} \, dx$.

(c) Hence deduce that $\displaystyle\int_0^{\pi} \frac{2x^3 - 3\pi x^2}{(1 + \sin x)^2} \, dx = -\frac{2\pi^3}{3}$.

50. Suppose $\displaystyle\int_{a+1}^{a+b} \left(\frac{b}{x-a} + a - x \right)^{2n} dx$ where $a, b \in \mathbb{R}$ and $n \in \mathbb{N}$.

By using the substitution $u = \dfrac{b}{x - a} + a - x$ show that

$$\int_{a+1}^{a+b} \left(\frac{b}{x-a} + a - x \right)^{2n} dx = \frac{(b-1)^{2n+1}}{2n+1}.$$

Endnotes

1. Named after the English mathematician Joseph Wolstenholme (1829–1891).
2. The integral is named after the French mathematician Joseph-Alfred Serret (1819–1885) who first evaluated such an integral in 1844. See J.-A. Serret, 'Sur l'intégrale $\displaystyle\int_0^1 \frac{\ell(1+x)}{1+x^2} \, dx$', *Journal de Mathématiques Pures et Appliquées* **9**, 436 (1844).

17

Integrating Inverse Functions

Anyone who has taken a calculus course at high school or college has read the instruction: 'Evaluate the following integral'. For many students, the words fill them with dread, for others they bring a shiver of excited anticipation. For both groups the reason is the same: integration is hard.

— Jonathan Borwein and Keith Devlin, *The Computer as Crucible: An Introduction to Experimental Mathematics*

An application of integration by parts (see Chapter 7) is to find integrals involving inverse functions such at the logarithmic function and the inverse trigonometric and hyperbolic functions. As a technique for integrating relatively simple expressions containing inverse functions this 'standard' method is more than adequate. However, for the integration of more complicated expressions containing inverse functions, use of the standard method can often prove difficult. To help overcome such difficulties, in this chapter we introduce a general method for finding and evaluating integrals for the inverse of the function, provided the function satisfies certain given conditions.

Theorem 17.1 (Definite integral of an inverse function). *Let f be a strictly monotonic function (meaning the function is either increasing or decreasing) with continuous derivative on the interval $[a, b]$. Then*

$$\boxed{\int_a^b f(x)\,dx + \int_{f(a)}^{f(b)} f^{-1}(x)\,dx = bf(b) - af(a)}$$

Proof Since f is a strictly monotonic function on some interval, it has an inverse f^{-1} on this interval. Integrating the function f by parts one

233

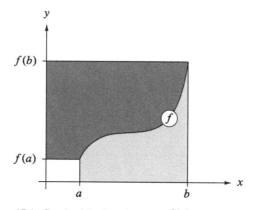

Figure 17.1. Graph of the function $y = f(x)$ and various areas.

has

$$\int_a^b f(x)\, dx = x f(x)\Big|_a^b - \int_a^b x f'(x)\, dx$$

$$= b f(b) - a f(a) - \int_a^b x f'(x)\, dx.$$

Set $y = f(x)$, $dy = f'(x)\, dx$ and $x = f^{-1}(y)$. For the limits of integration, when $x = b$, $y = f(b)$ and when $x = a$, $y = f(a)$. Thus

$$\int_a^b f(x)\, dx = b f(b) - a f(a) - \int_{f(a)}^{f(b)} f^{-1}(y)\, dy,$$

or

$$\int_a^b f(x)\, dx + \int_{f(a)}^{f(b)} f^{-1}(x)\, dx = b f(b) - a f(a),$$

on changing the dummy variable back to x and rearranging, as required to prove.   ∎

The result can be given a very simple geometric meaning. Consider the graph of the function $y = f(x)$ between $x = a$ and $x = b$ shown in Figure 17.1.
  If

$$b \cdot f(b) = \text{area of big rectangle}, \quad a \cdot f(a) = \text{area of small rectangle},$$

$$\int_a^b f(x)\, dx = \text{area} \quad \text{and} \quad \int_{f(a)}^{f(b)} f^{-1}(x)\, dx = \text{area} \ \blacksquare$$

from the figure, in terms of areas, one can immediately see that

area $\blacksquare$ = area of big rectangle − area of small rectangle − area $\blacksquare$

or

$$\int_a^b f(x)\,dx = bf(b) - af(a) - \int_{f(a)}^{f(b)} f^{-1}(x)\,dx,$$

as expected.

Example 17.1 As a simple example, we apply the result directly to the inverse sine function on the interval $[0, 1]$.

Solution As is well known, the inverse sine function on the interval $[0, 1]$ has an inverse, namely the sine function on the interval $[0, \frac{\pi}{2}]$. So from the result

$$\int_a^b f(x)\,dx + \int_{f(a)}^{f(b)} f^{-1}(x)\,dx = bf(b) - af(a),$$

setting $f^{-1}(x) = \sin^{-1} x$ so that $f(x) = \sin x$ and $a = 0, b = \frac{\pi}{2}$ giving $f(a) = f(0) = 0$ and $f(b) = f(\frac{\pi}{2}) = 1$, we see that

$$\int_0^1 \sin^{-1} x\,dx = \frac{\pi}{2} - \int_0^{\frac{\pi}{2}} \sin x\,dx = \frac{\pi}{2} - 1. \qquad \blacktriangleright$$

Example 17.2 Evaluate $\int_0^2 \left(\sqrt{1 + x^3} + \sqrt[3]{x^2 + 2x} \right) dx.$

Solution Let $f(x) = \sqrt{1 + x^3}$. Since $f'(x) = \dfrac{x^2}{\sqrt[3]{1 + x^3}} > 0$ for $x \in [0, 2]$ f is strictly increasing and will therefore have an inverse on this interval. On finding this inverse we have $f^{-1}(x) = \sqrt[3]{x^2 - 1}$. If we now set $a = 0$ so that $f(a) = f(0) = 1$ and $b = 2$ so that $f(b) = f(2) = 3$, on applying the result

$$\int_a^b f(x)\,dx + \int_{f(a)}^{f(b)} f^{-1}(x)\,dx = bf(b) - af(a),$$

we have

$$\int_0^2 \sqrt{1 + x^3}\,dx + \int_1^3 \sqrt[3]{x^2 - 1}\,dx = 2 \cdot 3 - 0 \cdot 1 = 6.$$

In the second integral, setting $x = u + 1$, we have $dx = du$ while the limits of integration are $x = 1, u = 0$ and $x = 3, u = 2$. Thus

$$\int_0^2 \sqrt{1 + x^3}\,dx + \int_0^2 \sqrt[3]{(u + 1)^2 - 1}\,du = 6,$$

or

$$\int_0^2 \left(\sqrt{1 + x^3} + \sqrt[3]{x^2 + 2x} \right) dx = 6,$$

where the dummy variable appearing in the second integral has been changed back to x. ▶

A result for the indefinite integral of an inverse function can also be readily found, which we give as a theorem in the next result.

Theorem 17.2 (Indefinite integral of an inverse function). *Let f be a* strictly monotonic *function with continuous derivative on some given interval. Then*

$$\boxed{\int f(x)\, dx = xf(x) - \int f^{-1}(u)\, du \quad \text{where} \quad u = f(x)}$$

This result is often written more conveniently as

$$\int^x f(u)\, du = xf(x) - \int^{f(x)} f^{-1}(u)du.$$

The upper limits of x and $f(x)$ appearing in the integrals are used to remind one that the variable needs to be changed back to x after the integration with respect to u has been performed.

Proof Since f is a strictly monotonic function on some interval, it has an inverse given by f^{-1}. Integrating the function f by parts, one has

$$\int f(x)\, dx = xf(x) - \int xf'(x)\, dx.$$

As $f^{-1}(f(x)) = x$, the integral on the right can be written as

$$\int f(x)\, dx = xf(x) - \int f^{-1}(f(x))f'(x)\, dx.$$

Setting $u = f(x), du = f'(x)\, dx$ gives

$$\int f(x)\, dx = xf(x) - \int f^{-1}(u)\, du,$$

as required to prove. ∎

Example 17.3 Find $\int (\cos^{-1} x)^2\, dx$.

Solution Let $f(x) = (\cos^{-1} x)^2$. Since f is a strictly decreasing function on its domain, its inverse will exist. On finding this inverse we have

$f^{-1}(x) = \cos \sqrt{x}$. Using the result for the indefinite integral of an inverse function, namely

$$\int f(x)\,dx = xf(x) - \int f^{-1}(u)\,du \quad \text{where } u = f(x),$$

one has

$$\int (\cos^{-1} x)^2\,dx = x(\cos^{-1} x)^2 - \int \cos \sqrt{u}\,du,$$

where $u = f(x) = (\cos^{-1} x)^2$. In the integral appearing to the right, if we let $u = t^2$ then $du = 2t\,dt$. So for this integral we have

$$\int \cos \sqrt{u}\,du = 2 \int t \cos t\,dt = 2t \sin t - 2 \int \sin t\,dt \quad \text{(by parts)}$$

$$= 2t \sin t + 2 \cos t + C = 2\sqrt{u} \sin \sqrt{u} + 2 \cos \sqrt{u} + C.$$

But as $u = (\cos^{-1} x)^2$ one has

$$\int (\cos^{-1} x)^2\,dx = x(\cos^{-1} x)^2 - 2 \sin(\cos^{-1} x) \cos^{-1} x$$

$$- 2 \cos(\cos^{-1} x) + C$$

$$= x(\cos^{-1} x)^2 - 2\sqrt{1 - x^2} \cos^{-1} x - 2x + C. \quad \blacktriangleright$$

Note in this case the integral could have just as conveniently been found by applying integration by parts directly.

Exercises for Chapter 17

✠ Warm-up

1. By applying directly the result for the definite integral of an inverse function to the inverse tangent function on the interval $[0, 1]$, show that

$$\int_0^1 \tan^{-1} x\,dx = \frac{\pi}{4} - \frac{\ln 2}{2}.$$

✠ Practice questions

2. Evaluate

(a) $\displaystyle \int_0^1 \left(\sqrt[3]{1 - x^7} - \sqrt[7]{1 - x^3} \right) dx$

(b) $\displaystyle \int_0^1 \left[\sin^3 \left(\frac{\pi x}{2} \right) + \frac{2}{\pi} \sin^{-1} \left(\sqrt[3]{x} \right) \right] dx$

3. Suppose that f is a strictly monotonic, continuous function on the interval $[0, 1]$.

If $f(0) = 0, f(1) = 1$, and $\displaystyle\int_0^1 f(x)\,dx = \frac{1}{3}$, find the value of $\displaystyle\int_0^1 f^{-1}(x)\,dx$.

4. Suppose that g is a continuous function that is strictly monotonic on the interval $[1, 5]$.

If $g(1) = 0, g(5) = 10$, and $\displaystyle\int_1^5 g(x)\,dx = 7$, find the value of $\displaystyle\int_0^{10} g^{-1}(x)\,dx$.

5. If $n \in \mathbb{N}$, show that $\displaystyle\int_0^1 {}^{n+1}\!\sqrt{1-x^n}\,dx = \int_0^1 \sqrt[n]{1-x^{n+1}}\,dx$.

6. (a) By reversing the roles of f and f^{-1}, show that the result for the definite integral of an inverse function can be written as

$$\int_a^b f^{-1}(x)\,dx = bf^{-1}(b) - af^{-1}(a) - \int_{f^{-1}(a)}^{f^{-1}(b)} f(x)\,dx.$$

(b) By reversing the roles of f and f^{-1}, show that the result for the indefinite integral of an inverse function can be written as

$$\int^x f^{-1}(u)\,du = xf^{-1}(x) - \int^{f^{-1}(x)} f(u)\,du.$$

7. Let $f : [0, 1] \to [0, 1]$ be a continuous, monotonically increasing function.

(a) Show that $\displaystyle\int_0^1 f^{-1}(x)\,dx = 1 - \int_0^1 f(x)\,dx$.

(b) Give a simple, graphical illustration of this result.

8. Suppose that $f(x) = \sin^{-1}\left(\sqrt{\dfrac{a}{a+x}}\right)$ where $a > 0$.

On the interval $[0, a]$, notice that f increases monotonically as the argument of the inverse sine function runs from $\dfrac{\pi}{4}$ to $\dfrac{\pi}{2}$. Thus f has an inverse on this interval.

(a) Show that the inverse of f is given by $f^{-1}(x) = a \cot^2 x$.

(b) Using the result for the definite integral of an inverse function, show that

$$\int_0^a \sin^{-1}\left(\sqrt{\frac{a}{a+x}}\right) dx = a.$$

9. Let $f(x) = x^3 + 3x + 4$.

(a) Show that f is an increasing function on the interval $[-1, 1]$ and therefore has an inverse on this interval.

(b) Denoting the inverse of f by f^{-1}, show that

$$\int_{-1}^1 f(x)\,dx + \int_0^4 f^{-1}(x)\,dx = \frac{23}{4}.$$

10. (a) Let $f(x) = x^3 - 2x^2 + 5$. Find the value of $\displaystyle\int_{37}^{149} f^{-1}(x)\,dx$.

(b) Let $g(x) = x + \cos x - 1$. Find the value of $\displaystyle\int_0^{2\pi} g^{-1}(x)\,dx$.

✠ Extension Questions and Challenge problems

11. Evaluate $\displaystyle\int_0^1 \left(\frac{x^x}{(1-x)^{1-x}} - \frac{(1-x)^{1-x}}{x^x}\right) dx$.

12. Suppose f is a strictly monotonic, continuous function on the interval $[a, b]$ $(b > a)$.

(a) If a and b are *fixed points*, that is, if $f(a) = a$ and $f(b) = b$, show that

$$\int_a^b \left(f(x) + f^{-1}(x)\right) dx = b^2 - a^2.$$

(b) Let $f(x) = x + \sin x$ on the interval $[0, \pi]$.

   (i) Show that f is a monotonically increasing function on the interval $[0, \pi]$ and $x = 0$ and $x = \pi$ are fixed points for the function.

   (ii) Even though no explicit formula for the inverse of f can be found, using the result given in (a) it is still possible to find the value for the definite integral $\displaystyle\int_0^\pi f^{-1}(x)\,dx$. Find its value.

13. It is known that $\displaystyle\int_0^{\frac{\pi}{4}} \ln(\cos x)\,dx = \frac{G}{2} - \frac{\pi}{4}\ln 2$.

Here G is a mathematical constant known as *Catalan's constant* (its value, correct to nine decimal places, is $G = 0.915\,965\,594$). Use this result to show that

$$\int_0^1 (\tan^{-1} x)^2 \, dx = \frac{\pi^2}{16} + \frac{\pi}{4} \ln 2 - G.$$

14. In this question an inequality known as *Young's inequality* will be proved. It is named after the English mathematician William Henry Young (1863–1942), who first proved the result in 1912.[1]

(a) Suppose that f is a continuous increasing function with $f(0) = 0$. If $a, b > 0$ then Young's inequality states that

$$\int_0^a f(x) \, dx + \int_0^b f^{-1}(x) \, dx \geqslant ab,$$

with equality holding only if $b = f(a)$.

To prove this result it is easiest if the three cases of $f(a) > b$, $f(a) < b$, and $f(a) = b$ are considered separately. For the first case of $f(a) > b$, start by writing the left-hand side of Young's inequality as

$$\int_0^a f(x) \, dx + \int_0^b f^{-1}(x) \, dx = \int_0^{f^{-1}(b)} f(x) \, dx$$
$$+ \int_{f^{-1}(b)}^a f(x) \, dx + \int_0^b f^{-1}(x) \, dx,$$

and apply the result of Exercise 6 (a) with a set equal to zero in the above expression before using one of the integral comparison properties. A similar thing can then be done for $f(a) < b$ while the case of $f(a) = b$ will follow immediately from either of the previous two cases.

(b) Illustrate Young's inequality graphically using separate graphs showing the separate cases when $f(a) > b$, $f(a) < b$, and $f(a) = b$.

(c) Young's inequality is so general that many interesting inequalities can be derived from it.

(i) When $f(x) = f^{-1}(x) = x$, show that the following inequality is obtained:

$$\sqrt{ab} \leqslant \frac{a + b}{2}.$$

This is a very famous and important inequality known as the *arithmetic mean–geometric mean inequality* (AM–GM inequality) for two positive numbers.

(ii) Show that if the particular function of $f(x) = x^{p-1}$ for $p > 1$ is taken in Young's inequality, one obtains

$$ab \leqslant \frac{a^p}{p} + \frac{b^q}{q} \quad \text{for} \quad p, q > 1 \quad \text{such that} \quad \frac{1}{p} + \frac{1}{q} = 1,$$

with equality occurring when $a^p = b^q$.

Endnote

1. W. H. Young, 'On the multiplication of successions of Fourier constants,' *Proceedings of the Royal Society A* **87**(596), 331–339 (1912).

18

Reduction Formulae

Human science fragments everything in order to understand it.
— Leo Tolstoy, *War and Peace*

A very useful technique of integration that is often employed comes from establishing what is known as a *reduction formula*. For expressions containing an integer parameter that is usually in the form of a power of one of the elementary functions, a reduction formula involves a *recurrence relation* in terms of the parameter that is reduced in a step-wise manner. Any of the common integration techniques discussed so far, be it integration using a substitution, integration by parts, integration using a trigonometric or hyperbolic substitution, integration using the method of partial fractions, and so on, can be used to reduce the original integral to an integral of the same or a very similar type with a lower integer parameter. This 'smaller' integral can be further reduced in a similar manner until one finally arrives at an integral that can be relatively easily found.

So what does the reduction technique actually do to the integral? It takes an integral of the form

$$I_n = \int f(x, n) \, dx,$$

where n is the parameter and reduces it to an integral of the form

$$I_k = \int f(x, k) \, dx,$$

where $k < n$. In many instances reduction formulae are established using integration by parts, but as we have already mentioned, that method is by no means the only way of creating reduction formulae. To better understand the process of establishing a reduction formula from a given integral, as well as typical applications to which such formulae can be put, we now consider examples that make use of a variety of different techniques.

Example 18.1 Suppose that $\int \sin^n x \, dx$, for $n \geqslant 0$.

(a) Show $\int \sin^n x \, dx = -\dfrac{1}{n} \cos x \sin^{n-1} x + \dfrac{n-1}{n} \int \sin^{n-2} x \, dx$.

(b) Hence find $\int \sin^5 x \, dx$.

Solution

(a) We start by rewriting the integrand as

$$\int \sin^n x \, dx = \int \sin x \cdot \sin^{n-1} x \, dx.$$

Integrating by parts, one has

$$\int \sin^n x \, dx = -\cos x \cdot \sin^{n-1} x$$

$$- \int (-\cos x) \cdot (n-1) \sin^{n-2} x \cdot \cos x \, dx$$

$$= -\cos x \sin^{n-1} x + (n-1) \int \cos^2 x \sin^{n-2} x \, dx$$

$$= -\cos x \sin^{n-1} x + (n-1) \int (1 - \sin^2 x) \sin^{n-2} x \, dx$$

$$= -\cos x \sin^{n-1} x + (n-1) \int \sin^{n-2} x \, dx$$

$$- (n-1) \int \sin^n x \, dx.$$

After collecting terms and rearranging, one finally has

$$\int \sin^n x \, dx = -\frac{1}{n} \cos x \sin^{n-1} x + \frac{n-1}{n} \int \sin^{n-2} x \, dx,$$

as required to show.

(b) Setting $n = 5$ in the reduction formula found in (a), we have

$$\int \sin^5 x \, dx = -\frac{1}{5} \cos x \sin^4 x + \frac{4}{5} \int \sin^3 x \, dx.$$

To find the last integral we set $n = 3$ in the reduction formula. Thus

$$\int \sin^3 x \, dx = -\frac{1}{3} \cos x \sin^2 x + \frac{2}{3} \int \sin x \, dx$$

$$= -\frac{1}{3} \cos x \sin^2 x - \frac{2}{3} \cos x + C.$$

So finally we have

$$\int \sin^5 x \, dx = -\frac{1}{5} \cos x \sin^4 x - \frac{4}{15} \cos x \sin^2 x - \frac{8}{15} \cos x + C. \quad \blacktriangleright$$

Example 18.2 Suppose that I_n is defined by $I_n = \int_0^{\pi/4} \tan^n x \, dx$, whenever $n \geqslant 0$.

(a) Find the value for I_0.

(b) Show that $I_n = \dfrac{1}{n-1} - I_{n-2}$ for all $n \geqslant 2$.

(c) Using the result in (b), evaluate $\int_0^{\pi/4} \tan^4 x \, dx$.

Solution

(a) When $n = 0$ we have

$$I_0 = \int_0^{\pi/4} (\tan x)^0 dx = \int_0^{\pi/4} dx = x \Big|_0^{\pi/4} = \frac{\pi}{4}.$$

(b) In this part of the question we will need to make use of the identity $\tan^2 x = \sec^2 x - 1$.

$$\begin{aligned}
I_n &= \int_0^{\pi/4} \tan^n x \, dx = \int_0^{\pi/4} \tan^{n-2} x \tan^2 x \, dx \\
&= \int_0^{\pi/4} \tan^{n-2} x (\sec^2 x - 1) dx \\
&= \int_0^{\pi/4} \tan^{n-2} x \sec^2 x \, dx - \int_0^{\pi/4} \tan^{n-2} x \, dx \\
&= \int_0^{\pi/4} \tan^{n-2} x \sec^2 x \, dx - I_{n-2}.
\end{aligned}$$

Now evaluating the integral using a substitution. If we let

$$u = \tan x, \qquad du = \sec^2 x \, dx$$

while for the limits of integration, $x = 0, u = 0$ and $x = \pi/4, u = 1$. Thus

$$\int_0^{\pi/4} \tan^{n-2} x \sec^2 x \, dx = \int_0^1 u^{n-2} du = \left[\frac{u^{n-1}}{n-1} \right]_0^1 = \frac{1}{n-1}.$$

So finally we have

$$I_n = \frac{1}{n-1} - I_{n-2},$$

as required to show.

(c) Substituting $n = 4$ into the reduction formula found in (b) we have

$$\int_0^{\pi/4} \tan^4 x \, dx = I_4 = \frac{1}{4-1} - I_2 = \frac{1}{3} - I_2.$$

So we see that we need to find I_2 first. Now

$$I_2 = \frac{1}{2-1} - I_0 = 1 - \frac{\pi}{4}.$$

Hence

$$\int_0^{\pi/4} \tan^4 x \, dx = \frac{1}{3} - \left(1 - \frac{\pi}{4}\right) = \frac{\pi}{4} - \frac{2}{3}. \qquad \blacktriangleright$$

Example 18.3 Suppose that I_n is defined by $I_n = \int_0^1 \frac{dx}{(1+x^2)^n}$ for $n \in \mathbb{N}$.

(a) Find the value for I_1.

(b) Show that $I_{n+1} = \frac{1}{n2^{n+1}} + \frac{2n-1}{2n} I_n$, for all $n \geqslant 1$.

(c) Using the result in (b), evaluate $\int_0^1 \frac{dx}{(1+x^2)^3} \, dx$.

Solution

(a) Setting $n = 1$ we have

$$I_1 = \int_0^1 \frac{dx}{1+x^2} = \tan^{-1} x \Big|_0^1 = \frac{\pi}{4}.$$

(b) Integrating by parts we have

$$I_n = \left[\frac{x}{(1+x^2)^n}\right]_0^1 + 2n \int_0^1 \frac{x^2}{(1+x^2)^{n+1}} \, dx$$

$$= \frac{1}{2^n} + 2n \int_0^1 \frac{(1+x^2) - 1}{(1+x^2)^{n+1}} \, dx$$

$$= \frac{1}{2^n} + 2n \int_0^1 \frac{dx}{(1+x^2)^n} - 2n \int_0^1 \frac{dx}{(1+x^2)^{n+1}}$$

$$= \frac{1}{2^n} + 2n I_n - 2n I_{n+1}.$$

Rearranging we have

$$I_{n+1} = \frac{1}{n2^{n+1}} + \frac{2n-1}{2n} I_n,$$

as required to show.

(c) To find the required integral, set $n = 2$ in the reduction formula. Thus

$$I_3 = \frac{1}{16} + \frac{3}{4} I_2.$$

Now in order to find I_2, set $n = 1$ in the reduction formula. Thus

$$I_2 = \frac{1}{4} + \frac{1}{2} I_1 = \frac{1}{4} + \frac{\pi}{8},$$

and for the integral we have

$$\int_0^1 \frac{dx}{(1+x^2)^3} = \frac{1}{16} + \frac{3}{4}\left(\frac{1}{4} + \frac{\pi}{8}\right) = \frac{1}{4} + \frac{3\pi}{32}.$$ ▶

Example 18.4 Let $I_n = \displaystyle\int_0^\pi \frac{1 - \cos nx}{1 - \cos x} dx$ for $n = 0, 1, 2, \ldots$

(a) Show that $I_{n+2} - 2I_{n+1} + I_n = 0$.
(b) Evaluate I_0 and I_1.
(c) Using induction, prove that $I_n = n\pi$ for all $n = 0, 1, 2, \ldots$

Solution

(a) We show this result by direct substitution.

$$\begin{aligned}
I_{n+2} - 2I_{n+1} + I_n &= \int_0^\pi \frac{(1 - \cos(n+2)x)}{1 - \cos x} dx \\
&\quad - 2\int_0^\pi \frac{(1 - \cos(n+1)x)}{1 - \cos x} dx \\
&\quad + \int_0^\pi \frac{(1 - \cos nx)}{1 - \cos x} dx \\
&= \int_0^\pi \frac{2\cos(n+1)x - \cos(n+2)x - \cos nx}{1 - \cos x} dx.
\end{aligned}$$

Now using the following sum-to-product result of

$$\cos\theta + \cos\varphi = 2\cos\left(\frac{\theta + \varphi}{2}\right)\cos\left(\frac{\theta - \varphi}{2}\right),$$

on setting $\theta = nx + 2x$ and $\varphi = nx$, we have

$$\cos(n+2)x + \cos nx = 2\cos(n+1)x \cos x.$$

Thus

$$I_{n+2} - 2I_{n+1} + I_n = \int_0^\pi \frac{2\cos(n+1)x - 2\cos(n+1)x\cos x}{1 - \cos x}\,dx$$

$$= \int_0^\pi \frac{2\cos(n+1)x \cdot (1 - \cos x)}{1 - \cos x}\,dx$$

$$= \int_0^\pi 2\cos(n+1)x\,dx = \frac{2}{n+1}\sin(n+1)x\Big|_0^\pi = 0,$$

as required to show.   ▶

(b) Setting $n = 0$ we have

$$I_0 = \int_0^\pi \frac{1 - \cos(0 \cdot x)}{1 - \cos x}\,dx = 0,$$

and setting $n = 1$ we have

$$I_1 = \int_0^\pi \frac{1 - \cos x}{1 - \cos x}\,dx = \int_0^\pi dx = \pi.$$

(c) When $n = 0$, $I_0 = 0 \cdot \pi = 0$ and when $n = 1$, $I_1 = 1 \cdot \pi = \pi$, which as seen from part (b) are both true. Assume the statement is true for $n = k - 1$ and $n = k$, that is,

$$I_{k-1} = (k-1)\pi \quad \text{and} \quad I_k = k\pi,$$

respectively. Now prove the statement is true for $n = k + 1$. From the reduction formula proved in (a), setting $n = k - 1$ one has

$$I_{k+1} - 2I_k + I_{k-1} = 0.$$

Thus

$$I_{k+1} = 2I_k - I_{k-1} = 2k\pi - (k-1)\pi = (k+1)\pi.$$

So the statement is true for $n = k + 1$. So by induction the statement must be true for $n = 0$ and all positive integers n. Hence proven.   ▶

Example 18.5 Let $I_n = \int_0^1 x^n \sqrt{1-x}\,dx$ for $n = 0, 1, 2, \ldots$

(a) Show that $I_n = \dfrac{2n}{2n+3} I_{n-1}$.

(b) Evaluate I_0.

(c) Hence deduce that $I_n = \dfrac{2^{2n+2} n!(n+1)!}{(2n+3)!}$.

Solution

(a) Integrating by parts we have

$$I_n = x^n \cdot -\frac{2}{3}(1-x)^{\frac{3}{2}}\Big|_0^1 + \frac{2n}{3}\int_0^1 x^{n-1}(1-x)^{\frac{2}{3}}\,dx$$

$$= \frac{2n}{3}\int_0^1 x^{n-1}(1-x)\sqrt{1-x}\,dx$$

$$= \frac{2n}{3}\int_0^1 x^{n-1}\sqrt{1-x}\,dx - \frac{2n}{3}\int_0^1 x^n\sqrt{1-x}\,dx$$

$$= \frac{2n}{3}I_{n-1} - \frac{2n}{3}I_n,$$

which, after collecting like terms and rearranging, gives

$$I_n = \frac{2n}{2n+3}I_{n-1},$$

as required to show.

(b) Setting $n = 0$ we have

$$I_0 = \int_0^1 \sqrt{1-x}\,dx.$$

Now let $x = \sin^2\theta, dx = 2\sin\theta\cos\theta\,d\theta$ while for the limits of integration, when $x = 0, \theta = 0$ while for $x = 1, \theta = \pi/2$. So we have

$$I_0 = 2\int_0^{\frac{\pi}{2}} \sin\theta\cos^2\theta\,d\theta = \frac{2}{3}\cos^3\theta\Big|_0^{\frac{\pi}{2}} = \frac{2}{3}.$$

(c) From the reduction formula for I_n, repeated application gives

$$I_n = \frac{2n}{2n+3}I_{n-1}$$

$$= \frac{2n}{2n+3}\cdot\frac{2n-2}{2n+1}\cdot I_{n-2}$$

$$\vdots$$

$$= \frac{2n}{2n+3}\cdot\frac{2n-2}{2n+1}\cdot\frac{2n-4}{2n-1}\cdots\frac{4}{7}\cdot\frac{2}{5}\cdot I_0.$$

Factoring out a factor of two in each term appearing in the numerator, as there are n of these terms, we have

$$
\begin{aligned}
I_n &= \frac{2^n \left[n(n-1)\cdots 2\cdot 1\right]}{(2n+3)(2n+1)\cdots 7\cdot 5} \cdot \frac{2}{3} \\[2mm]
&= \frac{2^{n+1} n!}{(2n+3)(2n+1)\cdots 7\cdot 5\cdot 3} \\[2mm]
&= \frac{2^{n+1} n!(2n+2)2n(2n-2)\cdots 6\cdot 4\cdot 2}{(2n+3)(2n+2)(2n+1)\cdots 5\cdot 4\cdot 3\cdot 2} \\[2mm]
&= \frac{2^{n+1} n! 2^{n+1}(n+1)!}{(2n+3)!} \\[2mm]
&= \frac{2^{2n+2} n!(n+1)!}{(2n+3)!},
\end{aligned}
$$

as required to show. ▶

For our last example, while a reduction formula is not explicitly asked for, we give an application that makes use of one as part of the solution to a particular problem.

Example 18.6 Evaluate $\displaystyle\int_0^{\frac{\pi}{6}} \frac{1 - \sin^8 x}{\cos^2 x}\, dx.$

Solution

$$
\begin{aligned}
\int_0^{\frac{\pi}{6}} \frac{1 - \sin^8 x}{\cos^2 x}\, dx &= \int_0^{\frac{\pi}{6}} \frac{1 - \sin^8 x}{1 - \sin^2 x}\, dx = \int_0^{\frac{\pi}{6}} \frac{\sin^8 x - 1}{\sin^2 x - 1}\, dx \\[2mm]
&= \int_0^{\frac{\pi}{6}} \frac{(\sin^2 x - 1)(\sin^6 x + \sin^4 x + \sin^2 x + 1)}{\sin^2 x - 1}\, dx \\[2mm]
&= \int_0^{\frac{\pi}{6}} (\sin^6 x + \sin^4 x + \sin^2 x + 1)\, dx.
\end{aligned}
$$

Now consider $I_n = \displaystyle\int_0^{\frac{\pi}{6}} \sin^n x\, dx$ where $n = 0, 1, 2, \ldots$ So we have

$$
\begin{aligned}
I_n &= \int_0^{\frac{\pi}{6}} \sin x \sin^{n-1} x\, dx \\[2mm]
&= -\cos x \sin^{n-1} x \Big|_0^{\frac{\pi}{6}} + (n-1)\int_0^{\frac{\pi}{6}} \cos^2 x \sin^{n-2} x\, dx,
\end{aligned}
$$

after integrating by parts. Evaluating the first term and rewriting the cosine squared term in the integrand in terms of sine squared, one has

$$I_n = -\frac{\sqrt{3}}{2^n} + (n-1)\int_0^{\frac{\pi}{6}} (1 - \sin^2 x)\sin^{n-2} x\, dx$$

$$= -\frac{\sqrt{3}}{2^n} + (n-1)\int_0^{\frac{\pi}{6}} \sin^{n-2} x\, dx - (n-1)\int_0^{\frac{\pi}{6}} \sin^n x\, dx$$

$$= -\frac{\sqrt{3}}{2^n} + (n-1)I_{n-2} - (n-1)I_n$$

$$\Rightarrow I_n = -\frac{\sqrt{3}}{n2^n} + \frac{n-1}{n}I_{n-2}$$

$$n = 0: I_0 = \int_0^{\frac{\pi}{6}} dx = \frac{\pi}{6}$$

$$n = 2: I_2 = -\frac{\sqrt{3}}{2 \cdot 2^2} + \frac{1}{2}\cdot I_0 = -\frac{\sqrt{3}}{8} + \frac{\pi}{12}$$

$$n = 4: I_4 = -\frac{\sqrt{3}}{4 \cdot 2^4} + \frac{3}{4}\cdot I_2 = -\frac{\sqrt{3}}{64} + \frac{3}{4}\left(-\frac{\sqrt{3}}{8} + \frac{\pi}{12}\right)$$

$$= -\frac{7\sqrt{3}}{64} + \frac{\pi}{16}$$

$$n = 6: I_6 = -\frac{\sqrt{3}}{6 \cdot 2^6} + \frac{5}{6}\cdot I_4 = -\frac{\sqrt{3}}{384} + \frac{5}{6}\left(-\frac{7\sqrt{3}}{64} + \frac{\pi}{16}\right)$$

$$= -\frac{3\sqrt{3}}{32} + \frac{5\pi}{96}.$$

So one has

$$\int_0^{\frac{\pi}{6}} \frac{1 - \sin^8 x}{\cos^2 x}\, dx = \int_0^{\frac{\pi}{6}} (\sin^6 x + \sin^4 x + \sin^2 x + 1)\, dx$$

$$= \int_0^{\frac{\pi}{6}} \sin^6 x\, dx + \int_0^{\frac{\pi}{6}} \sin^4 x\, dx$$

$$+ \int_0^{\frac{\pi}{6}} \sin^2 x\, dx + \int_0^{\frac{\pi}{6}} dx$$

$$= I_6 + I_4 + I_2 + I_0,$$

which we see are just the four integrals we already found using our reduction formula for I_n. Thus

$$\int_0^{\frac{\pi}{6}} \frac{1 - \sin^8 x}{\cos^2 x} \, dx = \left[-\frac{3\sqrt{3}}{32} + \frac{5\pi}{96} \right] + \left[-\frac{7\sqrt{3}}{64} + \frac{\pi}{16} \right]$$

$$+ \left[-\frac{\sqrt{3}}{8} + \frac{\pi}{12} \right] + \frac{\pi}{6}$$

$$= \frac{35\pi}{96} - \frac{21\sqrt{3}}{64}. \qquad \blacktriangleright$$

Exercises for Chapter 18

✠ **Warm-up**

1. The reduction formula

$$\int \operatorname{cosec}^n x \, dx = -\frac{\cot x \cdot \operatorname{cosec}^{n-2} x}{n-1} + \frac{n-2}{n-1} \int \operatorname{cosec}^{n-2} x \, dx,$$

is valid for all $n \geqslant 2$. Use this to find the integrals

$$\int \operatorname{cosec}^3 x \, dx \text{ and } \int \operatorname{cosec}^4 x \, dx.$$

✠ **Practice questions**

2. (a) If $\int \cos^n dx, n = 0, 1, 2, \ldots$ show that

$$\int \cos^n dx = \frac{1}{n} \sin x \cdot \cos^{n-1} x + \frac{n-1}{n} \int \cos^{n-2} x \, dx,$$

where $n = 2, 3, 4, \ldots$

(b) Hence find $\int \cos^4 x \, dx.$

(c) There is no reason why the result found in (a) cannot be applied to negative integers as well. By rearranging the result in (a), show that

$$\int \cos^{n-2} x \, dx = -\frac{1}{n-1} \sin x \cos^{n-1} x + \frac{n}{n-1} \int \cos^n x \, dx,$$

where $n = -1, -2, -3, \ldots$

(d) Hence find $\int \cos^{-3} x \, dx$ and $\int \cos^{-4} x \, dx.$

3. Suppose that $I_n = \displaystyle\int_0^1 x^n e^{-x}\, dx$.

    (a) Show that $I_n = n I_{n-1} - \dfrac{1}{e}$.

    (b) Hence evaluate $\displaystyle\int_0^1 x^3 e^{-x}\, dx$.

4. Suppose $I_n = \displaystyle\int_0^1 \sinh^n x\, dx, n \in \mathbb{N}$.

    (a) Show that $n I_n + (n-1) I_n = \cosh(1) \sinh^{n-1}(1)$.

    (b) Hence evaluate $\displaystyle\int_0^1 \sinh^3 x\, dx$.

5. (a) Show that

$$\int \sec^n x\, dx = \frac{\tan x \cdot \sec^{n-2} x}{n-1} + \frac{n-2}{n-1} \int \sec^{n-2} x\, dx, n \geqslant 2.$$

    (b) Hence find $\displaystyle\int \sec^4 x\, dx$.

6. Consider the integral $\displaystyle\int_0^1 (1-x^4)^n\, dx, n \geqslant 0$.

    (a) Show that $\displaystyle\int_0^1 (1-x^4)^n\, dx = \frac{4n}{4n+1} \int_0^1 (1-x^4)^{n-1}\, dx$,
    where $n \geqslant 1$.

    (b) Hence evaluate $\displaystyle\int_0^1 (1-x^4)^3\, dx$.

7. Suppose $I_n = \displaystyle\int_0^{\frac{\pi}{2}} x^n \cos x\, dx$ for $n \in \mathbb{Z}^+$.

    (a) Show that $I_n = \left(\dfrac{\pi}{2}\right)^n - n(n-1) I_{n-2}$ for $n \geqslant 2$.

    (b) Hence evaluate $\displaystyle\int_0^{\frac{\pi}{2}} x^6 \cos x\, dx$.

8. Let $I_{2n+1} = \displaystyle\int_0^1 x^{2n+1} e^{x^2}\, dx, n \in \mathbb{Z}^+$.

    (a) Show that $I_{2n+1} = \dfrac{e}{2} - n I_{2n-1}$.

    (b) Hence evaluate $\displaystyle\int_0^1 x^5 e^{x^2}\, dx$.

9. (a) Evaluate $\displaystyle\int_0^x \operatorname{sech} t \, dt$.

(b) If $\displaystyle J_n = \int_0^x \operatorname{sech}^n t \, dt$ for $n \in \mathbb{N}$, show that

$$J_n = \frac{1}{n-1} \tanh x \operatorname{sech}^{n-2} x + \frac{n-2}{n-1} J_{n-2}, n \geqslant 2.$$

(c) Hence find expressions for J_3 and J_4.

10. Suppose that $\displaystyle I_n = \int_0^\pi \frac{\cos nx}{5 - 4\cos x} dx$ for $n \geqslant 0$.

(a) By calculating directly, find expressions for I_0 and I_1.

(b) Show that $I_{n+1} + I_{n-1} - \frac{5}{2} I_n = 0$ for all $n \geqslant 1$.

(c) Hence find expressions for I_2 and I_3.

(d) Based on the expressions found for $I_0, I_1, I_2,$ and I_3 conjecture a closed-form expression for I_n where $n \in \mathbb{Z}^+$.

(e) Use induction to prove the conjecture you gave in (d).

11. Suppose $\displaystyle I_n = \int_0^{\frac{\pi}{2}} \frac{\sin^2 nx}{\sin x} dx$ for $n \in \mathbb{Z}^+$.

(a) Evaluate I_0.

(b) Prove that $\cos 2nx - \cos(2n + 2)x = 2\sin(2n + 1)x \sin x$.

(c) Show that $\displaystyle I_{n+1} - I_n = \frac{1}{2n + 1}$.

(d) Hence deduce that $\displaystyle \int_0^{\frac{\pi}{2}} \frac{\sin^2 nx}{\sin x} dx = \sum_{k=1}^n \frac{1}{2k - 1}$.

12. Let $\displaystyle I_k = \int_0^1 x^k (1 - x)^{n-k} \, dx$ where $k = 0, 1, 2, \ldots, n$ such that $n \geqslant k$.

(a) Find I_0.

(b) Show that $\displaystyle I_k = \frac{k}{n - k + 1} I_{k-1}$.

(c) Deduce that $\displaystyle I_k = \left[\binom{n}{k}(n + 1) \right]^{-1}$.

(d) Hence evaluate $\displaystyle \int_0^1 \binom{212}{8} x^8 (1 - x)^{204} \, dx$.

13. Suppose $\displaystyle S_n = \int (\sin^{-1} x)^n \, dx$ where $n \in \mathbb{N}$.

(a) For $n \geqslant 2$, show that

$$S_n = x(\sin^{-1} x)^n + n\sqrt{1 - x^2}(\sin^{-1} x)^{n-1} - n(n - 1)I_{n-2}.$$

(b) Hence deduce that $\displaystyle\int_0^1 (\sin^{-1} x)^3 \, dx = 6 - 3\pi + \frac{\pi^3}{8}$.

14. Let $V_n = \displaystyle\int_0^1 x(1 - x^3)^n \, dx$ where $n \in \mathbb{Z}^+$. Show that

$$V_n = \frac{3}{3n + 2} V_{n-1}.$$

15. (a) If $I_n = \displaystyle\int x^n \sin ax \, dx$ where $a \neq 0$ and $n \in \mathbb{Z}^+$, show that

$$I_n = -\frac{1}{a} x^n \cos ax + \frac{n}{a^2} x^{n-1} \sin ax - \frac{n(n-1)}{a^2} I_{n-2}.$$

(b) If $U_n = \displaystyle\int \frac{\cos ax}{x^n} \, dx$ where $a \in \mathbb{R}$ and $n = 3, 4, 5, \ldots$, show that

$$U_n = \frac{a \sin ax}{(n-1)(n-2)x^{n-2}} - \frac{\cos ax}{(n-1)x^{n-1}} - \frac{a^2}{(n-1)(n-2)} U_{n-2}.$$

16. Let $V_n(x) = \displaystyle\int_0^x \frac{u^n}{\sqrt{u^2 + a^2}} \, du, \; a \in \mathbb{R}, n \in \mathbb{Z}^+$.

(a) Show that $nV_n(x) = x^{n-1}\sqrt{x^2 + a^2} - (n-1)a^2 V_{n-2}(x)$, for $n \geqslant 2$.

(b) Using (a), evaluate $\displaystyle\int_0^2 \frac{x^5}{\sqrt{x^2 + 5}} \, dx$.

17. Let $\quad I_n = \displaystyle\int_0^\pi \frac{2 + 2\cos x - \cos(n-1)x - 2\cos nx - \cos(n+1)x}{1 - \cos 2x} \, dx$

for $n \in \mathbb{Z}^+$.

(a) Find values for the integral when $n = 0$ and $n = 1$.

(b) Show that $I_{n+1} = 2I_n - I_{n-1}$.

(c) Hence conclude that $I_{n+1} - I_n = I_n - I_{n-1}$ and that the terms in the sequence for I_n form an arithmetic progression.

(d) By finding the common difference between the terms in the arithmetic sequence, show that $I_n = n\pi$.

18. Let $I_n = \displaystyle\int_0^a (a^2 - x^2)^n \, dx$ such that $a \in \mathbb{R}$ and $n \in \mathbb{Z}^+$.

(a) Evaluate I_0.

(b) Show that $I_n = \dfrac{2na^2}{2n+1} I_{n-1}$.

(c) Hence deduce that $\displaystyle\int_0^a (a^2 - x^2)^n \, dx = \frac{(2^n n!)^2}{(2n+1)!} a^{2n+1}$.

(d) Using (c), evaluate $\int_0^1 (1 - x^2)^n \, dx$ where $n \in \mathbb{Z}^+$.

(e) By making use of the binomial theorem to find the integral given in (d) directly, show that

$$\sum_{k=0}^{n} \binom{n}{k} \frac{(-1)^n}{2n + 1} = \frac{(2^n n!)^2}{(2n + 1)!}.$$

19. (a) Let $u_n = \int_0^\pi \frac{\sin nx}{\sin x} \, dx$, $n \in \mathbb{Z}^+$.

    (i) Evaluate u_0 and u_1.

    (ii) Show that $u_{n+2} = u_n$ for $n \in \mathbb{Z}^+$.

    (iii) Hence deduce that $\displaystyle\int_0^\pi \frac{\sin nx}{\sin x} \, dx = \begin{cases} \pi, & n \text{ odd} \\ 0, & n \text{ even}. \end{cases}$

(b) Let $U_n = \int_0^\pi \frac{\sin^2 nx}{\sin^2 x} \, dx$, $n \in \mathbb{Z}^+$.

    (i) Using the result found for u_n, show that $U_n - U_{n-1} = \pi$.

    (ii) Hence deduce that $\displaystyle\int_0^\pi \frac{\sin^2 nx}{\sin^2 x} \, dx = n\pi$.

20. Suppose $J_n = \int_0^{\frac{\pi}{2}} \sin^n u \cos^n u \, du$ for $n \in \mathbb{N}$.

(a) Show that $J_n = \dfrac{n - 1}{4n} J_{n-2}$ for $n \geqslant 2$.

(b) Through the use of a suitable substitution, use the reduction formula given in (a) to evaluate $\int_0^2 (4 - x^2)^{\frac{3}{2}} x^4 \, dx$.

21. Suppose that $I_n = \int_0^\pi \frac{\sin(nx)}{\sin x} \, dx$, $n \in \mathbb{N}$.

(a) By using the following prosthaphaeresis formula for the trigonometric functions

$$\sin A - \sin B = 2 \sin\left(\frac{A - B}{2}\right) \cos\left(\frac{A + B}{2}\right),$$

show that $\sin nx - \sin(n - 2)x = 2 \sin x \cos(n - 1)x$.

(b) Hence show that $I_n = I_{n-2}$.

(c) Hence deduce that $I_n = \displaystyle\int_0^\pi \frac{\sin(nx)}{\sin x} \, dx = \begin{cases} 0, & n \text{ even} \\ \pi, & n \text{ odd}. \end{cases}$

22. In this question we will find the value for the definite integral $\int_0^\pi \frac{\sin^2(nx)}{\sin^2 x} \, dx$, for all $n \in \mathbb{N}$.

(a) Find values for the integral when $n = 0$ and $n = 1$.

(b) If f is continuous and bounded on the interval $[0, 2a]$ where $a > 0$, show that

$$\int_0^{2a} f(x)\,dx = \int_0^a f(x)\,dx + \int_0^a f(2a - x)\,dx.$$

(c) Use (b) to show that $\displaystyle\int_0^\pi \frac{\sin^2(nx)}{\sin^2 x}\,dx = \int_0^\pi \frac{1 - \cos(nx)}{1 - \cos x}\,dx.$

(d) Let $\displaystyle I_n = \int_0^\pi \frac{1 - \cos(nx)}{1 - \cos x}\,dx$ where $n \in \mathbb{N}$.

Show that $I_{n+1} + I_{n-1} = 2I_n$.

(e) Hence conclude that $I_{n+1} - I_n = I_n - I_{n-1}$ and that the terms in the sequence for I_n form an arithmetic progression.

(f) By finding the common difference between the terms in the arithmetic sequence, show that

$$I_n = \int_0^\pi \frac{\sin^2(nx)}{\sin^2 x}\,dx = \int_0^\pi \frac{1 - \cos(nx)}{1 - \cos x}\,dx = n\pi.$$

✠ **Extension Questions and Challenge problems**

23. If $m, n \in \mathbb{N}$, show that $\displaystyle\int_0^1 x^m(1 - x)^n\,dx = \frac{n!}{(m + n + 1)!}.$

24. Let $\displaystyle I_n = \int_0^\pi \frac{\cos nx - \cos ny}{\cos x - \cos y}\,dx, \, y \in \mathbb{R}, n \in \mathbb{Z}^+.$

(a) Evaluate I_1 and I_2.

(b) Show that $I_{n+1} - 2I_n \cos y + I_{n-1} = 0, n \in \mathbb{N}$.

(c) Using induction on n, show that $I_n = \dfrac{\sin ny}{\sin y}\pi$ for $n \in \mathbb{N}$.

25. Suppose $\displaystyle I_n = \int_0^{\frac{\pi}{2}} \cos^n x\,dx$ where n is a non-negative integer.

(a) Show that $I_{n+1} = \dfrac{n}{n + 1} I_{n-1}.$

(b) Hence show that $(n + 1)I_{n+1}I_n$ is independent of n. Find its value.

(c) Prove that $I_{n+1} < I_n < I_{n-1}$.

(d) Hence show that $\displaystyle\lim_{n \to \infty} \frac{I_n}{I_{n+1}} = 1.$

(e) Hence deduce that

$$\lim_{n \to \infty} \left(\sqrt{n + 1} \int_0^{\frac{\pi}{2}} \cos^n x\,dx \right) = \sqrt{\frac{\pi}{2}}.$$

26 (Leibniz's series).

Let $I_n = \displaystyle\int_0^1 \frac{x^{2n}}{1+x^2}\, dx$ where $n \in \mathbb{Z}^+$.

(a) Show that $I_0 = \dfrac{\pi}{4}$.

(b) Show that $I_n + I_{n-1} = \dfrac{1}{2n-1}$ for $n \geqslant 1$.

(c) Hence deduce that

$$I_n = \frac{1}{2n-1} - \frac{1}{2n-3} + \cdots + (-1)^{n-2} \cdot \frac{1}{3} + (-1)^{n-1} \cdot 1 + (-1)^n \cdot \frac{\pi}{4}.$$

(d) Show that the expression for I_n given in (c) can be rewritten as

$$I_n = (-1)^n \left[\frac{\pi}{4} - \sum_{k=1}^{n} \frac{(-1)^{k+1}}{2k-1} \right].$$

(e) Show that $0 \leqslant \dfrac{x^{2n}}{1+x^2} \leqslant x^{2n}$ for all $n \geqslant 0$ and $0 \leqslant x \leqslant 1$.

(f) Hence deduce that $0 \leqslant \left| \dfrac{\pi}{4} - \displaystyle\sum_{k=1}^{n} \dfrac{(-1)^{k+1}}{2k-1} \right| \leqslant \dfrac{1}{2n+1}$.

(g) Hence show that as $n \to \infty$ one has $\displaystyle\sum_{k=1}^{\infty} \frac{(-1)^{k+1}}{2k-1} = \frac{\pi}{4}$.

This series is what is known as the *Leibniz series*, it being named after one of its independent co-discoverers, the German polymath Gottfried Wilhelm von Leibniz (1646–1716).[1]

27 (Wallis's product). Suppose that $I_n = \displaystyle\int_0^{\frac{\pi}{2}} \sin^n x \, dx, n \in \mathbb{N}$.

(a) Evaluate I_0 and I_1.

(b) Show that $I_{2n} = \dfrac{2n-1}{2n} I_{2n-2}$.

(c) Hence deduce that

$$I_{2n} = \frac{2n-1}{2n} \cdot \frac{2n-3}{2n-2} \cdots \frac{3}{4} \cdot \frac{1}{2} \cdot \frac{\pi}{2} = \frac{\pi}{2} \prod_{k=1}^{n} \frac{2k-1}{2k}.$$

(d) Show that $I_{2n+1} = \dfrac{2n}{2n+1} I_{2n-1}$.

(e) Hence deduce that

$$I_{2n+1} = \frac{2n}{2n+1} \cdot \frac{2n-2}{2n-1} \cdots \frac{4}{5} \cdot \frac{2}{3} \cdot 1 = \prod_{k=1}^{n} \frac{2k}{2k+1}.$$

(f) Show that $I_{2n+1} \leqslant I_{2n} \leqslant I_{2n-1}$.

(g) Hence show that $\dfrac{I_{2n}}{I_{2n+1}} \to 1$ as $n \to \infty$.

(h) Use the results from (c), (d), and (g) to show that

$$\frac{2}{\pi} = \lim_{n \to \infty} \prod_{k=1}^{n} \frac{2k-1}{2k} \cdot \frac{2k+1}{2k} = \frac{1}{2} \cdot \frac{3}{2} \cdot \frac{3}{4} \cdot \frac{5}{4} \cdot \frac{5}{6} \cdot \frac{7}{6} \cdots .$$

This product is known as *Wallis's product* after the English mathematician John Wallis (1616–1703) who first gave the result in 1655.

28 (The Basel problem). Suppose that

$$A_n = \int_0^{\frac{\pi}{2}} \cos^{2n} x \, dx \ \text{ and } \ B_n = \int_0^{\frac{\pi}{2}} x^2 \cos^{2n} x \, dx \ \text{ for } n \in \mathbb{Z}^+.$$

(a) Show that $A_0 = \dfrac{\pi}{2}$ and $B_0 = \dfrac{\pi^3}{24}$.

(b) Show that $A_n = \dfrac{2n-1}{2n} A_{n-1}$ for $n \geqslant 1$.

(c) By integrating A_n twice by parts, show that $A_n = (2n-1)n B_{n-1} - 2n^2 B_n$ for $n \geqslant 1$.

(d) From the results given in (b) and (c), deduce that $\dfrac{2B_{n-1}}{A_{n-1}} - \dfrac{2B_n}{A_n} = \dfrac{1}{n^2}$ for $n \geqslant 1$.

(e) Use the result given in (d) to show the following telescoping sum:

$$\sum_{k=1}^{n} \frac{1}{k^2} = \sum_{k=1}^{n} \left(\frac{2B_{k-1}}{A_{k-1}} - \frac{2B_k}{A_k} \right) = \frac{2B_0}{A_0} - \frac{2B_n}{A_n} .$$

(f) Hence deduce that for all $n \geqslant 1$, we have

$$\sum_{k=1}^{n} \frac{1}{k^2} = \frac{\pi^2}{6} - 2\frac{B_n}{A_n} .$$

(g) By observing that the linear function $2x/\pi$ coincides with $\sin x$ at the points $x = 0$ and $x = \pi/2$, and as $\sin x$ is concave up on the interval $[0, \pi/2]$, one has

$$\sin x \geqslant \frac{2x}{\pi} \quad \text{for all } 0 \leqslant x \leqslant \frac{\pi}{2}.$$

By replacing n with $n+1$ in the reduction formula given in (b), together with the above inequality for sine, use this to show that

$$B_n = \int_0^{\frac{\pi}{2}} x^2 \cos^{2n} x \, dx \leqslant \frac{\pi^2}{4} \cdot \frac{A_n}{2(n+1)} .$$

(h) Using the result in (f) together with the inequality in (g), show that

$$0 \leqslant \frac{\pi^2}{6} - \sum_{k=1}^{n} \frac{1}{k^2} \leqslant \frac{\pi^2}{4(n+1)}.$$

(i) From the result in (h), deduce that as $n \to \infty$ one has $\sum_{k=1}^{\infty} \frac{1}{k^2} = \frac{\pi^2}{6}$.

This identity is a well-known result known as the *Basel problem*.[2]

29 (Irrationality of e). Let $E_n = \int_0^1 x^n e^{-x} \, dx, n = 0, 1, 2, \ldots$

(a) Show that $E_0 = 1 - \frac{1}{e}$.

(b) Using integration by parts, show that $E_n = -\frac{1}{e} + n E_{n-1}$ for $n \geqslant 1$.

(c) Hence deduce that

$$E_n = -\frac{1}{e} [1 + n + n(n-1) + \cdots n!] + n! E_0.$$

(d) By considering the integrand in the definition for E_n, show that

$$0 < \int_0^1 x^n e^{-x} \, dx \leqslant \frac{1}{n+1}.$$

(e) From the result given in (c), substituting for the value of E_0 from (a) and dividing by $n!$, one has

$$\frac{E_n}{n!} = 1 - \frac{1}{e} \left(1 + \frac{1}{1!} + \frac{1}{2!} + \cdots = \frac{1}{n!} \right) = 1 - \frac{1}{e} \sum_{k=0}^{n} \frac{1}{k!}.$$

Using this result together with the result given in (d), show that as $n \to \infty$

$$e = \lim_{n \to \infty} \left(1 + \frac{1}{1!} + \frac{1}{2!} + \cdots + \frac{1}{n!} \right) = \sum_{n=0}^{\infty} \frac{1}{n!}.$$

(f) We will now show the number e is irrational. Let $e = p/q$ where p and q are positive integers that are relative prime (no common factors) and choose n such that $n \geqslant \max\{q, e\}$ (n is greater than or equal to the larger of the two, q or e).

From the expression given in (c), it can be rewritten as

$$n!(e-1) - (1 + n + n(n-1) + \cdots + n!) = e E_n.$$

Briefly explain why the expression on the left side must reduce to an integer.

(g) Show that the right side of the expression given in (f) is between zero and one for the choice made for n, which is a contradiction, implying the number e must be irrational.

30 (Irrationality of π). Suppose q is a positive integer. For each $n \in \mathbb{N}$ define the integral I_n by

$$I_n = \frac{q^{2n}}{n!} \int_{-\frac{\pi}{2}}^{\frac{\pi}{2}} \left(\frac{\pi^2}{4} - x^2 \right)^n \cos x \, dx.$$

(a) Show that $I_0 = 2$ and $I_1 = 4q^2$.

(b) Using integration by parts twice, show that

$$I_n = (4n - 2)q^2 I_{n-1} - q^4 \pi^2 I_{n-2}, \quad n \geqslant 2.$$

(c) Suppose that $\pi = p/q$ where p and q are positive integers that are relative prime. Using induction on n prove that I_n is an integer for every value of n.

(d) Observe that

$$0 < \left(\frac{\pi^2}{4} - x^2 \right)^n \cos x \leqslant \left(\frac{\pi^2}{4} \right)^n,$$

whenever $n \geqslant 0$ and $-\pi/2 < x < \pi/2$. Using this result, by considering the integrand in the definition of I_n, show that

$$0 < I_n < \frac{p}{q} \cdot \frac{(p^2/4)^n}{n!}.$$

(e) It can be shown that if $a > 0$ then $\lim_{n \to \infty} a^n / n! = 0$. Use this result together with the result given in (d) to show that as $n \to \infty$, then $0 < I_n < 1$.

(f) Based on the result given in (e), what can one conclude about the number π?

31. In this question we show how the integral of any rational function that is proper can always be expressed in terms of rational functions and logarithmic and inverse tangent functions.

Recall the partial fraction decomposition for any rational function that is proper can be expressed as a sum of linear and irreducible quadratic terms of the form (see page 145 of Chapter 11)

$$\frac{A}{(ax+b)^m},$$

and

$$\frac{Bx+C}{(\alpha x^2 + \beta x + \gamma)^n} = \frac{Bx}{(\alpha x^2 + \beta x + \gamma)^n} + \frac{C}{(\alpha x^2 + \beta x + \gamma)^n}.$$

Here $m, n \in \mathbb{N}$, $A, B, C \in \mathbb{R}$ while a, b, α, β and γ are constants such that $a, \alpha \neq 0$ and $\beta^2 < 4\alpha\gamma$ since the quadratic factor appearing in the denominator of the second term is irreducible.

(a) Integrating the linear term, show that

$$\int \frac{A}{(ax+b)^m} \, dx = \begin{cases} \dfrac{A}{a} \ln |ax+b| + C & m = 1 \\[2ex] -\dfrac{A}{a(m-1)} \dfrac{1}{(ax+b)^{m-1}} + C & m > 1. \end{cases}$$

This shows that the integral of all linear factors of a rational function yields rational functions and/or logarithmic functions depending on the value of the positive integer m.

(b) Now consider the expression of the irreducible quadratic term.

   (i) Show that

$$\int \frac{Bx+C}{(\alpha x^2 + \beta x + \gamma)^n} \, dx = \frac{1}{\alpha^n} \int \frac{Bx+C}{\left[\left(x + \frac{\beta}{2\alpha} \right)^2 + \left(\frac{\gamma}{\alpha} - \frac{\beta^2}{4\alpha^2} \right) \right]^n} \, dx.$$

   (ii) By employing the substitution $u = x + \dfrac{\beta}{2\alpha}$, show that the integral in (b)(i) can be rewritten as

$$\int \frac{Bx+C}{(\alpha x^2 + \beta x + \gamma)^n} \, dx = \frac{B}{\alpha^n} \int \frac{u}{(u^2 + k^2)^n} \, du$$

$$+ \frac{1}{\alpha^n} \left(C - \frac{B\beta}{2\alpha} \right) \int \frac{du}{(u^2 + k^2)^n}.$$

   Here $k^2 = \dfrac{\gamma}{\alpha} - \dfrac{\beta^2}{4\alpha^2} > 0$ since $\beta^2 < 4\alpha\gamma$.

   (iii) For the first of the integrals appearing to the right in (b)(ii), by letting $u^2 = t$, show that such a substitution reduces the integral to an integral of the form given in (a), an integral that has already been solved for.

(iv) For the second of the integrals appearing to the right in (b)(ii), writing it as

$$\int \frac{du}{(u^2 + k^2)^n} = \frac{1}{k^2} \int \frac{(k^2 + u^2) - u^2}{(u^2 + k^2)^n} du$$

$$= \frac{1}{k^2} \int \frac{du}{(u^2 + k^2)^{n-1}}$$

$$- \frac{1}{k^2} \int \frac{u^2}{(u^2 + k^2)^n} du,$$

and using integration by parts on the right-most integral, show the above integral can be converted into the following reduction formula

$$k^2 \int \frac{du}{(u^2 + k^2)^n} = \frac{1}{(2n - 2)} \frac{u}{(u^2 + k^2)^{n-1}}$$

$$+ \frac{2n - 3}{2n - 2} \int \frac{du}{(u^2 + k^2)^{n-1}},$$

where $n = 2, 3, \ldots$.

(v) The reduction formula gives the integral for all positive integer orders greater than or equal to two. Here one sees that all such integrals give rational functions except for the $n = 2$ term itself. Finding this integral explicitly, we see it is equal to an inverse tangent function as

$$\int \frac{du}{u^2 + k^2} = \frac{1}{k} \tan^{-1} \left(\frac{u}{k} \right) + C.$$

(c) From the results found for each of the integrals in (a) and (b) one is able to conclude that any rational function that is proper, when integrated, consists of only rational functions and logarithmic and inverse tangent functions.

32. The set of polynomials known as the *Legendre polynomials*,[3] denoted by $P_n(x)$ for $n = 0, 1, 2, \ldots$, are known to satisfy the following recurrence relation

$$(n + 1)P_{n+1}(x) - (2n + 1)xP_n(x) + nP_{n-1}(x) = 0, \quad n \geqslant 1,$$

and the integral relation known as an *orthogonality* condition

$$\int_{-1}^{1} P_m(x)P_n(x)\, dx = 0, \quad m \neq n.$$

In this question we will show that when $m = n$ one has

$$\int_{-1}^{1} [P_n(x)]^2 \, dx = \frac{2}{2n+1}.$$

Start by defining $A_n = \int_{-1}^{1} [P_n(x)]^2 \, dx$.

(a) If $P_0(x) = 1$ show that $A_0 = 2$.

(b) By replacing n with $n - 1$ in the recurrence relation for $P_n(x)$ and rearranging, show that

$$P_n(x) = \frac{2n-1}{n} x P_{n-1}(x) - \frac{n-1}{n} P_{n-2}(x).$$

(c) Using the result from (b), show that A_n can be expressed as

$$A_n = \frac{2n-1}{n} \int_{-1}^{1} x P_n(x) P_{n-1}(x) \, dx.$$

(d) By making the $x P_n(x)$ term the subject in the recurrence relation for $P_n(x)$, on substituting this expression into the integral for A_n given in (c), show that

$$A_n = \frac{2n-1}{2n+1} A_{n-1}, \quad n \geqslant 1.$$

(e) Hence deduce that $A_n = \int_{-1}^{1} [P_n(x)]^2 \, dx = \frac{2}{2n+1}$.

Endnotes

1. Other earlier independent co-discoverers of the formula were the Scottish mathematician James Gregory (1638–1675) who discovered it in 1667 and the Indian mathematician Kerla Gargya Nilakantha who discovered it circa 1500.

2. The *Basel problem* was first posed by the Italian mathematician Pietro Mengoli (1625–1686) in 1644. It remained an open problem (that is, unsolved) for 90 years, until the Swiss mathematician Leonhard Euler (1707–1783) made his first waves in the mathematical community by solving it. During his life, Euler would present three different solutions to the problem, which asks for an evaluation of the infinite series $\sum_{k=1}^{\infty} k^{-2}$. Euler's first solution to the problem in 1735 at the age of 28 brought him immediate fame within the mathematical community. The name of the problem itself comes from the location of the publisher of Jakob Bernoulli's text *Tractatus de Seriebus Infinitis*, published posthumously in 1713 in Basel, which first asked for a solution to the problem.

3. The Legendre polynomials are named after the French mathematician Adrien-Marie Legendre (1752–1833). They have very important applications in applied mathematics and physics.

19

Some Other Special
Techniques and Substitutions

An idea which can only be used once is a trick. If you can use it more
than once it becomes a method.

— George Pólya and Gábor Szegö,
Problems and Theorems in Analysis

In this last chapter on proper integrals before we move onto what are known as
improper integrals, we present a number of other special techniques one often
finds used to integrate a function, or other special types of substitutions. Each
takes advantage of the particular form taken by the integrand.

§ Product of an Exponential Term with the Sum of
a Function and Its Derivative

A useful method of integration presents itself if the integrand consists of a
product of an exponential term with the sum of a function and its derivative. If f and g are differentiable functions, recognising that from the product
rule

$$\frac{d}{dx}\left(e^{f(x)}g(x)\right) = e^{f(x)}\left[f'(x)g(x) + g'(x)\right],$$

one immediately has

$$\boxed{\int e^{f(x)}\left(f'(x)g(x) + g'(x)\right)dx = e^{f(x)}g(x) + C}$$

As we see in the examples that follow, the trick is to be able to correctly identify
the function g.

Example 19.1 Find $\displaystyle\int e^x \left(\frac{x-2}{x^3}\right) dx$.

Solution Since $\displaystyle\int e^x \left(\frac{x-2}{x^3}\right) dx = \int e^x \left(\frac{1}{x^2} - \frac{2}{x^3}\right) dx$, observing that $g(x) = 1/x^2$, $g'(x) = -2/x^3$, as $f(x) = x$ from the result

$$\int e^{f(x)}(f'(x)g(x) + g'(x))\, dx = e^{f(x)}g(x) + C,$$

it is immediate that

$$\int e^x \left(\frac{x-2}{x^3}\right) dx = \frac{e^x}{x^2} + C. \qquad \blacktriangleright$$

Example 19.2 Find $\displaystyle\int \frac{e^{x^2}(2x^3 - 2x^2 - 1)}{(x-1)^2}\, dx$.

Solution
$$\int \frac{e^{x^2}(2x^3 - 2x^2 - 1)}{(x-1)^2}\, dx = \int \frac{e^{x^2}[2x^2(x-1) - 1]}{(x-1)^2}\, dx$$

$$= \int e^{x^2} \left[\frac{2x^2}{x-1} - \frac{1}{(x-1)^2}\right] dx$$

$$= \int e^{x^2} \left[2x \cdot \frac{x}{x-1} - \frac{1}{(x-1)^2}\right] dx.$$

Here we have $f(x) = x^2$, $f'(x) = 2x$ and $g(x) = x/(x-1)$, giving $g'(x) = -1/(x-1)^2$ as needed. So from the result

$$\int e^{f(x)} \left(f'(x)g(x) + g'(x)\right) dx = e^{f(x)}g(x) + C,$$

one has

$$\int \frac{e^{x^2}(2x^3 - 2x^2 - 1)}{(x-1)^2}\, dx = \frac{xe^{x^2}}{x-1} + C. \qquad \blacktriangleright$$

§ Reverse Product Rule

It is apparent that the method given above is just a special case of the more general technique one can use to integrate a function known as the *reverse product rule*. If f and g are differentiable functions, from the product rule

$$\frac{d}{dx}[f(x) \cdot g(x)] = f(x) \cdot g'(x) + f'(x) \cdot g(x),$$

we see immediately that

$$\int [f(x) \cdot g'(x) + f'(x) \cdot g(x)] \, dx = \int [f(x) \cdot g(x)]' \, dx,$$

or

$$\boxed{\int [f(x) \cdot g'(x) + f'(x) \cdot g(x)] \, dx = f(x) \cdot g(x) + C}$$

Once again the trick to using this technique depends on being able to recognise that the function to be integrated is the derivative of a product.

Example 19.3 Find $\displaystyle\int e^x \left(\ln x + \frac{1}{x} \right) dx.$

Solution

$$\int e^x \left(\ln x + \frac{1}{x} \right) dx = \int (e^x \cdot \ln x)' \, dx = e^x \ln x + C. \qquad \blacktriangleright$$

Example 19.4 Find $\displaystyle\int \left(\sin(\ln x) + \cos(\ln x) \right) dx.$

Solution

$$\int \left(\sin(\ln x) + \cos(\ln x) \right) dx = \int \left(x \sin(\ln x) \right)' dx = x \sin(\ln x) + C. \quad \blacktriangleright$$

§ Reverse Quotient Rule

A *reverse quotient rule* may also be used to integrate a function. If u and v are differentiable functions, from the quotient rule

$$\left(\frac{u}{v} \right)' = \frac{u'v - v'u}{v^2},$$

it is immediate that

$$\boxed{\int \frac{u'v - v'u}{v^2} \, dx = \frac{u}{v} + C}$$

Integrands that consist of a quotient where the term appearing in the denominator is squared may be amenable to the method of the reverse quotient rule. As shall be seen, it is often more difficult to apply the method employing the reverse quotient rule than the method that employs the reverse product rule.

Example 19.5 Find $\displaystyle\int \frac{\sin x - x \cos x - \cos x}{\sin^2 x}\, dx.$

Solution The appearance of a squared term in the denominator of the integrand suggests that a reverse quotient rule method may work. Setting $v(x) = \sin x, v'(x) = \cos x$. For the term in the numerator we have

$$u'v - v'u = u' \cdot \sin x - u \cdot \cos x = \sin x - x \cos x - \cos x.$$

By inspection (this is the hard part!) one can see that if $u(x) = x + 1$, then $u'(x) = 1$, and the term in the numerator will follow. Thus

$$\int \frac{\sin x - x \cos x - \cos x}{\sin^2 x}\, dx = \int \left(\frac{x+1}{\sin x}\right)'\, dx = \frac{x+1}{\sin x} + C. \quad \blacktriangleright$$

Note that often you may find it easier to find the integral using a method other than the one that relies on the reverse quotient rule.

§ Symmetric Substitution

A *symmetric substitution* is a substitution of the form

$$\boxed{u = x^a \pm \frac{1}{x^a}, \quad a \neq 0}$$

The substitution is applied to integrals where terms in the integrand consist of a term in the form of the symmetric substitution for u, together with a term of the form

$$x^{a-1} \mp \frac{1}{x^{a+1}},$$

corresponding to the differential of the substitution itself. More often than not, one will be required to manipulate the integrand into such a form. As the following examples demonstrate, such a substitution proves to be surprisingly useful, particularly with certain types of integrals involving rational functions. As with any technique of integration, the ability to recognise and know when to use such a substitution is only gained after having seen and worked through many different types of examples.

Example 19.6 Find $\displaystyle\int \frac{x^2 - 1}{x^4 + 3x^2 + 1}\, dx.$

Solution In this case, as the integrand is a rational function of x, it suggests a symmetric substitution may just work (of course, remember that it is just as likely not to work).

$$\int \frac{x^2 - 1}{x^4 + 3x^2 + 1}\,dx = \int \frac{x^2 - 1}{x^2\left(x^2 + \frac{1}{x^2} + 3\right)}\,dx = \int \frac{1 - \frac{1}{x^2}}{x^2 + \frac{1}{x^2} + 3}\,dx$$

$$= \int \frac{1 - \frac{1}{x^2}}{\left(x + \frac{1}{x}\right)^2 + 1}\,dx.$$

With the integrand in this final form, the following symmetric substitution can be used:

$$u = x + \frac{1}{x},\, du = \left(1 - \frac{1}{x^2}\right)dx.$$

Thus

$$\int \frac{x^2 - 1}{x^4 + 3x^2 + 1}\,dx = \int \frac{du}{u^2 + 1} = \tan^{-1} u + C = \tan^{-1}\left(x + \frac{1}{x}\right) + C,$$

since $u = x + \frac{1}{x}$.   ▶

Example 19.7 Find $\displaystyle\int \frac{x^5 - x}{x^8 + 1}\,dx$.

Solution

$$\int \frac{x^5 - x}{x^8 + 1}\,dx = \int \frac{x^5 - x}{x^4\left(x^4 + \frac{1}{x^4}\right)}\,dx = \int \frac{x - \frac{1}{x^3}}{x^4 + \frac{1}{x^4}}\,dx$$

$$= \int \frac{x - \frac{1}{x^3}}{\left(x^2 + \frac{1}{x^2}\right)^2 - 2}\,dx.$$

With the integrand in this final form, the following symmetric substitution can be used:

$$u = x^2 + \frac{1}{x^2},\, du = 2\left(x - \frac{1}{x^3}\right)dx.$$

Thus

$$\int \frac{x^5 - x}{x^8 + 1}\,dx = \frac{1}{2} \int \frac{du}{u^2 - 2} = \frac{1}{2} \int \frac{du}{u^2 - (\sqrt{2})^2}$$

$$= \frac{1}{4\sqrt{2}} \ln\left|\frac{u - \sqrt{2}}{u + \sqrt{2}}\right| + C = \frac{1}{4\sqrt{2}} \ln\left|\frac{x^2 + \frac{1}{x^2} - \sqrt{2}}{x^2 + \frac{1}{x^2} + \sqrt{2}}\right| + C$$

$$= \frac{1}{4\sqrt{2}} \ln\left|\frac{x^4 - x^2\sqrt{2} + 1}{x^4 + x^2\sqrt{2} + 1}\right| + C.$$ ▶

Example 19.8 Find $\displaystyle\int \frac{x^2-1}{x\sqrt{x^4+1}}\,dx.$

Solution

$$\int \frac{x^2-1}{x\sqrt{x^4+1}}\,dx = \int \frac{x^2-1}{x\sqrt{x^2\left(x^2+\frac{1}{x^2}\right)}}\,dx = \int \frac{x^2-1}{x^2\sqrt{x^2+\frac{1}{x^2}}}\,dx$$

$$= \int \frac{1-\frac{1}{x^2}}{\sqrt{\left(x+\frac{1}{x}\right)^2-2}}\,dx.$$

With the integrand in this final form, the following symmetric substitution can be used:

$$u = x + \frac{1}{x}, du = \left(1 - \frac{1}{x^2}\right)dx.$$

Thus

$$\int \frac{x^2-1}{x\sqrt{x^4+1}}\,dx = \int \frac{du}{\sqrt{u^2-2}} = \int \frac{du}{\sqrt{u^2-(\sqrt{2})^2}}$$

$$= \sin^{-1}\left(\frac{u}{\sqrt{2}}\right) + C = \sin^{-1}\left(\frac{1}{\sqrt{2}}\left(x+\frac{1}{x}\right)\right) + C$$

$$= \sin^{-1}\left(\frac{x^2+1}{x\sqrt{2}}\right) + C. \qquad \blacktriangleright$$

Exercises for Chapter 19

✠ **Warm-up**

1. By applying the reverse quotient rule, find by inspection

$$\int \frac{\cos x - x \sin x}{(x \cos x)^2}\,dx.$$

✠ **Practice questions**

2. Find the following integrals using

$$\int e^{f(x)}[f'(x)g(x) + g'(x)]\,dx = e^{f(x)}g(x) + C.$$

(a) $\displaystyle\int \frac{xe^x}{(1+x)^2}\,dx$

(b) $\displaystyle\int \frac{e^x(x^2-2x+1)}{(1+x^2)^2}\,dx$

(c) $\int e^x \left(2\ln x + \dfrac{1}{x} + \dfrac{1}{x^2}\right) dx$ (d) $\int \dfrac{(x^2+1)e^x}{(x+1)^2} dx$

(e) $\int \dfrac{2-x^2}{(1-x)\sqrt{1-x^2}} e^x \, dx$ (f) $\int \left(1+x-\dfrac{1}{x}\right) e^{x+\frac{1}{x}} \, dx$

(g) $\int e^{\cos x} \dfrac{x\sin^3 x + \cos x}{\sin^2 x} dx$

3. Consider the integral $\int \left(\dfrac{1-\sin x}{1-\cos x}\right) e^x \, dx$.

   (a) Let $g(x) = \dfrac{\sin x}{\cos x - 1}$. Show that $g(x) + g'(x) = \dfrac{1-\sin x}{1-\cos x}$.

   (b) Hence find $\int \left(\dfrac{1-\sin x}{1-\cos x}\right) e^x \, dx$.

4. By applying the reverse product rule, find the following integrals.

   (a) $\int (x\sec^2 x + \tan x) \, dx$ (b) $\int (2\ln x + \ln^2 x) \, dx$

   (c) $\int x^x (\ln x + 1) \, dx$ (d) $\int (2x\tan^{-1} x + 1) \, dx$

   (e) $\int \left(\dfrac{1}{\ln x} + \ln(\ln x)\right) dx$ (f) $\int e^{\sin x} \cos(x - \cos x) \, dx$

5. (a) Using the reverse product rule, show that
   $$\int \sin(mx + x)\sin^{m-1} x \, dx = \dfrac{1}{m}\sin mx \sin^m x + C, \quad m \neq 0.$$

   (b) Hence find $\int \sin(101x)\sin^{99} x \, dx$.

6. Evaluate the following integral by first using a suitable substitution before applying the reverse product rule.
   $$\int_0^{\frac{\pi}{4}} \dfrac{(1-x^2)\ln(1+x^2) + (1+x^2) - (1-x^2)\ln(1-x^2)}{(1-x^4)(1+x^2)} \, dx.$$

7. Suppose that $g'(x) = f(x)$ for all x. Evaluate
   $$\int_1^e \dfrac{xe^x + 1}{x} f(e^x + \ln x) \, dx.$$

8. By applying the reverse quotient rule, find the following integrals.

(a) $\displaystyle\int \frac{4x^5 - 1}{(x^5 + x + 1)^2}\,dx$

(b) $\displaystyle\int \frac{4x^3 - 3x^2 - 2}{(x^3 + 1)^2}\,dx$

(c) $\displaystyle\int \frac{\ln x}{x^2(1 - \ln x)^2}\,dx$

(d) $\displaystyle\int \frac{\sin^2 x}{(x\cos x - \sin x)^2}\,dx$

(e) $\displaystyle\int \left(\frac{\tan^{-1} x}{x - \tan^{-1} x}\right)^2 dx$

(f) $\displaystyle\int \left(\frac{x^2 - 3x - \frac{1}{3}}{x^3 - x + 1}\right)^2 dx$

9. In this question we will find $\displaystyle\int \frac{x^2}{(x\sin x + \cos x)^2}\,dx$ using three different methods.

(a) Method I: Use the method of the reverse quotient rule.

(b) Method II: Use a substitution of $x = \tan u$ followed by a substitution of $t = \tan u - u$.

(c) Method III: By first observing that $\dfrac{d}{dx}(x\sin x + \cos x) = x\cos x$, use integration by parts.

10. Find the following integrals using an appropriate symmetric substitution.

(a) $\displaystyle\int \frac{x^2 - 1}{x^4 + x^2 + 1}\,dx$

(b) $\displaystyle\int \frac{1 + x^2}{(1 - x^2)\sqrt{1 + x^4}}\,dx$

(c) $\displaystyle\int \frac{dx}{x^4 + 2016x^2 + 1}$

(d) $\displaystyle\int \frac{x^2 - 1}{x\sqrt{1 + x^4}}\,dx$

(e) $\displaystyle\int \frac{x^4 + 1}{x^2\sqrt{x^4 - 1}}\,dx$

(f) $\displaystyle\int \frac{x^2 - 1}{x^2 + 1}\frac{1}{\sqrt{x^4 + 1}}\,dx$

(g) $\displaystyle\int \frac{1 - x^2}{x^2 - x + 1}\frac{dx}{\sqrt{x^4 + x^2 + 1}}$

(h) $\displaystyle\int \frac{x^2 + 1}{x^4 + 3x^3 + 3x^2 - 3x + 1}\,dx$

11. (a) Using a symmetric substitution, find $\displaystyle\int \frac{x^2+1}{x^4+1}\,dx$.

(b) Using a symmetric substitution, find $\displaystyle\int \frac{x^2}{x^4+1}\,dx$.

(c) Using (a) and (b), find $\displaystyle\int \frac{dx}{x^4+1}$.

12. By using a substitution of $u = x^2$ followed by a symmetric substitution, show that

$$\int \frac{2x(x^4+1)^2}{x^{12}+1}\,dx = \tan^{-1}\left(x^2 - \frac{1}{x^2}\right) + C.$$

13. Let $\displaystyle P_n = \int \frac{x^{3n-1}}{x^{4n}+1}\,dx, \ n \neq 0$.

(a) Using a symmetric substitution, show that

$$P_n = \frac{1}{4n\sqrt{2}}\left[2\tan^{-1}\left(\frac{x^{2n}-1}{x^n\sqrt{2}}\right) + \ln\left|\frac{x^{2n}-x^n\sqrt{2}+1}{x^{2n}+x^n\sqrt{2}+1}\right|\right] + C.$$

(b) Hence find $\displaystyle\int \frac{x^{17}}{x^{24}+1}\,dx$.

✠ **Extension questions and Challenge problems**

14. Suppose that $\displaystyle I(x) = \int_0^{2\pi} e^{x\cos\theta}\cos(x\sin\theta)\,d\theta$.

(a) Evaluate $I(0)$.

(b) Show that

$$\frac{d}{dx}\left[e^{x\cos\theta}\cos(x\sin\theta)\right] = \frac{1}{x}\frac{d}{d\theta}\left[e^{x\cos\theta}\sin(x\sin\theta)\right].$$

(c) If we assume that when $I(x)$ is differentiated with respect to x, the x-derivative can be moved under the integral sign of the definite integral as follows:

$$\frac{dI}{dx} = \int_0^{2\pi} \frac{d}{dx}\left[e^{x\cos\theta}\cos(x\sin\theta)\right]d\theta.$$

Using the above result together with the result given in (b), show that $\dfrac{dI}{dx} = 0$.

(d) Hence find an expression for $I(x)$ and deduce the value for the definite integral

$$\int_0^{2\pi} e^{\cos\theta} \cos(\sin\theta)\, d\theta.$$

15. The substitution $u = (1 - x)/(1 + x)$ is an example of what is known as a *self-similar* substitution.[1]

When solving for x, one finds $x = (1 - u)/(1 + u)$. As the functional forms for x and u are identical, each is equal to its own inverse, which is why such a substitution is said to be self-similar.

Find or evaluate the following integrals using a self-similar substitution.

(a) $\displaystyle\int \frac{dx}{(1 - x)\sqrt{1 - x^2}}$

(b) $\displaystyle\int \ln\left(\frac{1 + x}{1 - x}\right) \frac{dx}{1 - x^2}$

(c) $\displaystyle\int_0^1 \frac{\ln(1 + x)}{1 + x^2}\, dx$

(d) $\displaystyle\int_{-\frac{1}{3}}^{2} \frac{\tan^{-1} x}{x^2 + x + 2}\, dx$

(e) $\displaystyle\int_{-\frac{1}{2}}^{3} \frac{\ln(x + 2)}{(x + 1)\left[\ln(x + 3) + \ln(x + 2) - \ln(x + 1)\right]}\, dx$

16. Without using a partial fraction decomposition, find $\displaystyle\int \frac{dx}{(x^4 - 1)^2}$. At some stage a symmetric substitution should be used.

17. Find

(a) $\displaystyle\int \frac{1 - x}{(1 + x)\sqrt{x + x^2 + x^3}}\, dx$

(b) $\displaystyle\int \frac{(x + 1)^4(x - 1)^2\sqrt{x^4 + x^2 + 1}}{(x^2 + 1)^2(x^6 - 1)}\, dx$

18. Using the reverse quotient rule, or otherwise, find

$$\int \frac{x^2 + 20}{(x\sin x + 5\cos x)^2}\, dx.$$

Endnote

1. Seán M. Stewart, 'Finding some integrals using an interesting self-similar substitution', *The Mathematical Gazette*, **101**(550), 103–108 (2017).

20

Improper Integrals

So what we do when calculating integrals from a to infinity is we take
a look forward, we take a look into the abyss, and then we decide, no.
We don't want to deal with that.

— Anonymous

Almost all the definite integrals considered in the preceding chapters have been
proper integrals. Recall that the definite integral

$$\int_a^b f(x)\,dx,$$

is said to be proper if both (i) the interval of integration $[a, b]$ is bounded, and
(ii) in the interval $[a, b]$ the integrand f is defined at every point x, is bounded,
and has at most a finite number of discontinuities. At times it is convenient
to relax either of these boundedness conditions found in (i) or (ii). Integrals
that result when either or both of these boundedness conditions are relaxed are
known as *improper integrals* and have wide-ranging applications in areas such
as applied mathematics, physics, and engineering.

For convenience an improper integral can be said to be of one of two types.
When the interval of integration becomes unbounded, condition (i) fails and the
resulting integral is known as a *type I* improper integral. When the integrand
becomes unbounded, condition (ii) fails and the resulting integral is known as a
type II improper integral. Some improper integrals can be of both types, where
the interval of integration and the integrand are both unbounded.

Improper integrals of either type can be readily handled by extending
our definition of the proper definite integral. As the extension to the proper
definite integral definition depends on type, we now consider each of these
separately.

274

§ Type I Improper Integrals

An improper integral with an unbounded interval of integration is defined as follows.

Definition 20.1 (Improper integral of the first type)

(a) If $\displaystyle\int_a^\alpha f(x)\,dx$ exists for every number $\alpha \geqslant a$, then

$$\int_a^\infty f(x)\,dx = \lim_{\alpha\to\infty} \int_a^\alpha f(x)\,dx,$$

provided the limit exists. In this case f is said to be integrable over $[a, \infty)$ with the integral said to be *convergent*. If the limit does not exist, the integral is said to be *divergent*.

(b) If $\displaystyle\int_\alpha^b f(x)\,dx$ exists for every number $\alpha \leqslant b$, then

$$\int_{-\infty}^b f(x)\,dx = \lim_{\alpha\to-\infty} \int_\alpha^b f(x)\,dx,$$

provided the limit exists. In this case f is said to be integrable over $(-\infty, b]$ with the integral said to be *convergent*. If the limit does not exist, the integral is said to be *divergent*.

Just as with definite integrals, for convergent improper integrals the property of *linearity* holds.

Theorem 20.1 (Linearity for improper integrals). *If $\int_a^\infty f(x)\,dx$ and $\int_a^\infty g(x)\,dx$ both exist, then for any $\alpha, \beta \in \mathbb{R}$, $\int_a^\infty [\alpha f(x) \pm \beta g(x)]\,dx$ is convergent and converges to $\alpha \int_a^\infty f(x)\,dx \pm \beta \int_a^\infty g(x)\,dx$.*

Proof Since

$$\int_a^\infty [\alpha f(x) \pm \beta g(x)]\,dx = \lim_{b\to\infty} \int_a^b [\alpha f(x) \pm \beta g(x)]\,dx,$$

from linearity for the definite integral and the limit, one can write

$$\int_a^\infty [\alpha f(x) \pm \beta g(x)]\,dx = \alpha \lim_{b\to\infty} \int_a^b f(x)\,dx \pm \beta \lim_{b\to\infty} \int_a^b g(x)\,dx$$

$$= \alpha \int_a^\infty f(x)\,dx \pm \beta \int_a^\infty g(x)\,dx,$$

which converges as both the improper integrals for f and g converge. ∎

Example 20.1 Determine if the improper integral $\displaystyle\int_1^\infty \frac{dx}{x^2}$ converges or diverges. If it converges, find its value.

Solution

$$\int_1^\infty \frac{dx}{x^2} = \lim_{\alpha\to\infty} \int_1^\alpha x^{-2}dx = \lim_{\alpha\to\infty}\left[-\frac{1}{x}\right]_1^\alpha = \lim_{\alpha\to\infty}\left[-\frac{1}{\alpha}+1\right] = 1 < \infty.$$

Thus the improper integral converges and has a value equal to 1. ▶

Example 20.2 Determine if the improper integral $\displaystyle\int_0^\infty \frac{x}{1+x^2}\,dx$ converges or diverges. If it converges, find its value.

Solution

$$\int_0^\infty \frac{x}{1+x^2}\,dx = \tfrac{1}{2}\lim_{\alpha\to\infty}\int_0^\alpha \frac{2x}{1+x^2}\,dx = \tfrac{1}{2}\lim_{\alpha\to\infty}\left[\ln(1+x^2)\right]_0^\alpha$$
$$= \tfrac{1}{2}\lim_{\alpha\to\infty}\left[\ln(1+\alpha^2)-0\right] = \infty.$$

Thus the improper integral diverges to infinity and has no value. ▶

We now consider improper integrals whose interval of integration is the entire real line.

Definition 20.2 f is said to be integrable over $(-\infty,\infty)$, and hence converges, if f is integrable over both $(-\infty,c)$ and (c,∞). Here $c \in \mathbb{R}$. In this case we write

$$\int_{-\infty}^\infty f(x)\,dx = \int_{-\infty}^c f(x)\,dx + \int_c^\infty f(x)\,dx.$$

If f is not integrable on either of the intervals $(-\infty,c)$ or (c,∞) then the improper integral diverges.

For convenience one typically sets $c = 0$, though this is not necessary as the next example shows.

Example 20.3 Determine if the improper integral $\displaystyle\int_{-\infty}^\infty \frac{dx}{1+x^2}$ converges or diverges. If it converges, find its value.

Solution We begin by writing the improper integral as

$$\int_{-\infty}^\infty \frac{dx}{1+x^2} = \int_{-\infty}^c \frac{dx}{1+x^2} + \int_c^\infty \frac{dx}{1+x^2},$$

where $c \in \mathbb{R}$. Considering the first of these improper integrals we have

$$\int_{-\infty}^{c} \frac{dx}{1+x^2} = \lim_{\alpha \to -\infty} \int_{\alpha}^{c} \frac{dx}{1+x^2} = \lim_{\alpha \to -\infty} \left[\tan^{-1} x \right]_{\alpha}^{c}$$

$$= \lim_{\alpha \to -\infty} \left[\tan^{-1} c - \tan^{-1} \alpha \right]$$

$$= \tan^{-1} c - \left(-\frac{\pi}{2} \right) = \tan^{-1} c + \frac{\pi}{2}.$$

And considering the second of the improper integrals, we have

$$\int_{c}^{\infty} \frac{dx}{1+x^2} = \lim_{\alpha \to \infty} \int_{c}^{\alpha} \frac{dx}{1+x^2} = \lim_{\alpha \to \infty} \left[\tan^{-1} x \right]_{c}^{\alpha}$$

$$= \lim_{\alpha \to \infty} \left[\tan^{-1} \alpha - \tan^{-1} c \right] = \frac{\pi}{2} - \tan^{-1} c.$$

Thus

$$\int_{-\infty}^{\infty} \frac{dx}{1+x^2} = \left(\tan^{-1} c + \frac{\pi}{2} \right) + \left(\frac{\pi}{2} - \tan^{-1} c \right) = \pi.$$

So the improper integral converges and has a value equal to π. ▶

All the rules we have considered in this text for finding proper integrals may be extended so as to apply to improper integrals. Thus a change of variable, integration by parts, a t-substitution, and so on can all be used to evaluate a convergent improper integral.

Example 20.4 Evaluate $\displaystyle\int_{0}^{\infty} \frac{x \tan^{-1} x}{(1+x^2)^2} \, dx$.

Solution Let $x = 1/u$, then $dx = -1/u^2 \, du$, and the limits of integration become $x \to 0^+, u \to \infty$ and $x \to \infty, u \to 0^+$. Thus

$$\int_{0}^{\infty} \frac{x \tan^{-1} x}{(1+x^2)^2} \, dx = \int_{0}^{\infty} \frac{\frac{1}{u} \tan^{-1} \left(\frac{1}{u} \right)}{\left(1 + \frac{1}{u^2} \right)^2} \frac{du}{u^2} = \int_{0}^{\infty} \frac{u \tan^{-1} \left(\frac{1}{u} \right)}{(1+u^2)^2} \, du.$$

Now since $u > 0$, taking advantage of the following property for the inverse tangent function

$$\tan^{-1} u + \tan^{-1} \left(\frac{1}{u} \right) = \frac{\pi}{2},$$

we can rewrite the improper integral as

$$\int_{0}^{\infty} \frac{x \tan^{-1} x}{(1+x^2)^2} \, dx = \frac{\pi}{2} \int_{0}^{\infty} \frac{x}{(1+x^2)^2} \, dx - \int_{0}^{\infty} \frac{x \tan^{-1} x}{(1+x^2)^2} \, dx,$$

where the dummy variable u has reverted back to x, or

$$\int_0^\infty \frac{x \tan^{-1} x}{(1 + x^2)^2} \, dx = \frac{\pi}{4} \int_0^\infty \frac{x}{(1 + x^2)^2} \, dx.$$

The improper integral that now remains can be readily evaluated using a substitution of $u = 1 + x^2$. Here $du = 2x \, dx$ while the limits of integration become $x = 0, u = 1; x \to \infty, u \to \infty$. Thus

$$\int_0^\infty \frac{x \tan^{-1} x}{(1 + x^2)^2} \, dx = \frac{\pi}{8} \int_1^\infty \frac{du}{u^2} = \frac{\pi}{8} \lim_{\alpha \to \infty} \left[-\frac{1}{u} \right]_1^\alpha = \frac{\pi}{8} \lim_{\alpha \to \infty} \left[1 - \frac{1}{\alpha} \right] = \frac{\pi}{8}.$$

▶

§ Comparison Tests

Often it is difficult to determine the convergence or divergence of a given improper integral since it is not always possible to find a primitive for the integral. This is the case for many important improper integrals found to arise in applications. We would therefore like to have a way of testing for the convergence of an improper integral without having to explicitly evaluate it.

One can gain information about the convergence or divergence of an improper integral by making a comparison with other improper integrals of known behaviour. A number of tests are available for such a purpose. Here we introduce two *comparison* tests. While neither test gives the value for a convergent integral, each can be used to establish the convergence or divergence of an improper integral.

Theorem 20.2 (The direct comparison test for improper integrals). *Suppose f and g are integrable functions with $0 \leqslant f(x) \leqslant g(x)$ for all $x \geqslant a$.*

(i) If $\displaystyle\int_a^\infty g(x) \, dx$ converges, then $\displaystyle\int_a^\infty f(x) \, dx$ also converges.

(ii) If $\displaystyle\int_a^\infty f(x) \, dx$ diverges, then $\displaystyle\int_a^\infty g(x) \, dx$ also diverges.

Proof We give the proof for (i). The proof for (ii) is similar. Suppose f and g are integrable functions such that $0 \leqslant f(x) \leqslant g(x)$ for all $x \geqslant a$ and $\int_a^\infty g(x) \, dx$ converges. For $\alpha \geqslant a$, let

$$F(\alpha) = \int_a^\alpha f(x) \, dx \quad \text{and} \quad G(\alpha) = \int_a^\alpha g(x) \, dx.$$

From the given hypotheses, by the comparison property for the definite integral one has

$$0 \leqslant \int_a^\alpha f(x) \, dx \leqslant \int_a^\alpha g(x) \, dx \quad \text{or} \quad 0 \leqslant F(\alpha) \leqslant G(\alpha).$$

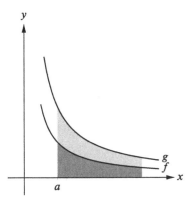

Figure 20.1. Geometric interpretation for the direct comparison test as corresponding to an area (either finite or infinite) beneath a curve.

Thus both of these functions are increasing as α increases. Furthermore, $F(\alpha)$ is bounded above by $\int_a^\infty g(x)\,dx$ as, from the hypothesis, the improper integral for g converges. That is to say $G(\alpha)$ tends to a limit L as $\alpha \to \infty$. It follows that

$$0 \leqslant F(\alpha) \leqslant G(\alpha) \leqslant L.$$

Since $F(\alpha)$ is increasing and bounded above by L it must also converge to a limit as $\alpha \to \infty$. Thus the improper integral $\int_a^\infty f(x)\,dx$ converges, as required to prove. ■

Thinking geometrically in terms of areas beneath the curves for f and g and the x-axis, the results for the direct comparison test for improper integrals are immediately obvious. Suppose $0 \leqslant f(x) \leqslant g(x)$ for all $x \geqslant a$. This we illustrate in Figure 20.1.

Clearly the area under the graph of g is greater than the area under the graph of f. So if the area under g is finite, so is the area under f, while if the area under f is infinite, so is the area under g. Put simply, a useful way to recall the results of the direct comparison test for improper integrals is as follows. If the BIG one converges, then so too does the SMALL one. If the SMALL one diverges, then so too does the BIG one.

The most difficult part of applying the direct comparison test is deciding which improper integral to compare with. Often properties of the functions found in the integrand may help you make that decision. Dominant term analysis on the integrand as x becomes large in an absolute sense may also prove useful.

Example 20.5 Determine if the improper integral $\displaystyle\int_1^\infty \frac{\cos^2 x}{x^4}\,dx$ converges or diverges.

Solution Since $0 \leqslant \cos^2 x \leqslant 1$ for all $x \geqslant 1$, as $x^4 > 0$ for all $x \geqslant 1$, one has

$$0 \leqslant \frac{\cos^2 x}{x^4} \leqslant \frac{1}{x^4}, \qquad \text{for all } x \geqslant 1.$$

So a comparison with the improper integral $\displaystyle\int_1^\infty \frac{dx}{x^4}$ will be made.

$$\int_1^\infty \frac{dx}{x^4} = \lim_{\alpha \to \infty} \int_1^\alpha x^{-4}\,dx = \lim_{\alpha \to \infty}\left[-\frac{1}{3x^3}\right]_1^\alpha$$

$$= \lim_{\alpha \to \infty}\left[-\frac{1}{3\alpha^3} + \frac{1}{3}\right] = \frac{1}{3} < \infty.$$

As the improper integral $\int_1^\infty \frac{dx}{x^4}$ converges, by the direct comparison test the improper integral $\int_1^\infty \frac{\cos^2 x}{x^4}\,dx$ also converges. ▶

Even though Example 20.5 shows that the improper integral converges, finding its value is far from easy as no primitive for $\int \frac{\cos^2 x}{x^4}\,dx$ in terms of the elementary functions exists.

Example 20.6 Determine if the improper integral $\displaystyle\int_1^\infty \frac{dx}{\sqrt{x^9 + 1}}$ converges or diverges.

Solution In deciding what improper integral the comparison should be made against, dominant term analysis on the function appearing in the integrand will be used. Dominant term analysis suggests

$$\frac{1}{\sqrt{x^9 + 1}} \to \frac{1}{\sqrt{x^9}},$$

when x is large. As $x^9 + 1 > x^9$ for $x \geqslant 1$, one has $\sqrt{x^9 + 1} > \sqrt{x^9}$ for $x \geqslant 1$, giving

$$\frac{1}{\sqrt{x^9 + 1}} < \frac{1}{\sqrt{x^9}}, \quad x \geqslant 1.$$

Thus

$$\int_1^\infty \frac{dx}{\sqrt{x^9 + 1}} < \int_1^\infty \frac{dx}{\sqrt{x^9}} = \lim_{\alpha \to \infty} \int_1^\alpha x^{-9/2} dx$$

$$= \lim_{\alpha \to \infty} \left[-\frac{2}{7\sqrt{x^7}} \right]_1^\alpha$$

$$= \lim_{\alpha \to \infty} \left[-\frac{2}{7\sqrt{\alpha^7}} + \frac{2}{7} \right] = \frac{2}{7} < \infty.$$

So by the direct comparison test the improper integral $\displaystyle\int_1^\infty \frac{dx}{\sqrt{x^9 + 1}} dx$ converges. ▶

Example 20.7 Determine if the improper integral $\displaystyle\int_2^\infty \frac{dx}{x^{5/3} - 1}$ converges or diverges.

Solution Dominant term analysis suggests that

$$\frac{1}{x^{5/3} - 1} \to \frac{1}{x^{5/3}},$$

when x is large. So we compare

$$\int_2^\infty \frac{1}{x^{3/2} - 1} dx \quad \text{with} \quad \int_2^\infty \frac{dx}{x^{3/2}}.$$

Since

$$\int_2^\infty \frac{dx}{x^{5/3}} = \lim_{\alpha \to \infty} \int_2^\alpha x^{-5/3} dx = \lim_{\alpha \to \infty} \left[-\frac{3}{2x^{2/3}} \right]_2^\alpha$$

$$= \lim_{\alpha \to \infty} \left[-\frac{3}{2\sqrt[3]{\alpha^2}} + \frac{3}{2\sqrt[3]{4}} \right] = \frac{3}{2\sqrt[3]{4}}.$$

Thus the improper integral $\int_2^\infty \frac{dx}{x^{5/3}}$ converges. But it is clear that

$$\frac{1}{x^{5/3} - 1} > \frac{1}{x^{5/3}}, \quad x \geqslant 2,$$

so we cannot immediately apply the direct comparison test. However, if we note that

$$x^{5/3} - 1 > x^{5/2} - \frac{1}{2}x^{5/3} = \frac{1}{2}x^{5/3}, \quad x \geqslant 2,$$

then

$$\frac{1}{x^{5/3} - 1} < \frac{2}{x^{5/3}}, \quad x \geqslant 2.$$

Now as

$$\int_2^\infty \frac{2}{x^{5/3}}\,dx = 2\int_2^\infty \frac{dx}{x^{5/3}},$$

which converges, by the direct comparison test the improper integral $\int_2^\infty \frac{dx}{x^{5/3}-1}$ also converges. ▶

As the previous example shows, dominant term analysis is not always straightforward and at times may not even be useful. An alternative approach to constructing a comparison to test for the convergence or divergence of improper integrals can, however, be done using what is known as the *limit comparison test*.

Theorem 20.3 (The limit comparison test for improper integrals). *Suppose* $a \in \mathbb{R}$ *and that* f *and* g *are non-negative and integrable on* $[a, \infty)$. *If*

$$\lim_{x\to\infty} \frac{f(x)}{g(x)} = L,$$

where $0 < L < \infty$, *then*

$$\int_a^\infty f(x)\,dx \quad and \quad \int_a^\infty g(x)\,dx,$$

both converge or both diverge.

The proof of this theorem is not given here. The main advantage of the limit comparison test over the direct comparison test is that it overcomes the difficulties often associated with finding an appropriate inequality needed to apply the latter test.

Example 20.8 Determine if the improper integral $\displaystyle\int_1^\infty \frac{dx}{\sqrt{1+x+x^2}}$ converges or diverges.

Solution Let $f(x) = \dfrac{1}{\sqrt{1+x+x^2}}$. We need to find a suitable function g to compare f with. From dominant term analysis we see that for large x

$$f(x) = \frac{1}{\sqrt{1+x+x^2}} \to \frac{1}{\sqrt{x^2}} = \frac{1}{x}.$$

So let $g(x) = 1/x$. Now

$$\int_1^\infty g(x)\,dx = \lim_{\alpha\to\infty}\int_1^\alpha \frac{dx}{x} = \lim_{\alpha\to\infty}[\ln x]_1^\alpha = \lim_{\alpha\to\infty}\ln\alpha = \infty.$$

Thus the improper integral $\int_1^\infty g(x)\,dx$ diverges. Next we consider the limit

$$\lim_{x\to\infty} \frac{f(x)}{g(x)} = \lim_{x\to\infty} \frac{x}{\sqrt{1+x+x^2}} = \lim_{x\to\infty} \sqrt{\frac{x^2}{1+x+x^2}}$$

$$= \lim_{x\to\infty} \sqrt{\frac{1}{\frac{1}{x^2}+\frac{1}{x}+1}} = 1.$$

Since this limit is a positive real number, from the limit comparison test we conclude the improper integral $\int_1^\infty f(x)\,dx$ diverges.   ▶

§ Absolute Convergence

In the preceding examples we have only considered integrands that were positive on their interval of integration. For integrands that become negative, the integrand needs to be at least bounded if the improper integral is to have any chance of converging; thus one needs to test for what is known as the *absolute convergence* of an improper integral. When the improper integral for $|f|$ converges, we say the improper integral for f is *absolutely convergent*.

Theorem 20.4 (Absolute convergence test). *Suppose f is integrable on any interval $[a,x]$ and $\int_a^\infty |f(x)|\,dx$ converges. Then so does $\int_a^\infty f(x)\,dx$.*

Proof From properties of the absolute value we know that

$$-|f(x)| \leqslant f(x) \leqslant |f(x)|,$$

for all $x \geqslant a$. Hence

$$0 \leqslant f(x) + |f(x)| \leqslant 2|f(x)|,$$

after adding the term $|f(x)|$ to both sides of the inequality. Since $\int_a^\infty |f(x)|\,dx$ converges, so does $\int_a^\infty 2|f(x)|\,dx$ by the linearity property for improper integrals, and therefore by the direct comparison test so does $\int_a^\infty [f(x) + |f(x)|]\,dx$. By the linearity property for improper integrals, since

$$\int_a^\infty f(x)\,dx = \int_a^\infty [(f(x)+|f(x)|)-|f(x)|]\,dx$$

$$= \int_a^\infty (f(x)+|f(x)|)\,dx - \int_a^\infty |f(x)|\,dx,$$

the improper integral on the left converges as both the improper integrals on the right converge, and completes the proof.   ∎

Theorem 20.4 tells us that absolute convergence implies convergence, though it does not necessarily hold true when reversed. That is, if the improper integral of f converges, that of $|f|$ may not converge at all.[1] As an example of how the test is applied, consider the following.

Example 20.9 Determine if the improper integral $\displaystyle\int_0^\infty \frac{\sin x}{1 + x^2}\, dx$ converges or diverges.

Solution Here the absolute convergence test in conjunction with the direct comparison test will be used. We begin by bounding the integrand.

$$\int_0^\infty \left| \frac{\sin x}{1 + x^2} \right| dx = \int_0^\infty \frac{|\sin x|}{1 + x^2}\, dx$$

$$\leqslant \int_0^\infty \frac{dx}{1 + x^2} = \lim_{\alpha \to \infty} \int_0^\alpha \frac{dx}{1 + x^2}$$

$$= \lim_{\alpha \to \infty} \left[\tan^{-1} x \right]_0^\alpha = \lim_{\alpha \to \infty} \tan^{-1} \alpha$$

$$= \frac{\pi}{2} < \infty.$$

Since $\int_0^\infty \left| \frac{\sin x}{1+x^2} \right| dx < \infty$ (and follows from the direct comparison test), then by the absolute convergence test $\int_0^\infty \frac{\sin x}{1+x^2}\, dx$ converges. ▶

§ Type II Improper Integrals

For the second type of improper integral the integrand becomes unbounded (blows up) at one or more points. This can occur if the integrand becomes infinite (unbounded) at either one or both of the end-points or at one or more points within the interval of integration.

Definition 20.3 (Improper integral of the second type)

(a) If f is continuous on $[a, b)$ and is unbounded (has an infinite discontinuity) at the end-point b, then

$$\int_a^b f(x)\, dx = \lim_{\alpha \to b^-} \int_a^\alpha f(x)\, dx,$$

if the limit exists. In this case f is said to be integrable over $[a, b]$ with the integral said to be *convergent*. If the limit does not exist, the integral is said to be *divergent*.

(b) If f is continuous on $(a, b]$ and is unbounded (has an infinite discontinuity) at the end-point a, then

$$\int_a^b f(x)\, dx = \lim_{\alpha \to a^+} \int_\alpha^b f(x)\, dx,$$

if the limit exists. In this case f is said to be integrable over $[a, b]$ with the integral said to be *convergent*. If the limit does not exist, the integral is said to be *divergent*.

(c) Suppose f is unbounded at point c within the interval $[a, b]$. f is said to be integrable over $[a, b]$, and hence converges, if f is integrable over both $[a, c]$ and $[c, b]$. In this case we write

$$\int_a^b f(x)\, dx = \int_a^c f(x)\, dx + \int_c^b f(x)\, dx.$$

If f is not integrable on either of the intervals $[a, c]$ or $[c, b]$ then the improper integral diverges.

From the definition of a type II improper integral we see that if the integrand has infinite discontinuities at both of its end-points or at one or more interior points within the interval of integration, one must first partition the interval of integration so that each integral has exactly one end-point infinite discontinuity.

Example 20.10 Determine if the improper integral $\displaystyle\int_0^1 \frac{dx}{1 - x}$ converges or diverges. If it converges, find its value.

Solution The integral is improper since the integrand becomes unbounded as $x \to 1^-$. Now

$$\int_0^1 \frac{dx}{1 - x} = \lim_{\alpha \to 1^-} \int_0^\alpha \frac{dx}{1 - x} = \lim_{\alpha \to 1^-} \left[-\ln(1 - x) \right]_0^\alpha$$

$$= -\lim_{\alpha \to 1^-} \ln(1 - \alpha) = \infty.$$

So the improper integral diverges to infinite and has no value. ▶

Example 20.11 Determine if the improper integral $\displaystyle\int_{-1}^1 \frac{dx}{\sqrt{1 + x}}$ converges or diverges. If it converges, find its value.

Solution The integral is improper since the integrand becomes unbounded as $x \to -1^+$. Now

$$\int_{-1}^{1} \frac{dx}{\sqrt{1+x}} = \lim_{\alpha \to -1+} \int_{\alpha}^{1} (1+x)^{-1/2} dx$$

$$= \lim_{\alpha \to -1+} \left[2\sqrt{1+x} \right]_{\alpha}^{1} = \lim_{\alpha \to -1+} \left[2\sqrt{2} - 2\sqrt{1+\alpha} \right]$$

$$= 2\sqrt{2} < \infty.$$

So the improper integral converges and has a value equal to $2\sqrt{2}$. ▶

The next example shows how important it is to always check that the integrand is bounded at all points within the interval of integration – a fact if overlooked will be to your peril!

Example 20.12 Determine if the improper integral $\displaystyle\int_{0}^{4} \frac{dx}{(x-3)^4}$ converges or diverges. If it converges, find its value.

Solution The integral is improper since the integrand becomes unbounded at the point $x = 3$. We therefore write the integral as

$$\int_{0}^{4} \frac{dx}{(x-3)^4} = \int_{0}^{3} \frac{dx}{(x-3)^4} + \int_{3}^{4} \frac{dx}{(x-3)^4}.$$

Consider the first of the improper integrals.

$$\int_{0}^{3} \frac{dx}{(x-3)^4} = \lim_{\alpha \to 3^-} \int_{0}^{\alpha} (x-3)^{-4} dx = \lim_{\alpha \to 3^-} \left[-\frac{1}{3(x-3)^3} \right]_{0}^{\alpha}$$

$$= \lim_{\alpha \to 3^-} \left[-\frac{1}{3(\alpha-3)^3} - \frac{1}{3^4} \right] = \infty.$$

So the improper integral diverges to infinite and has no value. ▶

Comparison tests for type II improper integrals analogous to those given for type I improper integrals exist. For each test the type I improper integral is simply replaced with a type II improper integral.

Example 20.13 Given that $x \tan x > \dfrac{4x - \pi}{\pi - 2x}$ for $x \in \left(0, \frac{\pi}{2}\right)$, use this to determine if the improper integral $\displaystyle\int_{0}^{\frac{\pi}{2}} x \tan x \, dx$ converges or diverges.

The inequality given above is proved in Exercise 27 on page 52 of Chapter 4.

Solution Observe the integral is improper since the integrand becomes unbounded at $x = \pi/2$. So

$$\int_0^{\frac{\pi}{2}} x \tan x \, dx > \int_0^{\frac{\pi}{2}} \frac{4x - \pi}{\pi - 2x} \, dx$$

$$= \int_0^{\frac{\pi}{2}} \left[-2 + \frac{\pi}{\pi - 2x} \right] dx$$

$$= \lim_{\alpha \to \frac{\pi}{2}^-} \int_0^{\alpha} \left[-2 + \frac{\pi}{\pi - 2x} \right] dx$$

$$= \lim_{\alpha \to \frac{\pi}{2}^-} \left[-2x - \frac{\pi}{2} \ln(\pi - 2x) \right]_0^{\alpha}$$

$$= \lim_{\alpha \to \frac{\pi}{2}^-} \left[-2\alpha - \frac{\pi}{2} \ln(\pi - 2\alpha) + \frac{\pi}{2} \ln(\pi) \right]$$

$$= \infty,$$

and diverges by the direct comparison test. ▶

Exercises for Chapter 20

✠ **Warm-ups**

1. Briefly explain why the following integrals are improper.

(a) $\displaystyle\int_1^{\infty} x^4 e^{-x^4} \, dx$

(b) $\displaystyle\int_0^{\frac{\pi}{2}} \sec x \, dx$

(c) $\displaystyle\int_0^2 \frac{x}{x^2 - 5x + 6} \, dx$

(d) $\displaystyle\int_1^{\infty} x^4 e^{-x^4} \, dx$

(e) $\displaystyle\int_0^{\infty} \frac{dx}{x - 1}$

(f) $\displaystyle\int_{-1}^1 \frac{dx}{(x^2 - 1)^4}$

2. If f and g are both integrable such that $\int_a^{\infty} f(x) \, dx$ and $\int_a^{\infty} g(x) \, dx$ both diverge, does $\int_a^{\infty} \big(f(x) + g(x)\big) \, dx$ diverge as well?

✠ **Practice questions**

3. Determine whether each of the following improper integrals converges or diverges. If it converges, find its value.

(a) $\displaystyle\int_1^\infty \frac{dx}{(3x+1)^2}$

(b) $\displaystyle\int_{-\infty}^1 3^x\,dx$

(c) $\displaystyle\int_0^\infty xe^{-5x}\,dx$

(d) $\displaystyle\int_{-\infty}^\infty \frac{x}{1+x^2}\,dx$

(e) $\displaystyle\int_0^3 \frac{dx}{\sqrt{x}}$

(f) $\displaystyle\int_{-1}^0 \frac{dx}{x^2}$

(g) $\displaystyle\int_{-2}^3 \frac{dx}{x+2}$

(h) $\displaystyle\int_{-1}^0 \frac{e^{1/x}}{x^3}\,dx$

(i) $\displaystyle\int_0^1 \frac{\ln x}{x^2}\,dx$

(j) $\displaystyle\int_e^\infty \frac{dx}{x\ln^2 x}$

(k) $\displaystyle\int_0^\infty \frac{\tan^{-1} x}{1+x^2}\,dx$

(l) $\displaystyle\int_0^{\frac{\pi}{2}} \cot x\,dx$

(m) $\displaystyle\int_{-1}^1 \frac{dx}{\sqrt{|x|}}$

(n) $\displaystyle\int_0^\infty \frac{dx}{2\sinh x + 1}$

(o) $\displaystyle\int_{-\infty}^\infty e^{-|x|}\,dx$

(p) $\displaystyle\int_{-\infty}^\infty \operatorname{sech} x\,dx$

(q) $\displaystyle\int_0^\infty \frac{dx}{(1+e^x)(1+e^{-x})}$

(r) $\displaystyle\int_0^1 \sqrt{\frac{1+x}{1-x}}\,dx$

(s) $\displaystyle\int_1^\infty \frac{dx}{(x+1)\sqrt{x^2+2x-2}}$

(t) $\displaystyle\int_0^\infty x\sin x\,dx$

(u) $\displaystyle\int_1^\infty \frac{\tan^{-1} x}{x^2}\,dx$

(v) $\displaystyle\int_1^\infty \frac{\ln x}{(1+x)^2}\,dx$

4. (a) Find $\displaystyle\int \frac{x}{(1+x)(1+x^2)}\,dx$.

(b) Hence deduce that $\displaystyle\int_0^\infty \frac{x}{(1+x)(1+x^2)}\,dx = \frac{\pi}{4}$.

5. The integral $\int_0^\infty \frac{dx}{\sqrt{x}(1+x)}$ is improper in two distinct ways: (i) the interval of integration is unbounded, and (ii) the integrand is unbounded.

If we rewrite the integral as $\int_0^1 \frac{dx}{\sqrt{x}(1+x)} + \int_1^\infty \frac{dx}{\sqrt{x}(1+x)}$, we have two improper integrals such that in the first the integrand is unbounded while in the second the interval is unbounded.

Show that each of these improper integrals converge, and by finding their respective values, find the value for the initial improper integral.

6. Use the direct comparison test to determine whether the following improper integrals converge or diverge.

(a) $\int_1^\infty \frac{\cos x}{1+x^2}\, dx$

(b) $\int_1^\infty \frac{dx}{x+e^x}$

(c) $\int_0^{\frac{\pi}{2}} \frac{dx}{x \sin x}$

(d) $\int_0^1 \sin\left(\frac{1}{x}\right) dx$

(e) $\int_0^\infty \frac{dx}{\sqrt{1+x^3}}$

(f) $\int_0^\infty x^n \sinh x\, dx, n \in \mathbb{N}$

(g) $\int_0^\infty \frac{\sin x}{1+\cos x+e^x}\, dx$

(h) $\int_0^\infty \frac{e^{\cos x}}{1+x^2}\, dx$

(i) $\int_0^\infty e^{-(x^4+x-4)}\, dx$

(j) $\int_1^\infty \frac{\ln x}{1+x^2}\, dx$

7. Use the limit comparison test to determine whether the following improper integrals converge or diverge.

(a) $\int_2^\infty \frac{x}{2x^3-1}\, dx$

(b) $\int_1^\infty \frac{dx}{\sqrt{1+x+x^4}}$

(c) $\int_2^\infty \frac{dx}{\sqrt[3]{1-x^4}}$

(d) $\int_0^1 \frac{\cos x}{x+\sqrt{x}}\, dx$

8. Find the value of a if $\int_1^\infty \left[\frac{2x^2+ax+a}{x(2x+a)} - 1\right] dx = 1$.

9. Consider the improper integral $\int_{2\pi}^\infty \frac{\sin^2 x}{x^2}\, dx$.

(a) Show that the improper integral converges.

(b) Hence show that $\int_{2\pi}^\infty \frac{\sin^2 x}{x^2}\, dx \leqslant \frac{1}{2\pi}$.

10. Consider the integral $\displaystyle\int_0^1 \frac{\sin x}{x^2}\, dx$.

(a) Briefly explain why the integral is improper.

(b) Show that for $0 < x < 1$, $\sin x > \dfrac{x}{2}$.

(b) By making use of the inequality given in (b), show that the improper integral diverges.

11. Consider the improper integral $\displaystyle\int_1^\infty \frac{dx}{x^p}$. Here $p \in \mathbb{R}$.

Improper integrals of this type are often call 'p-integrals'.

(a) Find the values of p for which the improper integral converges.

(b) Find the value of the improper integral for those values of p where it converges.

12. By using the substitution $x = 1/u$, show that $\displaystyle\int_0^\infty \frac{x \ln x}{(1 + x^2)^2}\, dx = 0$.

13. Suppose $f(x) = \dfrac{x^3}{e^{x/T} - 1}$ where $T > 0$.

If $\displaystyle\int_0^\infty \frac{u^3}{e^u - 1}\, du = \sigma$, where σ is a constant, show that $\displaystyle\int_0^\infty f(x)\, dx = \sigma T^n$

where n is a positive integer that needs to be determined.

14. Suppose $n = 0, 1, 2, \ldots$ Show that

(a) $\displaystyle\int_0^\infty t^n e^{-t}\, dt = n!$
  (b) $\displaystyle\int_0^1 (\ln x)^n\, dx = (-1)^n n!$

15. In this question we will show that the improper integral $\displaystyle\int_0^\infty \sin(x^2)\, dx$ converges.

(a) By writing the improper integral as

$$\int_0^\infty \sin(x^2)\, dx = \int_0^1 \sin(x^2)\, dx + \int_1^\infty \sin(x^2)\, dx,$$

briefly explain why the first of the integrals appearing on the right converges.

(b) For the second of the integrals appearing on the right in (a), by using the substitution $t = x^2$ followed by integration by parts, show that

$$\int_1^\infty \sin(x^2)\, dx = \frac{1}{2}\cos(1) - \frac{1}{2}\int_1^\infty \frac{\cos t}{t^{3/2}}\, dt.$$

(c) By applying the direct comparison test to the improper integral appearing in (b), show that it converges, thereby showing that $\int_0^\infty \sin(x^2)\,dx$ converges.

(d) Using this result, show that the improper integral $\int_0^\infty \dfrac{\sin x}{\sqrt{x}}\,dx$ converges.

16. (a) Show that $\displaystyle\int_0^\infty \dfrac{\ln x}{1+x^2}\,dx = 0$.

    (b) Hence evaluate $\displaystyle\int_0^\infty \dfrac{\ln x}{x^2+a^2}$. Here $a>0$.

17. (a) If $a>0$, show that

$$\int_0^a \frac{\sinh x}{\cosh 2x}\,dx = \frac{1}{\sqrt{2}}\coth^{-1}(\sqrt{2}) - \frac{1}{\sqrt{2}}\coth^{-1}(\sqrt{2}\cosh a)$$

and

$$\int_0^a \frac{\cosh x}{\cosh 2x}\,dx = \frac{1}{\sqrt{2}}\tan^{-1}(\sqrt{2}\sinh a).$$

    (b) Hence deduce that

$$\int_0^\infty \frac{\cosh x - \sinh x}{\cosh 2x}\,dx = \frac{\pi}{2\sqrt{2}} - \frac{1}{\sqrt{2}}\coth^{-1}(\sqrt{2}).$$

    (c) By making use of the substitution $u = e^x$ in the integral appearing in (b), use this to show that

$$\int_1^\infty \frac{dx}{1+x^4} = \frac{\pi}{4\sqrt{2}} - \frac{1}{2\sqrt{2}}\coth^{-1}(\sqrt{2}).$$

18. Using the result

$$\int_0^\infty f\left(x + \sqrt{1+x^2}\right)dx = \frac{1}{2}\int_1^\infty \left(1 + \frac{1}{x^2}\right)f(x)\,dx,$$

show that $\displaystyle\int_0^\infty \frac{dx}{(x+\sqrt{1+x^2})^3} = \frac{3}{8}$.

19. (a) Show that $\displaystyle\int_0^\infty \frac{dx}{a^2+x^2} = \frac{\pi}{2a},\, a>0$.

    (b) Hence show that $\displaystyle\int_0^\infty \frac{dx}{(a^2+x^2)^2} = \frac{\pi}{4a^3},\, a>0$.

20. (a) If $\alpha, \beta > 0$, show that

$$\int_0^\infty \frac{dx}{(x^2+\alpha^2)(x^2+\beta^2)} = \frac{\pi}{2\alpha\beta(\alpha+\beta)}.$$

(b) If $a > 0$, using the result in (a), find $\displaystyle\int_0^\infty \frac{dx}{x^4 + 2x^2 \cosh a + 1}$.

21. Consider the integral $\displaystyle\int_0^{\frac{\pi}{2}} \frac{d\theta}{3 + \sqrt{5}\cos 2\theta}$.

   (a) Though this integral is not an improper integral, show that it becomes improper when a substitution of $u = \tan\theta$ is used.

   (b) Hence find the value of the integral.

22. (a) Show that $\displaystyle\int_{-\infty}^\infty \varphi(x^2)\,dx = 2\int_0^\infty \varphi(x^2)\,dx$ provided the improper integral $\displaystyle\int_0^\infty \varphi(x^2)\,dx$ converges.

   (b) Show that if $\displaystyle\int_0^\infty x\varphi(x^2)\,dx$ converges, then $\displaystyle\int_{-\infty}^\infty x\varphi(x^2)\,dx = 0$.

23. Let $J_n = \displaystyle\int_0^\infty \operatorname{sech}^n x\,dx$, where $n \in \mathbb{N}$.

   (a) Show that $J_{n+2} = \dfrac{n}{n+1} J_n$.

   (b) Hence find the value of J_4.

24. Find the value of the improper integral $\displaystyle\int_1^\infty \frac{dx}{(x+a)\sqrt{x^2-1}}$ for all values of $a > -1$.

25. (a) If f is a continuous function on $[0, 1]$, show that

$$\int_0^\pi x f(\sin x)\,dx = \pi \int_0^{\frac{\pi}{2}} f(\sin x)\,dx.$$

   (b) Using the result in (a), show that $\displaystyle\int_0^\pi \frac{x}{\varphi - \cos^2 x}\,dx = \frac{\pi^2}{2}$.
   Here $\varphi = (1 + \sqrt{5})/2$ is the *golden ratio*.

   (c) Hence deduce that if $n \in \mathbb{N}$, then $\displaystyle\int_0^{n\pi} \frac{x}{\varphi - \cos^2 x}\,dx = \frac{\pi^2 n^2}{2}$.

26. (a) Using a symmetric substitution (see page 267 of Chapter 19) show that

$$\int_0^\infty \frac{1 + x^2}{1 + 4x^2 + x^4}\,dx = \frac{\pi}{\sqrt{6}}.$$

   (b) Hence show that

$$\int_0^\infty \frac{dx}{1 + 4x^2 + x^4} = \int_0^\infty \frac{x^2}{1 + 4x^2 + x^4}\,dx = \frac{\pi}{2\sqrt{6}}.$$

(c) Using a substitution of $x = 1/u$ together with the result found in (a), show that

$$\int_0^\infty \frac{(1 + x^2) \tan^{-1}(x^2)}{1 + 4x^2 + x^4} \, dx = \frac{\pi^2}{4\sqrt{6}}.$$

27. Suppose $I = \int_0^\infty \frac{(2x)^4}{(1 - x^2 + x^4)^3} \, dx$ and $J = \int_0^\infty \frac{(2x)^4}{(1 - x^2 + x^4)^4} \, dx$.

(a) If f is an integrable function such that $\int_{-\infty}^\infty f(x) \, dx$ exists, show that

$$\int_{-\infty}^\infty f(x) \, dx = \int_{-\infty}^\infty f\left(x - \frac{1}{x}\right) \, dx.$$

(b) By making use of the substitution $x = 1/u$, together with the result in (a), show that $I = 3\pi$.

(c) By considering $J - I$, show that $J = I$.

28. Consider the improper integral $\int_0^1 \frac{\ln x}{1 - x} \, dx$.

(a) By recalling the infinite sum formula for a geometric series, namely

$$\frac{1}{1 - x} = \sum_{n=0}^\infty x^n, \quad |x| < 1,$$

if this term in the integrand is replaced with the above infinite sum, and assuming the order of the summation and integration can be interchanged, one can write $\int_0^1 \frac{\ln x}{1-x} \, dx = \sum_{n=0}^\infty \int_0^1 x^n \ln x \, dx$.

Using integration by parts, show that $\int_0^1 \frac{\ln x}{1 - x} \, dx = -\sum_{n=1}^\infty \frac{1}{n^2}$.

(b) Recalling that the infinite sum found in (a) is the Basel problem (see Exercise 28 on page 258 of Chapter 18), deduce that

$$\int_0^1 \frac{\ln x}{1 - x} \, dx = -\frac{\pi^2}{6}.$$

29. If $n \in \mathbb{N}$, show that

$$\int_0^\infty \frac{dx}{(1 + x^2)^n} = \frac{2n - 3}{2n - 2} \cdot \frac{2n - 5}{2n - 4} \cdots \frac{3}{4} \cdot \frac{1}{2} \cdot \frac{\pi}{2}.$$

30. Let $\int_0^\infty f(x + \sqrt{1 + x^2}) \, dx$ where f is any function for which the improper integral exists.

(a) Show that $\int_0^\infty f(x + \sqrt{1 + x^2}) \, dx = \frac{1}{2} \int_1^\infty \left(1 + \frac{1}{x^2}\right) f(x) \, dx.$

(b) Using (a), evaluate $\displaystyle\int_0^\infty \frac{dx}{2x^2 + 1 + 2x\sqrt{x^2 + 1}}$.

(c) Using (a) together with the substitution $x = \tan\theta$, evaluate

$$\int_0^{\frac{\pi}{2}} \frac{d\theta}{(1 + \sin\theta)^3}.$$

31. (a) Show that $\displaystyle\int_1^\infty \frac{\cos x}{x^2}\, dx$ converges.

(b) Hence show that $\displaystyle\int_1^\infty \frac{\sin x}{x}\, dx$ converges.

32. (a) Show that $\displaystyle\int_0^{\frac{\pi}{2}} \ln(\sin x)\, dx = -\frac{\pi}{2}\ln 2$.

(b) If f is an integrable function, show that

$$\int_0^1 \frac{f(x)}{\sqrt{1 - x^2}}\, dx = \int_0^{\frac{\pi}{2}} f(\sin x)\, dx.$$

(c) Hence deduce that $\displaystyle\int_0^1 \frac{\ln x}{\sqrt{1 - x^2}}\, dx = -\frac{\pi}{2}\ln 2$.

33. Let $\displaystyle I_n = \int_0^\infty \frac{dx}{(1 + x^2)^n}$ where $n \in \mathbb{N}$.

(a) Show that $\displaystyle I_n - I_{n+1} = \frac{1}{2n}I_n$.

(b) Hence deduce that $\displaystyle I_{n+1} = \frac{(2n)!\pi}{2^{2n+1}(n!)^2}$.

(c) If $\displaystyle\int_0^\infty f\left[\left(x - \frac{1}{x}\right)^2\right] dx$ where f is any function for which the improper integral exists, show that

$$\int_0^\infty f(x^2)\, dx = \int_0^\infty \frac{1}{x^2} f\left[\left(x - \frac{1}{x}\right)^2\right] dx.$$

(d) Hence evaluate $\displaystyle\int_0^\infty \frac{x^{2n-2}}{(x^4 - x^2 + 1)^n}$ where $n \in \mathbb{N}$.

34. Let $\displaystyle I_{m,n} = \int_0^\infty \frac{x^m}{(1 + x)^{m+n}}\, dx$ where m and $n - 1$ are positive integers.

(a) Show that $\displaystyle I_{m,n} = \frac{m}{m + n - 1}I_{m-1,n}$.

(b) Hence deduce that $\displaystyle I_{m,n} = \frac{m!(n - 2)!}{(m + n - 1)!}$, $m \geqslant 0, n \geqslant 2$.

35. If $a^2 + b^2 = c^2$ such that $b > 0$, show that

$$\int_{-\infty}^{\infty} \frac{b}{1 + 2ax + c^2x^2} \cdot \frac{dx}{1 - 2ax + c^2x^2} = \frac{\pi}{2}.$$

36. Show that $\displaystyle\int_0^1 \frac{\cos(3x^2 + 2x + 1)}{\sqrt{x} + \sqrt[3]{x}}\, dx \leqslant 1.$

37. Show

(a) $\displaystyle\int_a^b \frac{dx}{\sqrt{(b-x)(x-a)}} = \pi$, where $b > a$

(b) $\displaystyle\int_1^{\infty} \frac{dx}{(x^2 + a^2)^{\frac{3}{2}}} = \frac{1}{a^2}\left(1 - \frac{1}{\sqrt{1 + a^2}}\right)$, $a > 0$

(c) $\displaystyle\int_0^{\frac{\pi}{2}} x \cot x\, dx = \frac{\pi}{2} \ln 2$

(d) $\displaystyle\int_0^1 \frac{\ln x}{\sqrt{x - x^2}}\, dx = -2\pi \ln 2$

(e) $\displaystyle\int_0^{\infty} \ln\left(x + \frac{1}{x}\right) \frac{dx}{1 + x^2} = \pi \ln 2$

(f) $\displaystyle\int_{-\infty}^{\infty} \frac{dx}{a \sinh x + b \cosh x} = \frac{\pi}{\sqrt{b^2 - a^2}}$, $0 < a < b$

(g) $\displaystyle\int_{-1}^1 \tan^{-1}\left(\frac{1}{\sqrt{1 - x^2}}\right) dx = (2 - \sqrt{2})\pi$

(h) $\displaystyle\int_0^{\infty} \frac{dx}{(1 + x)(\ln^2 x + 1)} = \frac{\pi}{2}$

(i) $\displaystyle\int_{-\infty}^{\infty} \frac{dx}{x^4 + x^2 + 1} = \frac{\pi}{\sqrt{3}}$

(j) $\displaystyle\int_1^{\infty} \frac{dx}{e^{x+1} + e^{3-x}} = \frac{\pi}{4e^2}$

38. (a) Consider the improper integral $\displaystyle\int_0^{\infty} \frac{\sqrt{x}}{1 + x^2}\, dx$. By using the substitution $x = \tan\theta$, show that

$$\int_0^{\infty} \frac{\sqrt{x}}{1 + x^2}\, dx = \int_0^{\frac{\pi}{2}} \sqrt{\tan\theta}\, d\theta = \frac{\pi}{\sqrt{2}}.$$

(b) If f is an integrable function, show that

$$\int_0^{\infty} f(x)\, dx = \int_0^1 \left[f(x) + \frac{1}{x^2} f\left(\frac{1}{x}\right)\right] dx.$$

(c) Let $f(x) = \dfrac{\sqrt{x}}{1 + x^2}$. Show that

$$f(x) + \frac{1}{x^2} f\left(\frac{1}{x}\right) = \frac{1 + x}{\sqrt{x}(1 + x^2)}.$$

(d) Hence find the value of $\displaystyle\int_0^1 \frac{1 + x}{\sqrt{x}(1 + x^2)}\, dx$.

39. Find the values of p for which the improper integral $\displaystyle\int_1^\infty \frac{x^{p-1}}{1 + x}\, dx$ converges.

✠ Extension questions and Challenge problems

40. Using the result $e^x > 1 + x$ for all $x \in \mathbb{R}$, where required, show that $\displaystyle\int_0^1 \frac{x^{s-1}}{e^x - 1}\, dx$ converges for all $s > 1$.

41. Let $P(x)$ be a polynomial of degree n with real coefficients and $\alpha > 0$.

(a) Show that

$$\int_0^\alpha e^{-x} P(x)\, dx = -e^{-x} P(x)\Big|_0^\alpha - e^{-x} P'(x)\Big|_0^\alpha - \cdots - e^{-x} P^{(n)}(x)\Big|_0^\alpha.$$

(b) Hence show that $\displaystyle\int_0^\infty e^{-x} P(x)\, dx = \sum_{k=0}^n P^{(k)}(0)$.

42 (The Laplace transform). For any suitable function f, the *Laplace transform* of the function $f(t)$ is the function $F(s)$ defined by

$$\mathcal{L}\{f(t)\} = F(s) = \int_0^\infty e^{-st} f(t)\, dt, \quad s > 0.$$

(a) Show that

  (i) $\mathcal{L}\{1\} = \dfrac{1}{s}$ (ii) $\mathcal{L}\{\sin t\} = \dfrac{1}{s^2 + 1}$

(b) If $a, \alpha > 0$, show that

  (i) $\mathcal{L}\{e^{-\alpha t} f(t)\} = F(s + \alpha)$ (ii) $\mathcal{L}\{f(at)\} = \dfrac{1}{a} F\left(\dfrac{s}{a}\right)$

(c) Use the results given in (a)(ii) and (b)(ii) to find $\mathcal{L}\{\sin(at)\}$.

(d) Show that $\mathcal{L}\{f'(t)\} = sF(s) - f(0)$.

(e) Use the results given in (a)(ii) and (d) to find $\mathcal{L}\{\cos t\}$.

(f) Hence find $\mathcal{L}\{e^{-2t}\cos t\}$.

43. Evaluate $\displaystyle\int_0^{\frac{\pi}{2}} \operatorname{cosec}^{-1}(\sqrt{1+\cot x})\,dx$.

Endnote

1. In such cases we say the improper integral is *conditionally convergent*, but we do not consider these cases any further here.

21
Two Important Improper Integrals

'Do you know what a mathematician is?', Lord Kelvin once asked a class. He stepped to the blackboard and wrote $\int_{-\infty}^{\infty} e^{-x^2}\, dx = \sqrt{\pi}$. Putting his finger on what he had written, he turned to the class. 'A mathematician is one to whom that is as obvious as that twice two makes four is to you.'

— *Lord Kelvin*

In this our final chapter we present evaluations for two important improper integrals that arise in applications. Each evaluation is made using elementary means based on ideas at a level not beyond those encountered and used throughout this text. The two improper integrals to be considered are

$$\int_0^\infty e^{-x^2}\, dx \quad \text{and} \quad \int_0^\infty \frac{\sin x}{x}\, dx,$$

which are known as the the *Gaussian integral*[1] and the *Dirichlet integral*[2], respectively. What makes the evaluation of these integrals particularly troublesome is that in neither case do the corresponding indefinite integrals admit expressions in elementary terms. By a remarkable stroke of luck, values in terms of known mathematical constants can, however, be found for each of these two improper integrals.

§ The Gaussian Integral

We start by showing the Gaussian integral converges. We write it as

$$\int_0^\infty e^{-x^2}\, dx = \int_0^1 e^{-x^2}\, dx + \int_1^\infty e^{-x^2}\, dx.$$

298

For the first of the integrals on the right, as e^{-x^2} is continuous for all $x \in [0, 1]$ the integral exists and is Riemann integrable. For the second integral appearing on the right, as $x^2 > x$ for $x > 1$, multiplying both sides of the inequality by negative one and exponentiating, as the exponential function is monotonic, the inequality is equivalent to $e^{-x^2} < e^{-x}$ for all $x > 1$. Since

$$\int_1^\infty e^{-x} \, dx = \left[-e^{-x} \right]_1^\infty = \frac{1}{e} < \infty$$

converges, by the direct comparison test $\int_1^\infty e^{-x^2} \, dx$ also converges, thereby showing that the Gaussian integral converges.

In evaluating the Gaussian integral the tools to be used are reduction formulae, two inequalities, and the squeeze theorem for limits. While the approach to be used is quite involved,[3] importantly it does not rely on techniques beyond the scope of this text.[4] Let

$$j_n = \int_0^1 (1 - x^2)^n \, dx \quad \text{and} \quad k_n = \int_0^\infty \frac{dx}{(1 + x^2)^n},$$

where $n \in \mathbb{N}$. In the first of these integrals, if we let $x \sin \theta, dx = \cos \theta \, d\theta$ while for the limits of integration, when $x = 0, \theta = 0$ and when $x = 1, \theta = \pi/2$. Thus

$$j_n = \int_0^{\frac{\pi}{2}} (1 - \sin^2 \theta)^n \cos \theta \, d\theta = \int_0^{\frac{\pi}{2}} \cos^{2n+1} \theta \, d\theta = I_{2n+1},$$

where $I_n = \int_0^{\frac{\pi}{2}} \cos^n \theta \, d\theta$.

Similarly, for the second of the integrals, if we let $x = \tan \theta, dx = \sec^2 \theta \, d\theta$ while for the limits of integration, when $x = 0, \theta = 0$ and when $x \to \infty, \theta \to \pi/2$. Thus

$$k_n = \int_0^{\frac{\pi}{2}} \frac{\sec^2 \theta}{(1 + \tan^2 \theta)^n} \, d\theta = \int_0^{\frac{\pi}{2}} \cos^{2n-2} \theta \, d\theta = I_{2n-2}.$$

Now we find a reduction formula for I_n. Writing

$$I_n = \int_0^{\frac{\pi}{2}} \cos^n \theta \, d\theta = \int_0^{\frac{\pi}{2}} \cos^{n-1} \theta \cos \theta \, d\theta,$$

integrating by parts gives

$$I_n = \sin\theta \cdot \cos^{n-1}\theta \Big|_0^{\frac{\pi}{2}} + (n-1)\int_0^{\frac{\pi}{2}} \cos^{n-2}\theta \sin^2\theta \, d\theta$$

$$= (n-1)\int_0^{\frac{\pi}{2}} \cos^{n-2}\theta(1-\cos^2\theta)\, d\theta$$

$$= (n-1)\int_0^{\frac{\pi}{2}} \cos^{n-2}\theta \, d\theta - (n-1)\int_0^{\frac{\pi}{2}} \cos^n\theta \, d\theta$$

$$= (n-1)I_{n-2} - (n-1)I_n,$$

or

$$I_n = \frac{n-1}{n}I_{n-2},$$

after rearranging. We also note that

$$I_0 = \int_0^{\frac{\pi}{2}} d\theta = \frac{\pi}{2} \quad \text{and} \quad I_1 = \int_0^{\frac{\pi}{2}} \cos\theta \, d\theta = 1.$$

Now shifting the index in the reduction formula for I_n by $n \mapsto 2n$ we have

$$I_{2n} = \frac{2n-1}{2n}I_{2n-2} = \frac{2n-1}{2n} \cdot \frac{2n-3}{2n-2} \cdots \frac{3}{4} \cdot \frac{1}{2} \cdot I_0,$$

and by $n \mapsto 2n+1$ we have

$$I_{2n+1} = \frac{2n}{2n+1}I_{2n-1} = \frac{2n}{2n+1} \cdot \frac{2n-2}{2n-1} \cdots \frac{4}{4} \cdot \frac{2}{3} \cdot I_1.$$

Taking their product and using the values found for I_0 and I_1, we have

$$I_{2n}I_{2n+1} = \frac{1}{2n+1} \cdot \frac{\pi}{2},$$

for all $n = 0, 1, \ldots$. Shifting the index in the product by $n \mapsto n-1$ gives

$$I_{2n-2}I_{2n-1} = \frac{1}{2n-1} \cdot \frac{\pi}{2},$$

for all $n = 1, 2, \ldots$.

Putting aside these reduction formulae for the moment, we now develop some inequalities that will be needed. The first is

$$1 + x^2 \leqslant e^{-x^2}, \quad \text{for } 0 \leqslant x \leqslant 1.$$

To prove this, consider the function $f(x) = 1 - x^2 - e^{-x^2}$. $f'(x) = 2x(e^{-x^2} - 1) \leqslant 0$ for all $x \in [0, 1]$ since $e^{-x^2} \leqslant 1$. Thus f is a decreasing function and $f(x) \leqslant f(0) = 0$, and the required inequality follows.

The second inequality we need is

$$e^{-x^2} \leqslant \frac{1}{1 + x^2}, \text{ for } x \geqslant 0.$$

To prove this inequality, consider the function $g(x) = e^{-x^2} - 1/(1 + x^2)$. Now,

$$g'(x) = \frac{(1 + x^2)e^{-x^2} - 1}{1 + x^2} = \frac{h(x)}{1 + x^2},$$

where $h(x) = (1 + x^2)e^{-x^2} - 1$. As $1 + x^2 > 0$ for all x, we need only consider the behaviour of the function h. As $h'(x) = -2x^3 e^{-x^2} \leqslant 0$ for all $x \geqslant 0$, h is a decreasing function. This implies g is a decreasing function and $g(x) \leqslant g(0) = 0$, and the required inequality follows.

If we now integrate the nth (positive) power of these inequalities, the first from 0 to 1, the second from 0 to ∞ we have, respectively,

$$\int_0^1 (1 - x^2)^n \, dx \leqslant \int_0^1 e^{-nx^2} \, dx:$$

and

$$\int_0^\infty e^{-nx^2} \, dx \leqslant \int_0^\infty \frac{dx}{(1 + x^2)^n}.$$

Noting that $e^{-nx^2} > 0$ for all x,

$$\int_0^\infty e^{-nx^2} \, dx = \int_0^1 e^{-nx^2} \, dx + \int_1^\infty e^{-nx^2} \, dx > \int_0^1 e^{-nx^2} \, dx,$$

the two inequalities can be combined

$$\int_0^1 (1 - x^2)^n \, dx \leqslant \int_0^1 e^{-nx^2} \, dx < \int_0^\infty e^{-nx^2} \, dx \leqslant \int_0^\infty \frac{dx}{(1 + x^2)^n},$$

or as

$$I_{2n+1} \leqslant \int_0^1 e^{-nx^2} \, dx < I_{2n-2},$$

in terms of the reduction formula I_n. Under the change of variable $x \mapsto x/\sqrt{n}$ this becomes

$$\sqrt{n} I_{2n+1} \leqslant \int_0^{\sqrt{n}} e^{-x^2} \, dx < \sqrt{n} I_{2n-2}. \tag{$*$}$$

Finally we find the ratio between the reduction formulae in the limit as $n \to \infty$. Since $\cos \theta < 1$ for $\theta \in (0, \pi/2)$, on repeatedly multiplying this inequality by the positive term $\cos \theta$ we find for the integrals

$$I_{2n+1} < I_{2n} < I_{2n-1},$$

or

$$1 < \frac{I_{2n}}{I_{2n+1}} < \frac{I_{2n-1}}{I_{2n+1}},$$

since I_n is positive for all n. But it was shown that $I_{2n+1} = \frac{2n}{2n+1} I_{2n-1}$. Thus the above inequality becomes

$$1 < \frac{I_{2n}}{I_{2n+1}} < \frac{2n+1}{2n} = 1 - \frac{1}{2n}.$$

As $n \to \infty$,

$$1 < \lim_{n \to \infty} \frac{I_{2n}}{I_{2n+1}} < 1,$$

so from the squeeze theorem for limits we have

$$\lim_{n \to \infty} \frac{I_{2n}}{I_{2n+1}} = 1.$$

Furthermore, since $I_{2n+1} I_{2n} = \frac{1}{2n+1} \cdot \frac{\pi}{2}$, this limit can be written as

$$\lim_{n \to \infty} (2n+1) \frac{2}{\pi} I_{2n}^2 = 1 \quad \text{or} \quad \lim_{n \to \infty} \sqrt{2n+1} I_{2n} = \sqrt{\frac{\pi}{2}}.$$

A change of the variable in the limit to $n \mapsto n - 1$ and $n \mapsto n + 1/2$ gives, respectively,

$$\lim_{n \to \infty} \sqrt{2n-1} I_{2n-2} = \sqrt{\frac{\pi}{2}} \quad \text{and} \quad \lim_{n \to \infty} \sqrt{2n+2} I_{2n+1} = \sqrt{\frac{\pi}{2}}.$$

In the limit, as $n \to \infty$ $(*)$ becomes

$$\lim_{n \to \infty} \sqrt{n} I_{2n+1} < \lim_{n \to \infty} \int_0^{\sqrt{n}} e^{-x^2} \, dx < \lim_{n \to \infty} \sqrt{n} I_{2n-2}. \qquad (**)$$

The two limits appearing in (∗∗) can be found. Here

$$\lim_{n\to\infty} \sqrt{n}\, I_{2n+1} = \lim_{n\to\infty} \frac{\sqrt{n}}{\sqrt{2n+2}} \cdot \sqrt{2n+2}\, I_{2n+1}$$

$$= \lim_{n\to\infty} \frac{\sqrt{n}}{\sqrt{2n+2}} \cdot \lim_{n\to\infty} \sqrt{2n+2}\, I_{2n+1}$$

$$= \lim_{n\to\infty} \sqrt{\frac{n}{2n+2}} \cdot \sqrt{\frac{\pi}{2}}$$

$$= \sqrt{\frac{\pi}{2}} \cdot \lim_{n\to\infty} \sqrt{\frac{1}{2+2/n}}$$

$$= \sqrt{\frac{\pi}{2}} \cdot \frac{1}{\sqrt{2}} = \frac{\sqrt{\pi}}{2},$$

and

$$\lim_{n\to\infty} \sqrt{n}\, I_{2n-2} = \lim_{n\to\infty} \frac{\sqrt{n}}{\sqrt{2n-1}} \cdot \sqrt{2n-1}\, I_{2n-2}$$

$$= \lim_{n\to\infty} \frac{\sqrt{n}}{\sqrt{2n-1}} \cdot \lim_{n\to\infty} \sqrt{2n-1}\, I_{2n-2}$$

$$= \lim_{n\to\infty} \sqrt{\frac{n}{2n-1}} \cdot \sqrt{\frac{\pi}{2}}$$

$$= \sqrt{\frac{\pi}{2}} \cdot \lim_{n\to\infty} \sqrt{\frac{1}{2-1/n}}$$

$$= \sqrt{\frac{\pi}{2}} \cdot \frac{1}{\sqrt{2}} = \frac{\sqrt{\pi}}{2}.$$

Thus (∗∗) reduces to

$$\frac{\sqrt{\pi}}{2} < \lim_{n\to\infty} \int_0^{\sqrt{n}} e^{-x^2}\, dx < \frac{\sqrt{\pi}}{2}.$$

Recognising

$$\lim_{n\to\infty} \int_0^{\sqrt{n}} e^{-x^2}\, dx = \int_0^{\infty} e^{-x^2}\, dx,$$

by the squeeze theorem the value for the Gaussian integral is

$$\boxed{\int_0^{\infty} e^{-x^2}\, dx = \frac{\sqrt{\pi}}{2}}$$

This is a truly remarkable result. Not only can a value for the Gaussian integral be found, but the unexpected appearance of π makes it all the more astounding.

Using this result, values for certain types of improper integrals related to the Gaussian integral can be found by reducing the integral to the Gaussian integral. Consider the following example.

Example 21.1 By reducing the integral to a Gaussian integral, find the value of $\int_1^\infty \dfrac{e^{1-x}}{\sqrt{x-1}}\, dx$.

Solution Let $x - 1 = u^2, dx = 2u\, du$ while for the limits of integration we have $x = 1, u = 0$ and $x \to \infty, u \to \infty$. Thus

$$\int_1^\infty \frac{e^{1-x}}{\sqrt{x-1}}\, dx = \int_0^\infty \frac{e^{-u^2}}{u} \cdot 2u\, du = 2 \int_0^\infty e^{-u^2}\, du = 2 \cdot \frac{\sqrt{\pi}}{2} = \sqrt{\pi}.$$

▶

§ The Dirichlet Integral

The second of our important improper integrals is the Dirichlet integral. While there are many, mostly advanced, methods that can be used to evaluate this integral the approach to be used here is via the *Dirichlet kernel*.[5] It is considered to be largely elementary in that it does not require anything beyond the scope of the material presented in this text.

You may recall in Example 15.10 on page 198 of Chapter 15 it was shown that the Dirichlet kernel can be expressed as a finite sum as follows:

$$\frac{\sin(n + \frac{1}{2})x}{2\sin(\frac{x}{2})} = \frac{1}{2} + \sum_{k=1}^n \cos kx.$$

At $x = 0$ the left-hand side is understood to correspond to its limiting value of $n + \frac{1}{2}$ as $x \to 0$. The integral of the Dirichlet kernel between the limits 0 to π, will be needed. It is

$$\int_0^\pi \frac{\sin(n + \frac{1}{2})x}{2\sin(\frac{x}{2})}\, dx = \int_0^\pi \left[\frac{1}{2} + \sum_{k=1}^n \cos kx \right] dx$$

$$= \left[\frac{x}{2} \right]_0^\pi + \sum_{k=1}^n \left[\frac{\sin kx}{k} \right]_0^\pi = \frac{\pi}{2}.$$

We first show the Dirichlet integral converges. We write the Dirichlet integral as

$$\int_0^\infty \frac{\sin x}{x}\, dx = \int_0^1 \frac{\sin x}{x}\, dx + \int_1^\infty \frac{\sin x}{x}\, dx. \qquad (\dagger)$$

For the first of the integrals, continuity of the function $x \mapsto \frac{\sin x}{x}$ is extended at zero by assigning it a value of one at $x = 0$. As the function in the integrand is now continuous for all x in the interval $[0, 1]$, the integral exists and is Riemann integrable.

For the second of the integrals, we integrate by parts:

$$\int_1^\infty \frac{\sin x}{x}\, dx = \lim_{\alpha \to \infty} \left[-\frac{\cos x}{x} \right]_1^\alpha - \int_1^\infty \frac{\cos x}{x^2}\, dx = \cos(1) - \int_1^\infty \frac{\cos x}{x^2}\, dx.$$

For the integral on the right

$$\int_1^\infty \left| \frac{\cos x}{x^2} \right| dx = \int_1^\infty \frac{|\cos x|}{x^2}$$

$$\leqslant \int_1^\infty \frac{dx}{x^2} = \lim_{\alpha \to \infty} \int_1^\alpha \frac{dx}{x^2}$$

$$= \lim_{\alpha \to \alpha} \left[-\frac{1}{x} \right]_1^\alpha = \lim_{\alpha \to \infty} \frac{1}{\alpha}$$

$$= 1 < \infty.$$

Since $\int_1^\infty \left| \frac{\cos x}{x^2} \right| dx < \infty$, by the absolute convergence test, $\int_1^\infty \frac{\cos x}{x^2}\, dx$ converges. Thus the second of the integrals in ($\dagger$) converges and shows that the Dirichlet integral converges.

We next turn our attention to considering the rather unusual function

$$f(t) = \frac{1}{t} - \frac{1}{2 \sin \frac{t}{2}}, \quad 0 < t \leqslant \pi.$$

At $t = 0$ the value for the function is understood as its limiting value as $t \to 0^+$ so in this case we have $f(0) = \lim_{t \to 0^+} f(t) = 0$; this means f is continuous for all $t \in [0, \pi]$. Its derivative f' can also be shown to be continuous on the interval $[0, \pi]$. Why we consider such a function at all will become clear in a moment's time. For now, consider the integral

$$\int_0^\pi \sin \left(\left(n + \frac{1}{2} \right) t \right) \left(\frac{1}{t} - \frac{1}{2 \sin \frac{t}{2}} \right) dt.$$

Integrating by parts, we have

$$\int_0^\pi \sin\left(\left(n+\frac{1}{2}\right)t\right)\left(\frac{1}{t}-\frac{1}{2\sin\frac{t}{2}}\right)dt = \int_0^\pi \sin\left(\left(n+\frac{1}{2}\right)t\right)\cdot f(t)\,dt$$

$$= \left[\frac{-1}{n+\frac{1}{2}}\cos\left(\left(n+\frac{1}{2}\right)t\right)\cdot f(t)\right]_0^\pi$$

$$+ \frac{1}{n+\frac{1}{2}}\int_0^\pi \cos\left(\left(n+\frac{1}{2}\right)t\right)\cdot f'(t)\,dt.$$

As $f(0) = 0$ and $\cos(n+\frac{1}{2})\pi = 0$, this integral reduces to

$$\int_0^\pi \sin\left(\left(n+\frac{1}{2}\right)t\right)\left(\frac{1}{t}-\frac{1}{2\sin\frac{t}{2}}\right)dt = \frac{1}{n+\frac{1}{2}}\int_0^\pi\cos\left(\left(n+\frac{1}{2}\right)t\right)\cdot f'(t)\,dt.$$

As $f'(t)$ is continuous on $[0, \pi]$, $|f'(t)|$ will be bounded by some constant $M > 0$. Thus $|f'(t)| \leqslant M$. So

$$\left|\int_0^\pi \cos\left(\left(n+\frac{1}{2}\right)t\right)f'(t)\,dt\right| \leqslant \int_0^\pi\left|\cos\left(\left(n+\frac{1}{2}\right)t\right)\right|\cdot|f'(t)|\,dt$$

$$\leqslant \int_0^\pi |f'(t)|\,dt,$$

since $|\cos(n+\frac{1}{2})t| \leqslant 1$ for all t. But as $|f'(t)| \leqslant M$ we have

$$\left|\int_0^\pi \cos\left(\left(n+\frac{1}{2}\right)t\right)f'(t)\,dt\right| \leqslant \int_0^\pi M\,dt = M\pi,$$

or after dividing both sides by the positive term $n+\frac{1}{2}$

$$\left|\frac{1}{1+\frac{1}{n}}\int_0^\pi \cos\left(\left(n+\frac{1}{2}\right)t\right)f'(t)\,dt\right| \leqslant \frac{M\pi}{n+\frac{1}{2}}.$$

In the limit as $n \to \infty$, we see that

$$\left|\frac{1}{1+\frac{1}{n}}\int_0^\pi \cos\left(\left(n+\frac{1}{2}\right)t\right)f'(t)\,dt\right| \to 0,$$

and can therefore conclude that[6]

$$\lim_{n\to\infty}\int_0^\pi \sin\left(\left(n+\frac{1}{2}\right)t\right)\left(\frac{1}{t}-\frac{1}{2\sin\frac{t}{2}}\right)dt = 0.$$

Now from the linearity of the definite integral, one has

$$\lim_{n\to\infty}\left[\int_0^\pi \frac{\sin\left(n+\frac{1}{2}\right)t}{t}\,dt - \int_0^\pi \frac{\sin\left(n+\frac{1}{2}\right)t}{2\sin\frac{t}{2}}\,dt\right] = 0.$$

Since the second integral was shown to have a value of $\frac{\pi}{2}$, we have

$$\lim_{n\to\infty}\left[\int_0^\pi \frac{\sin\left(n+\frac{1}{2}\right)t}{t}\,dt - \frac{\pi}{2}\right] = 0,$$

or

$$\lim_{n\to\infty}\int_0^\pi \frac{\sin\left(n+\frac{1}{2}\right)t}{t}\,dt = \frac{\pi}{2}.$$

Let $x = \left(n+\frac{1}{2}\right)t, dx = \left(n+\frac{1}{2}\right)dt$ while for the limits, when $t = 0, x = 0$ and when $t = \pi, x = \left(n+\frac{1}{2}\right)\pi$. Thus

$$\lim_{n\to\infty}\int_0^{\left(n+\frac{1}{2}\right)\pi} \frac{\sin x}{x}\,dx = \frac{\pi}{2},$$

or finally

$$\boxed{\int_0^\infty \frac{\sin x}{x}\,dx = \frac{\pi}{2}}$$

As was the case with the Gaussian integral, values for certain types of improper integrals related to the Dirichlet integral can be found by reducing the integral to the Dirichlet integral. Consider the following example.

Example 21.2 By reducing the integral to two Dirichlet integrals, find the value of $\int_0^\infty \frac{\sin^3 x}{x}\,dx$.

Solution Recalling that $4\sin^3 x = 3\sin x - \sin 3x$, the integral can be rewritten as

$$\int_0^\infty \frac{\sin^3 x}{x}\,dx = \frac{1}{4}\int_0^\infty \frac{3\sin x - \sin 3x}{x}\,dx$$

$$= \frac{3}{4}\int_0^\infty \frac{\sin x}{x}\,dx - \frac{1}{4}\int_0^\infty \frac{\sin 3x}{x}\,dx.$$

As the first integral is just the Dirichlet integral, it has a value equal to $\pi/2$. The second integral can be made equal to the Dirichlet integral. Setting $u = 3x, du = 3\,dx$ while the limits of integration remain unchanged, we have

$$\int_0^\infty \frac{\sin 3x}{x}\,dx = \int_0^\infty \frac{\sin u}{u}\,du = \frac{\pi}{2}.$$

Thus

$$\int_0^\infty \frac{\sin^3 x}{x} \, dx = \frac{3}{4} \cdot \frac{\pi}{2} - \frac{1}{4} \cdot \frac{\pi}{2} = \frac{\pi}{4}.$$ ▶

Exercises for Chapter 21

✠ **Practice questions**

1. Find the value for the following integrals by reducing each integral to the Gaussian integral.

   (a) $\displaystyle\int_{-\infty}^\infty e^{-x^2} \, dx$

   (b) $\displaystyle\int_{-\infty}^\infty e^{-x^2/2} \, dx$

   (c) $\displaystyle\int_0^\infty \frac{e^{-x}}{\sqrt{x}} \, dx$

   (d) $\displaystyle\int_0^\infty \frac{e^{-1/x^2}}{x^2} \, dx$

   (e) $\displaystyle\int_{-\infty}^\infty e^{-x^2+2x} \, dx$

   (f) $\displaystyle\int_{-\infty}^\infty e^{-x^2+8x-16} \, dx$

   (g) $\displaystyle\int_1^\infty \frac{e^{1-x}}{\sqrt{x-1}} \, dx$

   (h) $\displaystyle\int_0^\infty 2^{-x^2} \, dx$

   (i) $\displaystyle\int_0^\infty x^2 e^{-x^2} \, dx$

   (j) $\displaystyle\int_0^\infty x^4 e^{-x^2} \, dx$

   (k) $\displaystyle\int_0^1 \frac{dx}{\sqrt{-\ln x}}$

   (l) $\displaystyle\int_0^1 \sqrt{-\ln x} \, dx$

   (m) $\displaystyle\int_0^\infty (1-x) e^{-x^2} \, dx$

   (n) $\displaystyle\int_0^1 \left[\ln \left(\frac{1}{x} \right) \right]^{\frac{3}{2}} \, dx$

2. Find the value for the following integrals by reducing each integral to the Dirichlet integral.

   (a) $\displaystyle\int_{-\infty}^\infty \frac{\sin x}{x} \, dx$

   (b) $\displaystyle\int_0^\infty \frac{1}{x} \sin \left(\frac{1}{x} \right) \, dx$

   (c) $\displaystyle\int_0^\infty \frac{\sin(\pi x)}{\pi x} \, dx$

   (d) $\displaystyle\int_0^\infty \frac{\cos x - 1}{x^2} \, dx$

(e) $\displaystyle\int_0^\infty \frac{\sin x - \sin(x^2)}{x}\,dx$ \qquad (f) $\displaystyle\int_0^\infty \frac{\cos x \sin 5x}{x}\,dx$

3. (a) If $n \in \mathbb{N}$ find the value of the integral $\displaystyle\int_0^\infty \frac{\sin(x^n)}{x}\,dx$.

(b) If $a \neq 0$ find all values for the integral $\displaystyle\int_0^\infty \frac{\sin(ax)}{x}\,dx$.

4. Show that $\displaystyle\int_0^\infty e^{-\ln^2 x}\,dx = \sqrt[4]{e}\sqrt{\pi}$.

5. (a) If $a > 0$ show that $\displaystyle\int_0^\infty e^{-ax^2}\,dx = \frac{1}{2}\sqrt{\frac{\pi}{a}}$.

(b) If $a > 0$ and $b \in \mathbb{R}$ show that $\displaystyle\int_{-\infty}^\infty e^{-ax^2+bx}\,dx = e^{\frac{b^2}{4a}}\sqrt{\frac{\pi}{a}}$.

6. Show that $\displaystyle\int_0^\infty \frac{\cos 4x - \cos 6x}{x^2}\,dx = \pi$.

7. The *sine integral*[7] $\text{Si}\,(x)$ is defined by $\text{Si}\,(x) = \displaystyle\int_0^x \frac{\sin t}{t}\,dt$. It is found to have important applications in many areas of electrical engineering, including signal processing.

A function closely related to the sine integral is the function

$$\text{si}\,(x) = -\int_x^\infty \frac{\sin t}{t}\,dt.$$

Show that $\text{Si}\,(x) = \text{si}\,(x) + \frac{\pi}{2}$ and find values for $\text{Si}\,(0)$, $\text{Si}\,(\infty)$, $\text{Si}\,(-\infty)$, $\text{si}\,(0)$, $\text{si}\,(\infty)$, and $\text{si}\,(-\infty)$.

8. Find $\displaystyle\int_{-\infty}^\infty e^{-x^2}\cosh(2x)\,dx$.

9. In the theory of probability the *cumulative distribution function for the standard normal distribution* $\Phi(x)$ is defined by

$$\Phi(x) = \frac{1}{\sqrt{2\pi}}\int_{-\infty}^x e^{-t^2/2}\,dt.$$

A closely related function is the *error function* $\text{erf}(x)$ defined by

$$\text{erf}(x) = \frac{2}{\sqrt{\pi}}\int_0^x e^{-u^2}\,du.$$

(a) Using the substitution $u = -t/\sqrt{2}$, show that $\Phi(x)$ can be written in terms of the error function as

$$\Phi(x) = \frac{1}{2} + \frac{1}{2}\mathrm{erf}\left(\frac{x}{\sqrt{2}}\right).$$

(b) Show that the error function can be written in terms of the standard normal cumulative distribution function as $\mathrm{erf}(x) = 2\Phi(x\sqrt{2}) - 1$.

(c) Find values for $\mathrm{erf}(0), \mathrm{erf}(\infty), \mathrm{erf}(-\infty), \Phi(0)$, and $\Phi(\infty)$.

(d) A closely related function to the error function is the *complementary error function* $\mathrm{erfc}(x)$ defined by

$$\mathrm{erfc}(x) = \frac{2}{\sqrt{\pi}} \int_x^\infty e^{-u^2} \, du.$$

Show that $\mathrm{erfc}(x) = 1 - \mathrm{erf}(x)$ and hence find values for $\mathrm{erfc}(0), \mathrm{erfc}(\infty)$, and $\mathrm{erfc}(-\infty)$.

10. (a) By integrating the integral $\displaystyle\int_0^\infty \frac{\sin^2 x}{x^2} \, dx$ by parts, use this to show that $\displaystyle\int_0^\infty \frac{\sin^2 x}{x^2} \, dx = \frac{\pi}{2}$.

(b) By setting $x = 2u$ in the integral appearing in (a), use this to show that $\displaystyle\int_0^\infty \frac{\sin^4 x}{x^2} \, dx = \frac{\pi}{4}$.

(c) By integrating the integral appearing in (b) by parts, use this to show that $\displaystyle\int_0^\infty \frac{\sin^4 x}{x^4} \, dx = \frac{\pi}{3}$.

(d) Hence deduce that $\displaystyle\int_0^\infty \frac{\sin x \sin \frac{x}{3}}{x^2} \, dx = \frac{\pi}{6}$.

11. Suppose f is an integrable function such that $\int_{-\infty}^\infty f(x)\,dx$ exists.

(a) By writing

$$\int_{-\infty}^\infty f(x) \, dx = \int_{-\infty}^0 f(x) \, dx + \int_0^\infty f(x) \, dx,$$

and using a substitution of $x = -e^{-t}$ in the first of the integrals appearing on the right and a substitution of $x = e^t$ in the second of the integrals appearing on the right, show that

$$\int_{-\infty}^\infty f(x) \, dx = \int_{-\infty}^\infty f\left(x - \frac{1}{x}\right) dx.$$

(b) Using the result in (a), evaluate

(i) $\displaystyle\int_{-\infty}^{\infty} \exp\left[-\left(x - \frac{1}{x}\right)^2\right] dx$

(ii) $\displaystyle\int_{-\infty}^{\infty} \exp\left[-\frac{(x^2 - qx - 1)^2}{px^2}\right] dx, \; p > 0, q \in \mathbb{R}.$

✠ **Extension questions and Challenge problems**

12. The function si (x) was defined in Exercise 7. Using integration by parts, show that

$$\int_0^{\infty} \sin x \, \mathrm{si}\, x \, dx = -\frac{\pi}{4}.$$

13. The complementary error function erfc(x) was defined in Exercise 9.

(a) Show that $\displaystyle\frac{d}{dx}\mathrm{erfc}(x) = -\frac{2}{\sqrt{\pi}}e^{-x^2}$.

(b) Using l'Hôpital's rule for limits, show that $\displaystyle\lim_{x \to \infty} x \, \mathrm{erfc}(x) = 0$.

(c) Using integration by parts, show that $\displaystyle\int_0^{\infty} \mathrm{erfc}(x) \, dx = \frac{1}{\sqrt{\pi}}$.

14. Show that $\displaystyle\int_0^{\infty} \left(1 - x \sin\left(\frac{1}{x}\right)\right) dx = \frac{\pi}{4}$.

15. (a) Using properties for the hyperbolic functions, show that

$$\frac{\sinh^2(3\pi x)}{\sinh(\pi x)} = \frac{1}{2}e^{5\pi x} - \frac{1}{2}e^{-5\pi x} + \frac{1}{2}e^{3\pi x} - \frac{1}{2}e^{-3\pi x} + \frac{1}{2}e^{\pi x} - \frac{1}{2}e^{-\pi x}.$$

(b) If $a \in \mathbb{R}$, show that $\displaystyle\int_{-\infty}^{\infty} xe^{-\frac{\pi}{4}x^2 + ax} \, dx = \frac{4a}{\pi}e^{\frac{a^2}{\pi}}$.

(c) Using the results from (a) and (b), deduce that

$$\int_{-\infty}^{\infty} xe^{-\frac{\pi}{4}x^2} \frac{\sinh^2(3\pi x)}{\sinh(\pi x)} \, dx = 4(e^{\pi} + 3e^{9\pi} + 5e^{25\pi}).$$

16. Suppose $\displaystyle I = \int_{-\infty}^{\infty} \frac{1 - \cos x \cdot \cos 2x \cdots \cos nx}{x^2} \, dx$, where $n \in \mathbb{N}$.

(a) Show that $I_1 = \pi$, $I_2 = 2\pi$, and $I_3 = 3\pi$.

(b) Based on the results found in (a) it is tempting to conjecture $I_n = n\pi$ for all $n \in \mathbb{N}$. Show that this is not true by showing that $I_4 = 9\pi/2$.

Endnotes

1. The Gaussian integral, which is also known as the probability integral, is named in honour of the German mathematician and physicist Carl Friedrich Gauss (1777–1855), though it was first explicitly evaluated by the French mathematician Pierre-Simon Laplace (1749–1827) in his paper 'Mémoire sur la probabilité des causes par les évenments' of 1774. It is an important integral as it is often encountered in applications. For example, it plays a fundamental role in the theory of probability where it is related to the cumulative distribution function for the standard normal distribution (see Exercise 9).
2. The Dirichlet integral is named in honour of the nineteenth-century German mathematician Johann Peter Gustav Lejeune Dirichlet (1805–1859), though the result for the integral had been known to the great Swiss mathematician Leonhard Euler (1707–1783) by 1781.
3. The method we intend to follow seems to have been first presented by Michael Spivak as an exercise on page 329 of the first edition of his 1967 text *Calculus* (W. A. Benjamn, Menlo Park, California).
4. Many methods for finding the Gaussian integral are known, but almost all of these depend on one being familiar with more advanced techniques. For a selection of such methods see, for example, the article by Hirokazu Iwasawa, 'Gaussian integral puzzle', *The Mathematical Intelligencer*, **31**(3), 38–41 (2009).
5. We follow the method first suggested by Waclaw Kozakiewicz. See: Waclaw Kozakiewicz, 'A simple evaluation of an improper integral', *The American Mathematical Monthly*, **58**(3), 181–182 (1951).
6. In fact this result is a particular case of a more general result known as the *Riemann–Lebesgue lemma*, which states that if f is Riemann integrable on the interval $[a, b]$, then

$$\lim_{\lambda \to \infty} \int_a^b f(t) \sin(\lambda t) \, dt = 0.$$

   It is, respectively, named after the German and French mathematicians Georg Friedrich Bernhard Riemann (1826–1866) and Henri Léon Lebesgue (1875–1941).
7. The sine integral was first introduced by the Italian mathematician Lorenzo Mascheroni (1750– 1800) in 1790, though the value for the Dirichlet integral was already known to the great Swiss mathematician Leonhard Euler (1707–1783) by 1781.

Appendix A
Partial Fractions

The miraculous powers of modern calculation are due to three inventions: the Arabic Notation, Decimal Fractions and Logarithms.
— F. Cajori, *History of Mathematics*

In this appendix an overview for the method of partial fractions is given. In the same way that rational numbers are formed from the integers, *rational functions* may be constructed from polynomials. A rational function is an expression that can be written in the form

$$f(x) = \frac{P(x)}{Q(x)}.$$

Here $P(x)$ is a polynomial and $Q(x)$ is a nonzero polynomials of x.[1]

Two rational functions $P_1(x)/Q_1(x)$ and $P_2(x)/Q_2(x)$ are said to be equal, that is

$$\frac{P_1(x)}{Q_1(x)} = \frac{P_2(x)}{Q_2(x)},$$

if upon cross-multiplying, the two polynomials $P_1(x)Q_2(x)$ and $P_2(x)Q_1(x)$ are equal, that is $P_1(x)Q_2(x) = P_2(x)Q_1(x)$. Addition and multiplication of rational functions are defined in a natural manner using the same rules for fractions:

$$\frac{P(x)}{Q(x)} + \frac{A(x)}{B(x)} = \frac{P(x)B(x) + A(x)Q(x)}{Q(x)B(x)}, \quad \frac{P(x)}{Q(x)} \cdot \frac{A(x)}{B(x)} = \frac{P(x)A(x)}{Q(x)B(x)}.$$

Note that any polynomial $P(x)$ may be viewed as a rational function by thinking of it as $P(x)/1$ where $Q(x) = 1$.

313

Example A.1 Show that the following two rational function expressions

$$\frac{p(x)}{q(x)} = \frac{x+2}{x^2+x+1} \quad \text{and} \quad \frac{P(x)}{Q(x)} = \frac{2x^2-x-10}{2x^3-3x^2-3x-5},$$

are equal.

Solution For the two expressions to be equal we require $p(x)Q(x) = q(x)P(x)$. Now

$$p(x)Q(x) = (x+2)(2x^3-3x^2-3x-5) = (x+2)(2x-5)(x^2+x+1),$$

and

$$q(x)P(x) = (2x^2-x-10)(x^2+x+1) = (x+2)(2x-5)(x^2+x+1).$$

Thus $p(x)Q(x) = q(x)P(x)$ so the two expressions are equal. ▶

The method of partial fractions is a way of decomposing a rational function $P(x)/Q(x)$ into a sum of terms with denominators of degrees less than $Q(x)$ when a factorisation of $Q(x)$ is known. In cases where $P(x)/Q(x)$ is a rational function with real coefficients, when viewed as a real-valued function the method of partial fractions becomes a very important integration technique that can be used to find primitives for rational functions (see Chapter 11).

A rational function $P(x)/Q(x)$ is said to be *proper* if the degree of the numerator is less than the degree of the denominator, that is, deg P < deg Q. If deg $P \geqslant$ deg Q the rational function is said to be *improper*. Consider the rational function $A(x)/B(x)$ where deg $A \geqslant$ deg B. As it is improper the division algorithm for polynomials (polynomial long division) can be used to write $A(x)$ as

$$A(x) = B(x)Q(x) + R(x),$$

where $Q(x)$ is the quotient and $R(x)$ the remainder such that deg R < deg B. So any improper rational function can be written as

$$\frac{A(x)}{B(x)} = Q(x) + \frac{R(x)}{B(x)}.$$

Here $Q(x)$ is a polynomial while $R(x)/B(x)$ is a new rational function that is now proper. What this shows us is every rational function can be written as a sum of a polynomial and a proper rational function with the same denominator that appeared in the original improper fraction.

To give some definiteness to the idea we are trying to develop here, consider the following. For the partial fraction

$$\frac{2}{x+3} + \frac{1}{x-7},$$

a common denominator can be found by cross-multiplying, expanding, and simplifying. Doing so yields

$$\frac{2}{x+3} + \frac{1}{x-7} = \frac{3x-11}{x^2-4x-21}. \tag{A.1}$$

Algebraically this is a relatively simple operation to perform. The question is, if given the right-hand side of (A.1) to start with, is it possible to perform the operation in reverse in order to arrive at the left-hand side? The central challenge here is how to decompose any given proper fraction into a sum of *partial fractions* consisting of simpler (read, of lower degree) denominators? That it can be done, and in a unique way, is contained in the following theorem and is referred to as a *partial fraction decomposition*.

Theorem A.1 (Partial fraction decomposition). *Let $f(x)$ and $g(x)$ be polynomials over $\mathbb{R}[x]$ such that $\deg f < \deg g$ and where $g(x)$ can be written as a product of distinct linear and irreducible quadratic factors that may repeat, namely*

$$g(x) = \prod_{i=1}^{m}(a_i x + b_i)^{k_i} \cdot \prod_{j=1}^{n}(\alpha_j x^2 + \beta_j x + \gamma_j)^{s_j},$$

where $a_i, b_i, \alpha_j, \beta_j, \gamma_j \in \mathbb{R}$ and $k_i, s_j \in \mathbb{N}$; then there exists unique real constants A_{ij}, B_{ij}, and C_{ij} such that

$$\boxed{\frac{f(x)}{g(x)} = \sum_{i=1}^{m}\sum_{j=1}^{k_i}\frac{A_{ij}}{(a_i x + b_i)^j} + \sum_{i=1}^{n}\sum_{j=1}^{s_i}\frac{B_{ij}x + C_{ij}}{(\alpha_i x^2 + \beta_i x + \gamma_i)^j}}$$

As the proof of this theorem is rather technical, it will be omitted. The importance of the theorem is that it furnishes a method for actually finding the partial fraction decomposition of a rational expression provided the irreducible factorisation of the denominator is known; further, the uniqueness property shows that one never has to consider multiple solutions for the coefficients in the decomposition. One should also notice that, provided the fraction is proper, the $f(x)$ appearing in the numerator plays no role in how a partial fraction decomposition is formed.

Using the general expression for the partial fraction decomposition can be tedious. Streamlining the application of the final form found for the decomposition of a rational function into partial fractions is best done by handling one factor at a time for each of the factors appearing in the denominator of the rational function. Depending on whether the factors appearing in the denominator of the rational function are either linear or irreducible quadratic, and are repeated (contain multiplicities) or not (are distinct), at most only four separate cases need be considered. The final form for the partial fraction decomposition is then just the sum of the individual partial fractions found for each factor. We now consider how to find each of the four separate cases used in finding a partial fraction decomposition for a rational function.

Case I – $Q(x)$ is a product of a distinct linear factor

If $Q(x)$ can be written as a product of distinct linear factors, namely

$$Q(x) = (a_1 x + b_1)(a_2 x + b_2) \cdots (a_n x + b_n),$$

each linear factor can be written in the form $\frac{A}{ax+b}$ so that for the n distinct linear factors for the partial fraction decomposition one writes

$$\frac{P(x)}{Q(x)} = \frac{A_1}{a_1 x + b_1} + \frac{A_2}{a_2 x + b_2} + \cdots + \frac{A_n}{a_n x + b_n}.$$

Here $A_1, A_2, \ldots, A_n$ are constants to be determined. The number of these constants to be found is determined by the degree of $Q(x)$; observe that the numerator $P(x)$ has no bearing on the identity of the partial fraction decomposition formed.

Example A.2 Find the partial fraction decomposition for $\dfrac{3x - 11}{x^2 - 4x - 21}$.

Solution We have seen the answer to this question already as it just corresponds to (A.1). We now apply the procedure of partial fractions to show how the decomposition is achieved.

First observe that the degree of the numerator is less than the degree of the denominator so the rational function is proper. Factoring the denominator of the rational function as far as possible over the reals, we have

$$\frac{3x - 11}{x^2 - 4x - 21} = \frac{3x + 1}{(x + 3)(x - 7)}.$$

As the denominator contains two distinct linear factors, its partial fraction decomposition will be of the form

$$\frac{3x - 11}{x^2 - 4x - 21} = \frac{A}{x + 3} + \frac{B}{x - 7}.$$

The most general way to find the two unknown constants that is always guaranteed to work is to multiply through by the common denominator before equating for equal coefficients for x in the resulting polynomial equation. Doing so, we find

$$3x - 11 = A(x - 7) + B(x + 3) = (A + B)x + (-7A + 3B).$$

On equating equal coefficients for x we have

$$\text{coefficient for } x^1 : \quad 3 = A + B$$

$$\text{coefficient for } x^0 : \quad -11 = -7A + 3B.$$

Solving the pair of simultaneous equations we find $A = 2$ and $B = 1$, which gives

$$\frac{3x - 11}{x^2 - 4x - 21} = \frac{2}{x + 3} + \frac{1}{x - 7}$$

for the partial fraction decomposition, as expected. ▶

From this example the first thing you may have noticed is that the effort required in finding a partial fraction decomposition for a rational function is considerable compared to its reverse operation of summing together the partial fractions. The general method used in Example A.2 to find the unknown constants can, however, be considerably sped up by using appropriate substitutions for values of x.

Example A.3 Find the partial fraction decomposition for $\dfrac{3x - 11}{x^2 - 4x - 21}$ using the quicker method of substitution.

Solution As in Example A.2 we have

$$\frac{3x - 11}{x^2 - 4x - 21} = \frac{3x - 11}{(x + 3)(x - 7)} = \frac{A}{x + 3} + \frac{B}{x - 7},$$

which after multiplying through by $(x + 3)(x - 7)$ gives

$$3x - 11 = A(x - 7) + B(x + 3).$$

As this identity is true for all x, the values of A and B can be more easily determined by choosing suitable values for x. The values of x selected should lead to cancellation in one of the terms:

$$x = 7 : \quad 3(7) - 11 = B(7 + 3) \quad \Rightarrow B = 1$$

$$x = -3 : \quad 3(-3) - 11 = A(-3 - 7) \quad \Rightarrow A = 2$$

giving

$$\frac{3x - 11}{x^2 - 4x - 21} = \frac{2}{x + 3} + \frac{1}{x - 7},$$

as before. By employing the substitution method, the need to solve a pair of simultaneous equations has been avoided. ▶

Case II – $Q(x)$ contains some repeated linear factors

If the linear factor $(ax + b)$ appears in the factorisation of $Q(x)$ k-times, instead of writing the part of the partial fraction as $\dfrac{A}{(ax + b)}$ it is instead written as

$$\frac{A_1}{(ax + b)} + \frac{A_2}{(ax + b)^2} + \cdots + \frac{A_k}{(ax + b)^k}.$$

Example A.4 Find the partial fraction decomposition for $\dfrac{2x}{(x + 1)^2}$.

Solution First observe that the degree of the numerator is less than the denominator so the rational function is proper. As the denominator contains one distinct linear factor that is repeated, its partial fraction decomposition will be of the form

$$\frac{2x}{(x + 1)^2} = \frac{A}{x + 1} + \frac{B}{(x + 1)^2},$$

where A and B are two unknown constants to be determined. To find the constants we clear the denominator by multiplying through by $(x + 1)^2$ to obtain

$$2x + 1 = A(x + 1) + B.$$

Substituting obvious values for x, values for A and B can be quickly found.

$$x = -1 : \quad 2(-1) + 1 = B \qquad \Rightarrow B = -1$$

$$x = 0 : \quad 2(0) + 1 = A + B \Rightarrow A = 1 - B = 2.$$

Hence we obtain for the partial fraction decomposition

$$\frac{2x + 1}{(x + 1)^2} = \frac{2}{x + 1} - \frac{1}{(x + 1)^2}. \qquad ▶$$

Example A.5 Find the partial fraction decomposition for $\dfrac{x^2 + 6x + 11}{(x - 1)(x + 2)^2}$.

Solution First observe that the degree of the numerator is less than the denominator so the rational function is proper. As the denominator contains two linear factors, one of which is repeated, its partial fraction decomposition will be of the form

$$\frac{x^2 + 6x + 11}{(x - 1)(x + 2)^2} = \frac{A}{x - 1} + \frac{B}{x + 2} + \frac{C}{(x + 2)^2},$$

where A, B, and C are three unknown constants to be determined. To find the constants we multiply through by $(x - 1)(x + 2)^2$ to obtain

$$x^2 + 6x + 11 = A(x + 2)^2 + B(x - 1) + C(x - 1)(x + 2).$$

Substituting obvious values for x, values for A and B can be quickly found:

$$x = 1 : \quad (1)^2 + 6(1) + 11 = A(1 + 2)^2 \quad \Rightarrow A = 2$$

$$x = -2 : \quad (-2)^2 + 6(-2) + 11 = B(-2 - 1) \quad \Rightarrow B = -1.$$

To determine C, as A and B are already known, we can substitute any other value of x we wish. It is, however, best to choose a small integer to keep the arithmetic manageable. We choose $x = 0$.

$$x = 0 : \quad 11 = 4A - B - 2C \quad \Rightarrow C = 2A - \tfrac{1}{2}B - \tfrac{11}{2} = -1.$$

Hence we obtain for the partial fraction decomposition

$$\frac{x^2 + 6x + 11}{(x - 1)(x + 2)^2} = \frac{2}{x - 1} - \frac{1}{x + 2} - \frac{1}{(x + 2)^2}. \qquad \blacktriangleright$$

Case III – $Q(x)$ contains an irreducible quadratic factor that is not repeated

If the irreducible quadratic factor $(\alpha x^2 + \beta x + \gamma)$ appears in the factorisation of $Q(x)$, the form of its partial fraction decomposition will be

$$\frac{Ax + B}{\alpha x^2 + \beta x + \gamma}.$$

Example A.6 Find the partial fraction decomposition for $\dfrac{7}{(x + 2)(x^2 + 3)}$.

Solution First observe that the degree of the numerator is less than the denominator so the rational function is proper. As the denominator contains one distinct linear factor and one distinct irreducible quadratic factor, its partial fraction decomposition will be of the form

$$\frac{7}{(x + 2)(x^2 + 3)} = \frac{A}{x + 2} + \frac{Bx + C}{x^2 + 3},$$

where A, B, and C are three unknown constants to be determined. To find the constants by clearing the denominator by multiplying through by

$(x + 2)(x^2 + 3)$, one obtains

$$7 = A(x^2 + 3) + Bx(x + 2) + C(x + 2).$$

Substituting obvious values for x, values for A, B, and C can be readily found.

$$x = -2 : \; 7 = 7A \qquad\qquad \Rightarrow A = 1$$

$$x = 0 : \; 7 = 3A + 2C \qquad\quad \Rightarrow C = \tfrac{7}{2} - \tfrac{3}{2}A = 2$$

$$x = 1 : \; 7 = 4A + 3B + 3C \;\; \Rightarrow B = \tfrac{7}{3} - \tfrac{4}{3}A - C = -1.$$

Hence we obtain for the partial fraction decomposition

$$\frac{7}{(x + 2)(x^2 + 3)} = \frac{1}{x + 2} - \frac{x - 2}{x^2 + 3}. \qquad\blacktriangleright$$

Case IV – $Q(x)$ contains an irreducible quadratic factor that is repeated

If the irreducible quadratic factor $(\alpha x^2 + \beta x + \gamma)$ appears in the factorisation of $Q(x)$ k-times, instead of writing the part of the partial fraction as $\dfrac{Ax + B}{\alpha x^2 + \beta x + \gamma}$ it is instead written as

$$\frac{A_1 x + B_1}{\alpha x^2 + \beta x + \gamma} + \frac{A_2 x + B_2}{(\alpha x^2 + \beta x + \gamma)^2} + \cdots + \frac{A_k x + B_k}{(\alpha x^2 + \beta x + \gamma)^k}.$$

Example A.7 Find the partial fraction decomposition for $\dfrac{3x^4 + 5}{x(x^2 + 1)^2}$.

Solution First observe that the degree of the numerator is less than the denominator so the rational function is proper. As the denominator contains one distinct linear factor and one irreducible quadratic factor that is repeated, its partial fraction decomposition is of the form

$$\frac{3x^4 + 5}{x(x^2 + 1)^2} = \frac{A}{x} + \frac{Bx + C}{x^2 + 1} + \frac{Dx + E}{(x^2 + 1)^2}.$$

To find the five unknown constants the most direct method is to multiply through by the common denominator before equating for equal coefficients for x in the resulting polynomial equation found. Doing so, after some algebra, we find

$$3x^4 + 5 = (A + B)x^4 + Cx^3 + (2A + B + D)x^2 + (C + E)x + A.$$

On equating equal coefficients for x we have

$$\text{coefficient for } x^4 : \quad 3 = A + B$$

$$\text{coefficient for } x^3 : \quad 0 = C$$

$$\text{coefficient for } x^2 : \quad 0 = 2A + B + D$$

$$\text{coefficient for } x^1 : \quad 0 = C + D$$

$$\text{coefficient for } x^0 : \quad 5 = A.$$

As A and C are known, the other three constants can be quickly found. We have $A = 5, B = -2, C = 0, D = -8, E = 0$, and the required partial fraction decomposition is

$$\frac{3x^4 + 5}{x(x^2 + 1)^2} = \frac{5}{x} - \frac{2x}{x^2 + 1} - \frac{8x}{(x^2 + 1)^2}. \qquad \blacktriangleright$$

Example A.8 Write down the form of the partial fraction decomposition for the rational function

$$\frac{2x^3 - x^2 + 7}{(x + 5)(x - 1)^3(x^2 + x + 1)(x^2 + 4)^2}.$$

You are not required to evaluate any of the constants that appear in the expression for the partial fraction decomposition.

Solution First observe that the degree of the numerator is less than the denominator so the rational function is proper. As the denominator contains two linear factors, one of which is repeated, and two irreducible quadratic factors, one of which is repeated, its partial fraction decomposition is of the form

$$\frac{2x^3 - x^2 + 7}{(x + 5)(x - 1)^3(x^2 + x + 1)(x^2 + 4)^2} = \frac{A}{x + 5} + \frac{B}{x - 1} + \frac{C}{(x - 1)^2}$$

$$+ \frac{D}{(x - 1)^3} + \frac{Ex + F}{x^2 + x + 1} + \frac{Gx + H}{x^2 + 4} + \frac{Ix + J}{(x^2 + 4)^2},$$

where $A, B, \dots, J$ are real constants. $\qquad \blacktriangleright$

§ Heaviside Cover-Up Method

Finding a partial fraction decomposition for proper rational functions using the clearing of fractions method, while always guaranteed to work, can be very time consuming and slow. While the substitution method is faster, it is still often tedious to apply. For the case where the denominator contains distinct linear factors, a much faster way to find a partial fraction decomposition is to use what is known as the *Heaviside cover-up* method.[2]

If the denominator of a rational function can be factorised into n distinct linear factors, none of which are repeated, as we have seen one can write

$$\frac{P(x)}{Q(x)} = \frac{P(x)}{(x - a_1)(x - a_2) \cdots (x - a_n)} = \frac{A_1}{x - a_1} + \frac{A_2}{x - a_2} + \cdots + \frac{A_n}{x - a_n}.$$

$$(A.2)$$

To find A_1 cover up the $(x - a_1)$ factor appearing in the denominator of the left-hand side by pretending it is not there and evaluate the left-hand side at $x = a_1$. All other unknown constants are found in a similar way. For example, to find A_2 cover up the $(x - a_2)$ factor appearing in the denominator of the left-hand side and evaluate the left-hand side at $x = a_2$, and so on.

Example A.9 Using the Heaviside cover-up method, find the partial fraction decomposition for $\dfrac{1}{(x - 1)(x + 2)(x - 3)(x - 4)}$.

Solution First observe that the degree of the numerator is less than the denominator so the rational function is proper. As the denominator contains four distinct linear factors its partial fraction decomposition will be of the form

$$\frac{1}{(x - 1)(x + 2)(x - 3)(x + 4)} = \frac{A_1}{x - 1} + \frac{A_2}{x + 2} + \frac{A_3}{x - 3} + \frac{A_4}{x + 4}.$$

To find A_1, after the factor $(x - 1)$ on the left-hand side of the equation is covered up, evaluating at $x = 1$ gives

$$\frac{1}{(***)(x + 2)(x - 3)(x + 4)}\bigg|_{x=1} = A_1$$

$$\frac{1}{(***)(1 + 2)(1 - 3)(1 + 4)} = A_1$$

$$\Rightarrow A_1 = -\frac{1}{30}.$$

To find A_2, on the left-hand side of the equation we cover up the factor $(x + 2)$ before evaluating it at $x = -2$.

$$\frac{1}{(x - 1)(***)(x - 3)(x + 4)}\bigg|_{x=-2} = A_2 \quad \Rightarrow \quad A_2 = \frac{1}{30}.$$

To find A_3, on the left-hand side of the equation we cover up the factor $(x-3)$ before evaluating it at $x = 3$.

$$\frac{1}{(x-1)(x+2)(* * *)(x+4)}\bigg|_{x=3} = A_3 \Rightarrow A_3 = \frac{1}{70}.$$

Finally, to find A_4, on the left-hand side of the equation we cover up the factor $(x+4)$ before evaluating it at $x = -4$.

$$\frac{1}{(x-1)(x+2)(x-3)(* * *)}\bigg|_{x=-4} = A_4 \Rightarrow , A_4 = -\frac{1}{70}.$$

Thus

$$\frac{1}{(x-1)(x+2)(x-3)(x+4)} = -\frac{1}{30(x-1)} + \frac{1}{30(x+2)} + \frac{1}{70(x-3)}$$
$$- \frac{1}{70(x+4)}. \qquad \blacktriangleright$$

It is not too hard to see why the method works the way it does. Clearing the denominator of (A.2) we see that

$$P(x) = A_1(x-a_2)(x-a_3)\cdots(x-a_n) + A_2(x-a_1)(x-a_3)\cdots(x-a_n)$$
$$+ \cdots + A_n(x-a_1)(x-a_2)\cdots(x-a_{n-1}).$$

As this identity is true for all x on evaluating it at $x = a_1$ we see all terms on the right after the first is cancelled out, leaving

$$P(a_1) = A_1(a_1-a_2)(a_1-a_3)\cdots(a_1-a_n),$$

or

$$A_1 = \frac{P(a_1)}{(a_1-a_2)(a_1-a_3)\cdots(a_1-a_n)},$$

which is exactly the term one obtains from the cover-up process. Repeating the process by setting in turn $x = a_2, \ldots, a_n$, all other constants appearing in Equation (A.2) can be found.

§ Distinct Linear Factors – A Special Method

So far in our determination to find the unknown constants that appear in the partial fraction expression when a rational function is decomposed, purely algebraic methods have been used. We now consider one special case where methods from analysis, that is techniques from the calculus such as limits and differentiation, can be used to find the unknown constants.

We now show how if the denominator appearing in the expression for a rational function can be factored into n distinct linear factors, the n unknown coefficients in the partial fraction expansion can be determined by applying differentiation to the term appearing in the denominator.

Theorem A.2. *If $P(x)$ is a nonzero polynomial and $Q(x)$ is a polynomial that can be factored into n distinct linear factors,*

$$Q(x) = (x - \alpha_1)(x - \alpha_2) \cdots (x - \alpha_n),$$

such that $\deg P < n$, *the partial fraction decomposition for the rational function $P(x)/Q(x)$ is given by*

$$\boxed{\frac{P(x)}{Q(x)} = \sum_{k=1}^{n} \frac{P(\alpha_k)}{Q'(\alpha_k)} \frac{1}{x - \alpha_k}}$$

Proof We now give a simple proof of this result. Since $Q(x)$ consists of n distinct linear factors, using the product notation it can be written as

$$Q(x) = (x - \alpha_1)(x - \alpha_2) \cdots (x - \alpha_n) = \prod_{k=1}^{n}(x - \alpha_k),$$

and the partial fraction decomposition for the expression $P(x)/Q(x)$ can be written as

$$\frac{P(x)}{Q(x)} = \frac{A_1}{x - \alpha_1} + \frac{A_2}{x - \alpha_2} + \cdots + \frac{A_n}{x - \alpha_n} = \sum_{k=1}^{n} \frac{A_k}{x - \alpha_k}.$$

Taking the natural logarithm of $Q(x)$ and differentiating with respect to x, the logarithm turns products into sums

$$\frac{Q'(x)}{Q(x)} = \sum_{k=1}^{n} \frac{1}{x - \alpha_k}$$

$$\Rightarrow Q'(x) = \sum_{k=1}^{n} Q(x) \cdot \frac{1}{x - \alpha_k} = \sum_{k=1}^{n} \prod_{\substack{i=1 \\ i \neq k}}^{n} \frac{x - \alpha_i}{x - \alpha_k}.$$

Note the $i = k$ term in the product for $Q'(x)$ is not included as it cancels. Also, for $P(x)$ it can be rewritten as

$$P(x) = Q(x) \cdot \sum_{k=1}^{n} \frac{A_k}{x - \alpha_k} = \sum_{k=1}^{n} A_k \prod_{\substack{i=1 \\ i \neq k}}^{n} \frac{x - \alpha_i}{x - \alpha_k}.$$

Evaluating $P(x)$ and $Q'(x)$ at $x = \alpha_k$, except for the kth term, all terms in each sum cancel to give

$$Q'(\alpha_k) = \prod_{\substack{i=1 \\ i \neq k}}^{n} (\alpha_k - \alpha_i),$$

and

$$P(\alpha_k) = A_k \prod_{\substack{i=1 \\ i \neq k}}^{n} (\alpha_k - \alpha_i) = A_k Q'(\alpha_k).$$

Thus $A_k = P(\alpha_k)/Q'(\alpha_k)$ and the result follows. ∎

Example A.10 Use the formula given in Theorem A.2 to find the partial fraction decomposition for

$$\frac{6x^2}{(x+1)(x+2)(x+3)(x+4)}.$$

Solution In this example we have $P(x) = 6x^2$, $Q(x) = (x+1)(x+2)(x+3)(x+4)$ where $\alpha_1 = -1, \alpha_2 = -2, \alpha_3 = -3, \alpha_4 = -4$. So

$$P(-1) = 6(-1)^2 = 6$$
$$P(-2) = 6(-2)^2 = 24$$
$$P(-3) = 6(-3)^2 = 54$$
$$P(-4) = 6(-4)^2 = 96.$$

Finding $Q'(x)$ using the product rule we have

$$Q'(x) = (x+2)(x+3)(x+4) + (x+1)(x+3)(x+4)$$
$$+ (x+1)(x+2)(x+4) + (x+1)(x+2)(x+3).$$

Evaluating gives

$$Q'(-1) = (1)(2)(3) = 6$$
$$Q'(-2) = (-1)(1)(2) = -2$$
$$Q'(-3) = (-2)(-1)(1) = 2$$
$$Q'(-4) = (-3)(-2)(-1) = -6,$$

and the partial fraction decomposition becomes

$$\frac{6x^2}{(x+1)(x+2)(x+3)(x+4)} = \frac{P(-1)}{Q'(-1)(x+1)} + \frac{P(-2)}{Q'(-2)(x+2)}$$

$$+ \frac{P(-3)}{Q'(-3)(x+3)} + \frac{P(-4)}{Q'(-4)(x+4)}$$

$$= \frac{1}{x+1} - \frac{12}{x+2} + \frac{27}{x+3} - \frac{16}{x+4}. \quad \blacktriangleright$$

Remark: Finding a partial fraction decomposition using Theorem A.2 is usually quite long and slow compared to the Heaviside cover-up method and for this reason would not be used. Its real utility is as an important theoretical tool rather than as a practical algorithmic tool.

Exercises for Appendix A

✠ Warm-ups

1. Show that the following two rational function expressions

$$\frac{p(x)}{q(x)} = \frac{x-3}{2x^2+x+1} \quad \text{and} \quad \frac{P(x)}{Q(x)} = \frac{2x^2-5x-3}{4x^3+4x^2+3x+1}$$

are equal.

2. For the following proper rational functions write down the number of constants that would need to be found if each were decomposed into a sum of partial fractions.

(a) $\dfrac{x-1}{(x+2)(x-3)}$

(b) $\dfrac{x^2+1}{(x-1)^3(x+7)}$

(c) $\dfrac{x^2+1}{(x^2+9)^2(x-2)}$

(d) $\dfrac{1}{(x^2+3)(x^2+x+1)^2}$

✠ Practice questions

3. Write out the form of the partial fraction decomposition for the following rational functions. Do NOT attempt to find values for any of the unknown constants you give in the partial fraction sums.

(a) $\dfrac{2x}{(x+3)(3x+1)}$

(b) $\dfrac{1}{x^3+2x^2+x}$

(c) $\dfrac{2}{x^2 + 3x - 4}$ (d) $\dfrac{x^2}{(x-1)(x^2 + x + 1)}$

(e) $\dfrac{x^3}{x^4 - 1}$ (f) $\dfrac{x^4 + x^2 + 1}{(x^2 + 1)(x^2 + 4)^2}$

4. Resolve each of the following proper rational function expressions into partial fractions.

(a) $\dfrac{1}{(x-1)(x-2)}$ (b) $\dfrac{5x + 4}{x^2 + x - 2}$

(c) $\dfrac{7x - 1}{2x^2 - x - 1}$ (d) $\dfrac{25}{18x^3 - 9x^2 - 11x + 2}$

(e) $\dfrac{x^2 + 15x - 4}{(x+1)(x+2)(x-8)}$ (f) $\dfrac{48}{(x^2 - 1)(x^2 - 9)}$

(g) $\dfrac{1}{x^3 + 10x^2 + 25x}$ (h) $\dfrac{x}{(x+4)^3}$

(i) $\dfrac{11x + 18}{(x^2 - 4)(x + 3)}$ (j) $\dfrac{17x}{(2x+1)(x^2 + 4)}$

(k) $\dfrac{5}{(x+1)(x^2 + 4)^2}$ (l) $\dfrac{x^3 - 3x^2 + 4x - 6}{(x^2 + 1)(x^2 + 4)}$

(m) $\dfrac{1}{(x-1)(x^2 + x + 1)}$ (n) $\dfrac{2x - 1}{(x+1)(x^2 - x + 2)^2}$

(o) $\dfrac{x^5 - 4x^4 + 3x^2 - 2}{(x^2 - x + 2)^3}$ (p) $\dfrac{2}{(x^2 - x + 1)(x^2 + x + 1)}$

5. Resolve each of the following improper rational function expressions into partial fractions.

(a) $\dfrac{20x^3}{(2x+3)(x-1)}$ (b) $\dfrac{2x^3 - 27x^2 + 89x - 66}{x^2 - 12x + 27}$

(c) $\dfrac{x^3 - 2x}{x(x^2 + 4)}$ (d) $\dfrac{6x^4 - 4x^3 + 3x^2 - x - 2}{2x^2 + 1}$

6. If $f(x)$ is a polynomial of degree at most 3 and a, b, c, and d are distinct numbers, find expressions for the constants A, B, C, and D in terms of values for f and the numbers a, b, c, and d if

$$\frac{f(x)}{(x-a)(x-b)(x-c)(x-d)} = \frac{A}{x-a} + \frac{B}{x-b} + \frac{C}{x-c} + \frac{D}{x-d}.$$

7. In this question a partial fraction decomposition for $\dfrac{1}{x^4+1}$ will be found.

   (a) Find the four complex roots of the equation $x^4 + 1 = 0$.

   (b) Using (a), show that over the reals the term $x^4 + 1$ can be factored as

   $$x^4 + 1 = (x^2 + \sqrt{2}\,x + 1)(x^2 - \sqrt{2}\,x + 1).$$

   (c) Using the result in (b), show that the partial fraction decomposition for the rational function is

   $$\frac{1}{x^4+1} = \frac{x+\sqrt{2}}{2\sqrt{2}(x^2+\sqrt{2}\,x+1)} - \frac{x-\sqrt{2}}{2\sqrt{2}(x^2-\sqrt{2}\,x+1)}.$$

8. Consider the following rational function $f(x) = \frac{x^2+1}{x^3-x}$. By finding a partial fraction decomposition for f, use it to find $f^{(100)}(x)$, the 100th order derivative of f.

✠ **Extension questions and Challenge problems**

9. A simple and quick method that can be used to find a partial fraction decomposition when the denominator consists of a single repeated linear factor of the form $(ax + b)^n$ is to use the substitution $x = t/a - b$. The resulting expression for the rational function in terms of t is then readily resolved into partial fractions before back-substituting for x.

   As an example, consider the following rational function expression

   $$\frac{x^2 - 3x + 4}{(x-2)^3}.$$

   If we set $t = x - 2$, then $x = t + 2$, and on substituting into the rational function expression for x one has

   $$\frac{(t+2)^2 - 3(t+2) + 4}{t^3} = \frac{t^2 + t - 2}{t^3} = \frac{1}{t} + \frac{1}{t^2} - \frac{2}{t^3}.$$

   So it follows that the original partial function expression has

   $$\frac{x^2 - 3x + 4}{(x-2)^3} = \frac{1}{x-2} + \frac{1}{(x-2)^2} - \frac{2}{(x-2)^3},$$

   as its partial fraction decomposition.

By applying this method, find partial fraction decompositions for the following rational function expressions.

(a) $\dfrac{x+3}{(x-1)^2}$

(b) $\dfrac{x^2+7}{(x+1)^4}$

(c) $\dfrac{x^3+x^2-2x-3}{(x-3)^4}$

(d) $\dfrac{x^2-6x+5}{(x+2)^3}$

10. In addition to arithmetic and geometric series that have simple formulae for their sums, another type of series where simple formulae for their sums can be readily found are known as *telescoping* series. Here the series can be summed by first decomposing the summand appearing in the series into partial fractions.

For the following telescoping series, by finding a partial fraction decomposition for the summand first, find a simple expression for the sum of each series.

(a) $\displaystyle\sum_{k=1}^{n} \frac{1}{k^2+k}$

(b) $\displaystyle\sum_{k=1}^{n} \frac{1}{4k^2-1}$

(c) $\displaystyle\sum_{k=1}^{n} \frac{k^2-1}{k^4+k^2+1}$

A telescoping series is a special type of series where most of the terms in the series cancel out, leaving a very simple sum. If you have ever seen an old-fashioned telescope made of two or more tubes that slide inside one another, allowing it to be compactly stored, you will begin to understand where the series derives its name from. Like one of these old-fashioned telescopes, a telescoping series similarly 'collapses' down to a very simple sum.

11. Another special technique in finding a partial fraction decomposition applies to rational functions that contain a nonrepeated irreducible quadratic factor of the form (x^2+ax+b) in the denominator. It is due to Rear Admiral John P. Merrel (1846–1916), an officer who served in the United States Navy for over 40 years.[3] After clearing all denominators one replaces all x^2 terms by $-ax-b$ as often as is necessary and simplifies until at most only linear terms in x remain, allowing the unknown constants to be found by equating equal coefficients for x.

As an example of the method we find the partial fraction decomposition for

$$\frac{x^2 + 2}{(x + 1)(x^2 - x + 1)}.$$

As the denominator consists of a linear and an irreducible quadratic factor its partial fraction decomposition will be of the form

$$\frac{x^2 + 2}{(x + 1)(x^2 - x + 1)} = \frac{Ax + B}{x^2 - x + 1} + \frac{C}{x + 1}.$$

The constant C can be immediately found using the Heaviside cover-up method. Here $C = 1$. Next, clearing denominators one has

$$x^2 + 2 = A(x^2 + x) + B(x + 1) + C(x^2 - x + 1).$$

Replacing all x^2 that appear with $x - 1$ gives

$$(x - 1)^2 + 2 = A((x - 1) + x) + B(x + 1) + C((x - 1) - x + 1)$$

$$x + 1 = A(2x - 1) + B(x + 1)$$

$$x + 1 = (2A + B)x - A + B.$$

On equating equal coefficients for x we find $2A + B = 1$ and $-A + B = 1$, which when solved gives $A = 0$ and $B = 1$. Thus the required partial fraction decomposition is

$$\frac{x^2 + 2}{(x + 1)(x^2 - x + 1)} = \frac{1}{x^2 - x + 1} + \frac{1}{x + 1}.$$

By applying Merrel's method, find partial fraction decompositions for the following rational function expressions.

(a) $\dfrac{x^2 + 3}{(x + 1)^2(x^2 + x + 1)}$

(b) $\dfrac{x^3 - 8x^2 - 10x - 30}{(x^2 + x + 3)(x^2 + 2x + 5)}$

12. If $n \in \mathbb{N}$ show that the partial fraction decomposition for the rational function $\dfrac{1}{x^n(1 - x)}$ is given by

$$\frac{1}{x^n(1 - x)} = \sum_{k=1}^{n} \frac{1}{x^k} + \frac{1}{1 - x}.$$

13. Consider the partial function

$$f(x) = \frac{P(x)}{Q(x)},$$

where $P(x) = x^r$ and $Q(x) = (1 - x)(1 - 2x)(1 - 3x) \cdots (1 - rx)$. Here $r \in \mathbb{N}$.

(a) Since $\deg P = \deg Q$, f is improper. After performing polynomial long division it can be shown that

$$f(x) = \frac{(-1)^r}{r!} + \frac{x^r - \frac{1}{r!}(1 - x)(1 - 2x) \cdots (1 - rx)}{(1 - x)(1 - 2x) \cdots (1 - rx)}.$$

Prove this result by using induction on r.

(b) Using (a), show the partial fraction decomposition for f is given by

$$f(x) = \frac{1}{r!} \sum_{k=0}^{r} \binom{r}{k} \frac{(-1)^{r-k}}{1 - kx}.$$

Hint: Find an expression for the kth coefficient A_k in the partial fraction sum.

14. Let $f(x) = a_0 x^m + a_1 x^{m-1} + \cdots + a_m$ and $g(x) = b_0 x^n + b_1 x^{n-1} + \cdots + b_n$ where $\deg f \leqslant \deg g - 1$ and $a_0, b_0 \neq 0$. If $g(x)$ has n distinct roots $\alpha_1, \alpha_2, \ldots, \alpha_n$ such that the decomposition of $f(x)/g(x)$ into partial fractions is given by

$$\frac{f(x)}{g(x)} = \frac{A_1}{x - \alpha_1} + \frac{A_2}{x - \alpha_2} + \cdots + \frac{A_n}{x - \alpha_n},$$

show that

$$A_1 + A_2 + \cdots + A_n = \begin{cases} 0 & \text{if } \deg f < \deg g - 1 \\ \dfrac{a_0}{b_0} & \text{if } \deg f = \deg g - 1. \end{cases}$$

15. If $n \in \mathbb{N}$ and $a > 0$ prove

$$\frac{1}{x^{2n} + a^{2n}} = \frac{1}{n a^{2n-1}} \sum_{k=1}^{n} \frac{a - x \cos\left(\frac{(2k-1)\pi}{2n}\right)}{x^2 - 2ax \cos\left(\frac{(2k-1)\pi}{2n}\right) + a^2}.$$

Hint: Factor $x^{2n} + a^{2n}$ into $2n$ distinct linear factors over $\mathbb{C}$ before applying the result of Theorem A.2.

16. In this question we will show the logarithmic function $\ln : (0, \infty) \to \mathbb{R}$ is not a rational function. That is, there do not exist polynomials $p(x)$, $q(x)$ for $x > 0$ that do not contain any factors in common such that $q(x) \neq 0$ for all $x > 0$ and $\ln x = p(x)/q(x)$ for all $x > 0$.

(a) Suppose to the contrary that there are polynomials $p(x)$, $q(x)$ that are relatively prime (do not contain any nonconstant common factors) such that $q(x) \neq 0$ for all $x > 0$ and $\ln x = \frac{p(x)}{q(x)}$ for all $x > 0$. From $\ln x = p(x)/q(x)$, by taking the derivative of both sides of the equation, show that

$$q(x)^2 = x \left[p'(x)q(x) - p(x)q'(x) \right].$$

Note that both sides of this equation are polynomials as it must be true for all $x > 0$ is an identity of polynomials.

(b) From (a) observe that the polynomial x divides the polynomial $q(x)$. Now let $q(x) = x^k q_1(x)$ where $k \in \mathbb{N}$ and $q_1(x)$ is a polynomial in x that is not divisible by x, that is, $q_1(0) \neq 0$. Show that the polynomial equation in (a), in terms of $q_1(x)$, can be written as

$$k p(x) q_1(x) = x \left[p'(x)q_1(x) - p(x)q_1'(x) - x^{k-1}q_1(x)^2 \right].$$

(c) The polynomial identity given in (b) implies that the polynomial x divides the polynomial $p(x)$. This is a contradiction since $p(x)$ and $q(x)$ were assumed to have no nonconstant factors in common (they were assumed to be relatively prime), from which we conclude the logarithmic function is not a rational function.

Endnotes

1. The term 'rational function' is somewhat misleading. As $P(x)$ and $Q(x)$ are polynomials, the formal symbol x is an indeterminate so the fraction $P(x)/Q(x)$ is not a function at all, but is instead a formal expression in the same sense as polynomials are. The term, however, is an old one and is the conventional term used in connection with expressions of this type.
2. The method is named after the self-taught English electrical engineer, mathematician, and physicist Oliver Heaviside (1850–1925) who popularised the method.
3. L. S. Johnston, 'A note on partial fractions,' *The American Mathematical Monthly*, **43**(7), 413–414 (1936).

Appendix B

Answers to Selected Exercises

Chapter 1

1. In general the two are not equal. The first is equal to $f(x)$ while the second is equal to $f(x) - f(a)$.

2. (c) $\frac{1}{3}$

3. (c) $\frac{1}{4}$

4. (a) $\frac{9}{2}$ (b) 4 (c) 2 (d) π (e) $\frac{9}{2}\pi$ (f) 4

5. (a) $x \sin x$ (b) e^{x^2} (c) $\cos(x^2)$ (d) $x^2 - \cos x$

6. 16

8. $\sqrt{2}$

9. (a) $\dfrac{2}{\sqrt{\pi}} e^{-x^2}$

Chapter 2

1. (a) 35 (b) 4 (c) 16

2. (a) positive (b) positive (c) positive (d) negative
   (e) negative (f) negative (g) positive (h) negative

3. (a) 6 (b) 3 (c) 4

4. (a) $\frac{10}{3}$ (b) -15 (c) $\frac{143}{60}$

Chapter 3

1. (a) $x + C$ (b) $\frac{1}{a}\sin(ax) + C$

2. (a) $\frac{1}{4}x^4 + C$ (b) $\frac{2}{5}x^{5/2} + C$ (c) $-\frac{1}{12}x^{-4} + C$
   (d) $x^3 - x^2 + x + C$ (e) $\frac{2}{9}x^{9/2} + \frac{6}{5}x^{5/2} + 2\sqrt{x} + C$

333

(f) $\frac{4}{3}x^3 - 2x^2 + x + C$ (g) $\frac{1}{3}x^3 - 3x^2 + C$

(h) $\frac{3}{7}x^{7/3} + \frac{3}{4}x^{4/3} + C$ (i) $\frac{1}{3}x^3 - \frac{9}{2}x^2 + 18x + C$

(j) $\frac{1}{4}x^4 + x^{-1} + C$ (k) $\frac{1}{4}x^4 + 3x^3 + \frac{27}{2}x^2 + 27x + C$

(l) $\frac{2}{3}x\sqrt{x} + 2\sqrt{x} + C$

3. (a) $-\frac{50}{21}$ (b) $\frac{16}{15}$ (c) 0 (d) $\frac{44}{3}$ (e) $\frac{2}{35}$ (f) $\frac{33}{5}$

4. (a) $\frac{1}{2}\ln x + C$ (b) $x^3 + \ln x + C$

  (c) $e^{4x} + \cos 2x + C$ (d) $\frac{1}{\sqrt{3}}\tan^{-1}\left(\frac{x}{\sqrt{3}}\right) + C$

  (e) $-\frac{1}{3}\cos(3x + 2) + C$ (f) $\frac{1}{2}\tan^{-1}(2x) + C$

  (g) $2\tan\frac{x}{2} + 2\sin\frac{x}{2} + C$ (h) $\frac{1}{2}\sin^{-1} 2x + C$

  (i) $3\sec x - \frac{1}{3}\cos 3x + C$ (j) $\frac{1}{3}\tan^{-1}\frac{x}{3} + \frac{1}{3}\sin^{-1}\frac{x}{3} + C$

5. (a) $\frac{101}{30}$ (b) 0 (c) $\frac{2}{3}$

  (d) $\frac{1}{2}\cos(5) - \frac{1}{2}\sin(5)$ (e) $\frac{\pi}{8}$ (f) $\frac{5}{8}\pi$

  (g) $\frac{\pi}{6}$ (h) $\frac{\pi}{9}$

7. (b) $\frac{7}{64}$

8. (b) $\sqrt{2}\tan^{-1}\left(\dfrac{\sqrt{\tan x} - \sqrt{\cot x}}{\sqrt{2}}\right)$

9. (a) $x^6 - 6x^5 + 15x^4 - 20x^3 + 15x^2 - 6x + 1$ (b) $\frac{1}{7}$

10. (a) $\sec x + C$ (b) $-\operatorname{cosec} x + C$

11. $\frac{1}{4}x^4 - \frac{1}{4}$

12. (b) Yes

13. $\dfrac{12\,839}{73\,728}$

17. $f(x) = 2x^{11}$ and $k = -\frac{1}{7}$

Chapter 4

1. (a) -3 (b) 176

2. (a) $F(6) = 0, F(12) = -6$ (b) $x = 0, 6$

  (c) $x = 3$ is a local maximum while $x = 10$ is a local minimum.

3. (a) 1 (b) 3 (c) $\frac{20}{3}$

5. (a) $f''(1) = 3$ and $f'''(1) = 8$

7. (a) False   (b) True   (c) True   (d) False

9. 8

13. $\varphi(x) = x + \frac{1}{4}$

14. (a) $\frac{20}{3}$ (b) 4

15. $\frac{2}{3}$

17. 13

21. $1 + e^\pi + e^{2\pi} + e^{3\pi}$

25. There are no such functions.

26. (b) Equality holds when $f(x) = \lambda g(x), g(x) \neq 0$.

Chapter 5

1. $\frac{1}{2}x^2 + C$

2. (a) $x - 5\ln|x + 5| + C$ (b) $2x - 4\tan^{-1}\left(\frac{x}{2}\right) + C$

   (c) $\frac{1}{6}x^6 + \frac{1}{5}x^5 + \frac{1}{4}x^4 + \frac{1}{3}x^3 + \frac{1}{2}x^2 + x + C$

   (d) $\frac{8}{9}x\sqrt{2\sqrt{3\sqrt{x}}} + C$ (e) $-\ln|\cos x + \sin x| + C$

   (f) $\ln|x + 3| + C$ (g) $\frac{1}{2}\pi x + C$

   (h) $\frac{1}{2}x^2 - 3x + C$ (i) $x\cos 2 - \sin 2\ln|\sin x| + C$

   (j) $-\frac{1}{2}\cos x - \frac{1}{10}\cos 5x + C$

   (k) $\frac{1}{2^9}\left(\frac{5}{2}x - \frac{15}{16}\sin 4x + \frac{3}{16}\sin 8x - \frac{1}{48}\sin 12x\right) + C$

   (l) $-\frac{3}{4}\cos x + \frac{1}{12}\cos 3x + C$

   (m) $\ln|\cos(x + 5)| + \ln|\cos(x - 5)| - \frac{1}{2}\ln|\cos 2x| + C$

   (n) $\frac{1}{\ln 2 + 1}x2^{\ln x} + C$ (o) $x - \frac{1}{\sqrt{2}}\tan^{-1}\left(\frac{x}{\sqrt{2}}\right) + C$

   (p) $-\cot x + \tan x - 4x + C$ (q) $\frac{8}{17}\sqrt[8]{x}(x^2 + 17) + C$

   (r) $-\ln|\cos(x - 2)| + C$ (s) $\frac{1}{3}x^3 - \frac{1}{2}x^2 + x + C$

   (t) $-\frac{1}{3}\ln|\cos 3x| + \frac{1}{2}\ln|\sin 2x| + \ln|\sin x| + C$

3. (a) $\dfrac{1}{1+\pi}$ (b) $\dfrac{1}{\ln \pi}(\pi - 1)$ (c) $-\ln\left(\cos\frac{\pi}{8} - \sin\frac{\pi}{8}\right)$

   (d) $\dfrac{\sqrt{2}}{3}$ (e) $\frac{16}{3} - 8\ln(2)$ (f) $\frac{173}{192}$

   (g) $2 - \frac{5\pi}{8}$ (h) $\frac{1}{3} - \frac{2\sqrt{2}}{3} + \sqrt{3}$

4. (a) $-\cot x + C$ (b) $\frac{1}{2}x - \frac{1}{4}\sin 2x + C$

5. (a) $\frac{1}{3}\pi - \tan^{-1}(\sqrt{2})$ (b) $\frac{\pi}{6}$

6. (a) $\dfrac{1}{\ln a + 1}a^x e^x + C$ (b) $\dfrac{1}{\ln a + 1}x^{\ln a + 1} + C$

7. (b) $\tan\left(x + \frac{\pi}{4}\right) + C$

8. (b) $\tan\left(\frac{x}{2} - \frac{\pi}{4}\right) + C$

9. (b) $I = \dfrac{3\pi}{20} - \dfrac{\ln 3}{10}, J = \dfrac{3}{10}\ln 3 + \dfrac{\pi}{20}$

11. $2x\tan(6x^2) - \dfrac{1}{2\sqrt{x}}\tan(6\sqrt{x})$

12. $x = 0$ is a local maximum.

14. (a) $\dfrac{1}{2}\ln\left|\dfrac{1+x}{1-x}\right| + \displaystyle\sum_{k=1}^{n}\dfrac{x^{2k-1}}{2k-1} + C$

    (b) $-\tan^{-1}x + \displaystyle\sum_{k=1}^{n}\dfrac{(-1)^{k+1}x^{2k-1}}{2k-1} + C$

15. $e^2 - \frac{7}{8}$

16. $f(x) = 1 + \frac{5x}{3} - \frac{x^2}{2} + \frac{x^3}{3}$ and $g(x) = x(x-1) + \frac{5}{3}$

17. 0

19. (c) $n!x - n!\ln|e^x + \cos x + \sin x + P_n(x)| + C$

20. (b) $\ln\left|\dfrac{e^x\sin x + \cos x}{e^x\cos x - \sin x}\right|$

21. $-2015\ln|x| + \displaystyle\sum_{k=1}^{n}\dfrac{x^k}{k} + C$

22. (a) $\frac{2}{5}(x^6 + x^4 + x^{-1/4})^{\frac{5}{4}} + C$ (b) $2\sin x + x$

23. $\sqrt{3}$

24. $\frac{11}{4}$

Chapter 6

1. (a) $\frac{1}{2}\sin(2x) + C$ (b) $\frac{1}{2}\tan^2 x + C$ (c) $\frac{1}{4}(1 + x^2)^4 + C$

   (d) $e^{\sin x} + C$ (e) $-\frac{1}{6}\cos^6 x + C$ (f) $-2\cos\sqrt{x} + C$

2. (a) $\frac{2}{15}(5x + 7)^{\frac{3}{2}}$ (b) $-\frac{1}{3}e^{x^3} + C$ (c) $\frac{1}{2}\ln^2 x + C$

   (d) $-\frac{1}{2}(\tan^{-1} x)^2 + C$ (e) $\sqrt{1 + x^4} + C$ (f) $-\frac{1}{\ln x} + C$

   (g) $2e^{\sqrt{x}} + C$ (h) $-\frac{2}{5}\sqrt{1 - x^5} + C$ (i) $\frac{1}{3}\sec^3 x + C$

   (j) $\ln(2 - \cos x) + C$ (k) $\tan^{-1}(e^x) - \frac{1}{2}\ln(e^{2x} + 1) + C$

   (l) $-2\sqrt{x} + \ln\left|\frac{1+\sqrt{x}}{1-\sqrt{x}}\right| + C$

   (m) $-\frac{1}{15}\sqrt{25 - x^2}(3x^4 + 100x^2 + 5000) + C$

   (n) $2\sqrt{x} + C$ (o) $\frac{2}{27}x^{12}\left(\frac{x^3-3}{x^{11}}\right)^{\frac{3}{2}} + C$

   (p) $e^{\tan x} + C$ (q) $-\frac{1}{2}e^{\cos x^2} + C$

   (r) $\frac{1}{8\ln 3}\ln\left|\frac{3^x-8}{3^x}\right| + C$ (s) $\frac{3}{2}(\sin x - \cos x)^{\frac{2}{3}} + C$

   (t) $\frac{x[\ln 2]^{\ln x}}{1 + \ln(\ln 2)} + C$ (u) $\frac{7}{8}\sin 2x - \frac{1}{8}\sin 6x - \frac{7}{16}\cos 4x + \frac{1}{64}\cos 8x + C$

3. (a) 72 (b) $\sqrt{2} - 1$ (c) $\sqrt{3} - \frac{2\sqrt{2}}{3}$

   (d) $e^{e^e} - e$ (e) $2 - e - \ln 3 + \ln(1 + e)$

   (f) $\frac{\pi}{16}$ (g) $e^2 - 1$ (h) $\dfrac{2}{p(p + 1)}$

4. (a) 2 (b) 5

5. $k = \dfrac{\pi}{2} \pm 2n\pi$ where $n = 0, 1, 2, \ldots$

6. (a) 10 (b) 0

7. $\ln x + (\ln a - \ln b)\ln(\ln(bx)) + C$

8. (c) $\dfrac{1}{10\,302}$

9. $f_{\max}\left(\dfrac{\pi}{2}\right) = \dfrac{\pi}{8}$, $f_{\min}(0) = 0$

10. (a) 112 (b) 3 (c) $\frac{9}{2}$

11. (a) $I = \ln\left(\dfrac{\sin b}{\sin a}\right)$ (b) $J = \ln\left(\dfrac{\cos a}{\cos b}\right)$

12. $\sin^{-1} x - \dfrac{1}{\sqrt{1-x^2}} + C$

14. $a = e^e$

15. (a) $\ln(\ln x + x) + C$ (b) $-e^{\cot x} + C$

   (c) $\frac{1}{91}\sin^{13}(7x) - \frac{1}{105}\sin^{15}(7x) + C$

   (d) $\frac{2}{3}(1 + \sec x)^{\frac{3}{2}} + C$

17. 0

18. (a) $-\dfrac{2}{\sqrt{x + \ln x}} + C$

19. $\sin^{-1} x + \sqrt{1-x^2} + C$

20. $\frac{1}{2}\ln\left|(x-y)^2 - 1\right| + C$

21. (a) $\tan^{-1}\left(\dfrac{1-x}{\sqrt{1-x^2}}\right) - \dfrac{\pi}{4}$ (b) $-\frac{1}{2}\sin^{-1} x$

22. (b) $\tan^{-1}\left(x - \dfrac{1}{x}\right) + C$

23. $\displaystyle\int_0^1 x(x+1)^m \, dx = \begin{cases} \dfrac{m2^{m+1} + 1}{(m+1)(m+2)}, & m \neq -1, -2 \\ 1 - \ln 2, & m = -1 \\ \ln 2 - \frac{1}{2}, & m = -2 \end{cases}$

24. $\dfrac{1}{2}\tan^{-1}\left(\dfrac{e^x + 2e^{-x}}{2}\right) + C$

25. (b) $\dfrac{1}{3}\tan^{-1}\left(\dfrac{x^3 + 1}{3x}\right) + C$

26. (a) $\sqrt{x(x+2)} + 2\ln\left|\sqrt{x+2} + \sqrt{x}\right| + C$, and
      $\sqrt{x(x+2)} - 2\ln\left|\sqrt{x+2} + \sqrt{x}\right| + C$
   (b) $2\ln\left|\sqrt{x+2} + \sqrt{x}\right|$

27. $-\dfrac{1}{(1 + \sqrt{\tan x})^2} + \dfrac{2}{3(1 + \sqrt{\tan x})^3} + C$

28. (a) and (b) $\ln 2 - \frac{5}{6}$ (c) Yes

29. (a) $\frac{1}{2}\pi \ln \pi$ (b) $\frac{3}{4}\pi$

30. (a) 0

32. (a)(i) $(\sin^{-1} x)^2 + C_1$ (ii) $-(\cos^{-1} x)^2 + C_2$

36. (b) $\tan^{-1}(x^2 + x + 1) + C$

37. (b) $H_1 = 1$, $H_2 = \frac{3}{2}$, $H_3 = \frac{11}{6}$, $H_4 = \frac{25}{12}$, $H_5 = \frac{137}{60}$

39. (c) $I = \dfrac{1}{4\sqrt{2}} \ln \left| \dfrac{\sqrt{1 + x^4} + x\sqrt{2}}{\sqrt{1 + x^4} - x\sqrt{2}} \right| - \dfrac{1}{2\sqrt{2}} \tan^{-1} \left(\dfrac{\sqrt{1 + x^4}}{x\sqrt{2}} \right) + C$

$J = \dfrac{1}{8\sqrt{2}} \ln \left| \dfrac{\sqrt{1 + x^4} + x\sqrt{2}}{\sqrt{1 + x^4} - x\sqrt{2}} \right| + \dfrac{1}{4\sqrt{2}} \tan^{-1} \left(\dfrac{\sqrt{1 + x^4}}{x\sqrt{2}} \right) + C$

41. $6\sqrt{3}$

42. $\dfrac{\left(2x^{3n} + 3x^{2n} + 6x^n \right)^{\frac{n+1}{n}}}{6(n + 1)} + C$

44. $-2\tan^{-1} \left(\sqrt{\dfrac{1 - x}{x + 2}} \right) + C$

45. (e) $\dfrac{1}{n + 1}$

Chapter 7

1. (a) $\frac{1}{4}x^4 + C$ (b) $\frac{1}{6}x^5 + \frac{1}{2}C_1 x^2 + \frac{1}{3}C_2 x^3 + C_3$

   (c) They are not equal, which is to be expected since there is no such thing as a product rule for integrals.

2. The arbitrary constant of integration needs to be included.

3. (a) $\sin x - x \cos x + C$ (b) $\frac{1}{3}x^3 \ln x - \frac{1}{9}x^3 + C$

   (c) $x \sec x - \ln |\sec x + \tan x| + C$

   (d) $x \tan x + \ln |\cos x| + C$

   (e) $x - x \ln x + C$ (f) $2x(\ln x - 1) + C$

   (g) $x \tan^{-1} x - \frac{1}{2} \ln(1 + x^2) + C$

   (h) $x \ln^2 x - 2x \ln x + 2x + C$

   (i) $\frac{4}{7} \sin 4x \sin 3x + \frac{3}{7} \cos 4x \cos 3x + C$

   (j) $\frac{1}{2}x(\sin(\ln x) + \cos(\ln x)) + C$

(k) $\frac{1}{13}e^{-3x}(2\sin 2x - 3\cos 2x) + C$

(l) $e^x(x\ln x - 1) + C$ (m) $\ln(\ln x + 1) + C$

(n) $x\ln(1 + x^2) - 2x + 2\tan^{-1} x + C$

(o) $\frac{1}{2}x^2\ln(x^2 e^{x^2}) - \frac{1}{2}x^2 - \frac{1}{4}x^4 + C$

(p) $\frac{1}{2}e^{x^2}(x^2 - 1) + C$

4. (a) $\frac{1}{4}$ (b) $\frac{1}{2}(1 - \ln 2)$ (c) $\frac{\pi}{8}$

   (d) $\frac{2}{9}(1 + 2e^3)$ (e) $\frac{1}{2}(1 + e^{\pi/2})$ (f) $\ln 2 - 2 + \frac{\pi}{2}$

5. (a) $\ln 4 - \frac{7}{4}$ (b) 5

6. $x\ln(x^2 - 1) - 2x + \ln\left|\dfrac{1 + x}{1 - x}\right| + C$

9. $I_0 = \dfrac{\pi}{2}, I_1 = 1, I_2 = \dfrac{\pi^3 + 6\pi}{48}$

10. $2e^{\sqrt{x}}(\sqrt{x} - 1) + x + C$

11. 25

12. $\dfrac{\pi^2 + 4\pi - 8}{8(\pi + 2)}$

13. (b) $\dfrac{\pi x}{4} + \dfrac{1}{2}\left(\sqrt{1 - x^2} - x\cos^{-1} x\right) + C$

15. 4

16. (a) $F_1(x) = -(x + 1)e^{-x}, F_2(x) = -(x^2 + 2x + 2)e^{-x}$
    (b) $66k$, where k is a constant.

17. $5\pi + 2$

19. (b) $A = 1, B = -1$

20. (c) $\ln\sqrt{2} - \frac{1}{2} + \frac{\pi}{4}$

21. (b) $-\frac{\pi}{2}$

23. $a = e^{\frac{1}{n+1}}$

26. $\dfrac{1}{2}\left(\dfrac{1}{2} + \dfrac{1}{\pi + 2} - \gamma\right)$

27. (a) $3(x^2 - 2)\sin x - x(x^2 - 6)\cos x + C$

(b) $-\dfrac{1}{4}\dfrac{x^3}{(1+x)^4} - \dfrac{3}{12}\dfrac{x^2}{(1+x)^3} - \dfrac{6}{24}\dfrac{x}{(1+x)^2} - \dfrac{6}{24}\dfrac{1}{1+x} + C$

(c) $(5x^2 + 15)(3x + 2)^{4/5} - \frac{50}{27}x(3x + 2)^{9/5} + \frac{125}{567}(3x + 2)^{14/5} + C$

(d) $x\tan x + \ln|\cos x| + C$

(e) $\frac{1}{27}e^{3x}(9x^3 + 45x^2 + 69x + 31) + C$

28. (a) $(x - 1)\ln(1 + \sqrt{x}) - \frac{1}{2}x + \sqrt{x} + C$

(b) $x\ln|x + \sqrt{x}| - x + \sqrt{x} - \ln|\sqrt{x} + 1| + C$

(c) $x\ln|x + \sqrt{x^2 + 1}| - \sqrt{x^2 + 1} + C$

29. $\alpha = \dfrac{1}{2}$

30. (a) $x\tan x + \ln|\cos x| - \frac{1}{2}x^2 + C$ (b) $-x\cot x + \ln|\sin x| - \frac{1}{2}x^2 + C$

31. (a) $2\sqrt{x}\sin(\sqrt{x}) + 2\cos(\sqrt{x}) + C$ (b) $e^x\sin(e^x) + \cos(e^x) + C$

(c) $\frac{x^2}{5}(2\sin(\ln x) - \cos(\ln x)) + C$

32. $\frac{1}{4}\pi^2$

34. (a) $-\dfrac{1}{x + 1} + C$ (b) $\dfrac{e^x}{x + 1} + C$

35. $2\cos\sqrt{x}\left(-x^{5/2} + 20x^{3/2} - 120\right) + \sin\sqrt{x}\left(5x^2 - 60x + 120\sqrt{x}\right) + C$

36. $x = 2n\pi, n \in \mathbb{Z}$

37. $-\frac{1}{11}(f(0) + f(1))$

39. (a) $\sqrt{x^2 + 4} + C$ (b) $\frac{1}{3}(x^2 - 8)\sqrt{x^2 + 4} + C$

41. (a) $\frac{1}{2}x^2\ln\left|\frac{x+1}{x}\right| + \frac{1}{2}x - \frac{1}{2}\ln|1 + x| + C$ (b) $-x\cot x + C$

43. $-\frac{7}{8}k + \frac{\sqrt{2}}{4} - \frac{1}{162}$

45. $\dfrac{5051}{5050}$

47. (a) $-\frac{1}{4}u\cos^4 u + \frac{1}{16}\sin 2u + \frac{1}{128}\sin 4u + \frac{3}{32}u + C$ where $u = \tan^{-1} x$

(b) $-\frac{1}{4}v^2 + v + xuv - 2xu + u^2 + C$

where $u = \tan^{-1} x$ and $v = \ln(1 + x^2)$.

49. (a) $e - 1$ (b) $\frac{1}{2} - \frac{1}{8\sqrt{2}} \ln^2 2 - \frac{1}{4\sqrt{2}} \ln 2 - \frac{1}{2\sqrt{2}}$

51. $\frac{1}{2}$

54. (b)(ii) $\frac{1}{3}x^3 + x^2 + 4x + 8\ln(x - 2) + C$

Chapter 8

1. (a) $\tan^2 x + C$ (b) $-\cot x + C$ (c) $\sec x + C$ (d) $-\operatorname{cosec} x + C$

2. The second as it can be readily solved using a substitution of $u = \cos x$.

3. (a) $c_1 = \sin x + C$

   $$c_2 = \frac{1}{2}x + \frac{1}{4}\sin 2x + C$$

   $$c_3 = \sin x - \frac{1}{3}\sin^3 x + C$$

   $$c_4 = \frac{3x}{8} + \frac{1}{4}\sin 2x + \frac{1}{32}\sin 4x + C$$

   $$c_5 = \sin x - \frac{2}{3}\sin^2 x + \frac{1}{5}\sin^4 x + C$$

   (b) $t_1 = \ln|\sin x| + C$

   $$t_2 = -\cot x - x + C$$

   $$t_3 = -\ln|\sin x| - \frac{1}{2}\cot^2 x + C$$

   $$t_4 = -\frac{1}{3}\cot^3 x + \cot x + x + C$$

   $$t_5 = -\frac{1}{4}\cot^4 x + \frac{1}{2}\cot^2 x + \ln|\sin x| + C$$

6. (a) $\frac{1}{6}$ (b) $\frac{\pi}{32}$ (c) $\frac{5}{36} - \frac{1}{16}\ln 3$ (d) $\frac{14\sqrt{3}}{5}$

8. (a) $-\frac{1}{4}\cot^4 x - \frac{1}{6}\cot^6 x + C$ (b) $\frac{5}{12}$

9. (a) $-\ln|\operatorname{cosec} x + \cot x| + \cos x + C$
   (b) $\cos x + \sec x + C$
   (c) $-\frac{3}{2}x - \frac{1}{4}\sin 2x + \cot x + C$
   (d) $\frac{5}{3}\tan x - \cot x + \frac{1}{3}\tan x \sec^2 x + C$
   (e) $\tan x - \cot x + C$
   (f) $\sec x - \ln|\operatorname{cosec} x + \cot x|$

10. (a) $-\frac{1}{12}\cos 6x - \frac{1}{24}\cos 12x + C$

(b) $\frac{1}{4}\sin 2x - \frac{1}{24}\sin 12x + C$

(c) $\frac{1}{8}\sin 4x + \frac{1}{24}\sin 12x + C$

(d) $-\frac{1}{2}\cos x - \frac{1}{6}\cos 3x + C$

11. (c) $\frac{128}{315}$

12. (d) $\frac{1}{4^n}\sin(4^n x) + C$

Chapter 9

1. $\frac{1}{3}x^3 + C$

2. The arbitrary constant of integration needs to be included.

3. (b) $-\frac{1}{2}e^{-2x} + C$ and $\frac{1}{2}e^{2x} + C$ (c) $\frac{1}{2}e^{-2x} + C$

4. (a) $-\frac{1}{2}\coth(2x + 5) + C$ (b) $x\cosh x - \sinh x + C$

(c) $2\sinh\sqrt{1-x} + C$ (d) $\frac{1}{\ln 2}2^{\tanh x} + C$

(e) $\frac{1}{4}\sinh 2x - \frac{1}{2}x + C$ (f) $\frac{1}{2}\cos x\cosh x + \frac{1}{2}\sin x\sinh x + C$

(g) $-x\coth x + \ln|\sinh x| + C$ (h) $\frac{1}{4}\ln|\cosh 2x| + C$

(i) $\frac{3}{5}\sinh 2x\cosh 3x - \frac{2}{5}\cosh 2x\sinh 3x + C$

(j) $\frac{1}{2}\tan^{-1}(\sinh 2x) + C$

(k) $\frac{1}{2}\tan^{-1}\left(\dfrac{\tanh x}{2}\right) + C$ (l) $2\ln|\sinh\frac{x}{2}| + C$

5. (a) $\frac{4}{15}$ (b) $\ln\left(\frac{5}{4}\right)$ (c) $\cosh(1) - 1$

(d) $\frac{1}{4}e^2 - \frac{3}{4}$ (e) $\sinh(1)$ (f) $(2 - e^2 + e^4)/4$

6. (b) $x - 2\tan^{-1}x + C$

9. (a) $\cosh 2a\ln(\cosh(x + a)) - x\sinh 2a + C$

(b) $x\cosh 2a + \sinh 2a\ln|\sinh(x - a)| + C$

10. $\frac{1}{4}(2x^4 + 6x^2 + 3)\sinh 2x - 2x(2x^2 + 3)\cosh 2x + C$

11. (a) $\ln|\tanh x| + \frac{1}{2}\text{sech}^2 x + C$

(b) $\tanh x - \frac{1}{3}\tanh^3 x + C$

(c) $\frac{1}{4}\sinh 2x + \frac{1}{32}\sinh 4x + \frac{3}{8}x + C$

12. $\frac{1}{12}\cosh 3x - \frac{3}{4}\cosh x + C$

13. (a) $\frac{1}{7}\cosh^7 x - \frac{1}{5}\cosh^5 x + C$

(b) $\frac{1}{8}\cosh^8 x - \frac{1}{6}\cosh^6 x + C$

(c) $\frac{1}{64}\sinh 4x + \frac{1}{48}\sinh^3 2x - \frac{1}{16}x + C$

(d) $\frac{1}{2}\tan^{-1}(\sinh x) - \frac{1}{2}\operatorname{sech} x \tanh x + C$

14. (a) $\frac{1}{2n}e^{2nx} + C$ (b) $\frac{(-1)^n}{2}e^{2x} + C$

15. (a) $\frac{16}{3}\sinh^3\left(\frac{x}{2}\right) + 8\sinh\left(\frac{x}{2}\right) + C$ (b) $-2\ln\left|\sinh(\frac{x}{2})\right| + C$

17. $\sinh x + \frac{10}{3}\sinh^3 x + \frac{24}{5}\sinh^5 x + \frac{16}{7}\sinh^7 x + C$

18. (a) $\operatorname{sech} x$ (e) $-\frac{\pi}{2}$ and $\frac{\pi}{2}$

19. (c) $x + \sum_{k=1}^{n}\frac{1}{k}\sinh 2kx + C$ (d) $x + \sinh 2x + \frac{1}{2}\sinh 4x + \frac{1}{3}\sinh 6x + C$

Chapter 10

1. $\frac{\pi}{4}$

2. (a) $\frac{1}{2}\sqrt{25 + x^2} + \frac{25}{2}\ln|x + \sqrt{25 + x^2}| + C$

(b) $\frac{x}{2}\sqrt{x^2 - 3} - \frac{3}{2}\ln|\sqrt{x^2 - 3} + x| + C$

(c) $\dfrac{x}{\sqrt{1 - x^2}} + C$ (d) $\frac{9}{2}\sin^{-1}\left(\frac{x}{3}\right) - \frac{x}{2}\sqrt{9 - x^2} + C$

(e) $\ln|x + \sqrt{1 + x^2}| - \dfrac{\sqrt{1 + x^2}}{x} + C$ (f) $-\tan^{-1}\left(\dfrac{1}{\sqrt{x^2 - 1}}\right) + C$

(g) $\sqrt{4 - x^2} - 2\ln|\sqrt{4 - x^2} + 2| + 2\ln x + C$

(h) $\sqrt{x^2 - 1} + C$ (i) $\dfrac{\sqrt{4 - x^2}}{4x} + C$

(j) $\frac{1}{2}\tan^{-1}x - \dfrac{x}{2(x^2 + 1)} + C$ (k) $8\sin^{-1}\left(\frac{x}{8}\right) - \sqrt{64 - x^2} + C$

(l) $\ln|x + \sqrt{x^2 - 1}| - \cos^{-1}\left(\frac{1}{x}\right) + C$ (m) $\dfrac{x^2 + 2}{\sqrt{x^2 + 1}} + C$

(n) $\sqrt{x^2 - 1} + \frac{1}{3}(x^2 - 1)^{\frac{3}{2}} + C$ (o) $\frac{1}{3}(x^2 - 8)\sqrt{x^2 + 4} + C$

3. (a) $\frac{1}{8}(3 - \sqrt{5})$ (b) $\frac{3\sqrt{3}}{4}$

(c) $\frac{1}{\sqrt{2}}\ln(1 + \sqrt{2})$ (d) $\frac{\pi}{32}$

4. (a) $-\sqrt{5 - 4x - x^2} + \sin^{-1}\left(\dfrac{x + 2}{3}\right) + C$

   (b) $\frac{3}{2}(x - 3)\sqrt{6x - x^2 - 8} + \frac{3}{2}\sin^{-1}(x - 3) + \frac{1}{3}(6x - x^2 - 8)^{\frac{3}{2}} + C$

   (c) $-\dfrac{x + 1}{4\sqrt{2}\sqrt{x^2 + 2x - 1}} + C$

   (d) $\frac{1}{2}(x + 3)\sqrt{x^2 - 2x + 3} + C$

5. (a) $16\sqrt{x^2 - 16} + \frac{16}{3}(x^2 - 16)^{\frac{3}{2}} + C$

   (b) $\frac{1}{640}(8x^2 - 3)(9 + 16x^2)^{\frac{3}{2}} + C$

   (c) $\frac{1}{2}\ln\left|\dfrac{x}{\sqrt{4 - x^2} + 2}\right| + C$

6. (a) $\dfrac{x}{\sqrt{1 + x^2}} + C$

   (b) $\frac{1}{\sqrt{2}}\ln|x + 1| - \frac{1}{\sqrt{2}}\ln\left|\sqrt{2x^2 + 2} - x + 1\right| + C$

   (c) $\frac{2}{\sqrt{3}}\tan^{-1}\left(\dfrac{2 - x - 2\sqrt{1 - x^2}}{x\sqrt{3}}\right) + C$

   (d) $\sqrt{1 - x^2}(1 - \ln x) - \ln\left|\dfrac{1 + \sqrt{1 - x^2}}{x}\right| + C$

8. (a) and (b) $\dfrac{x}{4(x^2 + 2)} + \dfrac{1}{4\sqrt{2}}\tan^{-1}\left(\dfrac{x}{\sqrt{2}}\right) + C$

   (c) The trigonometric substitution used in (b).

9. (a) $\frac{1}{4}\pi^2$ (b) $\frac{1}{\sqrt{2}}\pi$

10. $\dfrac{2(1 + x^2)^{3/2}}{3x^3}\left[1 + \ln\left(\dfrac{x}{\sqrt{1 + x^2}}\right)\right] + C$

11. (a) $\frac{1}{2}\ln|\sec x + \tan x| + \frac{1}{2}\tan x \sec x + C$

   (b) $\frac{1}{3}\sec^3 x + C$

   (c) $\frac{1}{2}\tan x \sec x - \frac{1}{2}\ln|\sec x + \tan x| + C$

   (d) $\frac{1}{8}\tan x \sec x(1 + 2\tan^2 x) - \frac{1}{8}\ln|\tan x + \sec x| + C$

   (e) $-\ln|\operatorname{cosec} x + \cot x| + C$

   (f) $-\frac{1}{2}\cot x \operatorname{cosec} x + \frac{1}{2}\ln|\operatorname{cosec} x + \cot x| + C$

   (g) $\ln|\sec x + \tan x| - \operatorname{cosec} x + C$

   (h) $\sec x - \frac{1}{2}\cot x \operatorname{cosec} x - \frac{3}{2}\ln|\cot x + \operatorname{cosec} x| + C$

12. (a) $\frac{\pi}{2}$

Chapter 11

1. (a) $\ln|x^2 + x + 1| + C$ (b) $x - 3\ln|x + 3| + C$

   (c) $\frac{1}{2}\ln(x^2 + 1) + 7\tan^{-1} x + C$

2. (a) $A = 1, B = -1, C = 2$ (b) $\frac{2}{3}\tan^{-1}\left(\frac{x}{3}\right) + \ln\left(\dfrac{x + 2}{\sqrt{x^2 + 9}}\right) + C$

3. (a) $2\ln|x + 5| - \ln|x - 2| + C$

   (b) $2\ln|x + 1| - \dfrac{1}{x + 1} + C$

   (c) $\ln|x - 1| - \frac{1}{2}\ln|x^2 + 9| - \frac{1}{3}\tan^{-1}\left(\frac{x}{3}\right) + C$

   (d) $\ln|x + 2| - 2\tan^{-1} x + C$

   (e) $x + \frac{1}{4}\ln\left|\dfrac{x - 1}{x + 1}\right| - \frac{1}{2}\tan^{-1} x + C$

   (f) $\frac{1}{2}x^2 - x + 5\ln|x| - 4\ln|x + 1| + C$

   (g) $x + \ln|x^2 - x + 1| + \frac{2}{\sqrt{3}}\tan^{-1}\left(\dfrac{2x - 1}{\sqrt{3}}\right) + C$

   (h) $\dfrac{1}{4\sqrt{2}}\ln\left|\dfrac{1 + x\sqrt{2} + x^2}{1 - x\sqrt{2} + x^2}\right| + \dfrac{1}{2\sqrt{2}}\tan^{-1}\left(\dfrac{x\sqrt{2}}{1 - x^2}\right) + C$

   (i) $\frac{1}{2}\tan^{-1}\left(\frac{x}{2}\right) - \frac{1}{3}\tan^{-1}\left(\frac{x}{3}\right) + C$

   (j) $\frac{1}{2}x^2 - 18\ln|x^2 + 36| + 6\tan^{-1}\left(\frac{x}{6}\right) + C$

4. (a) $\frac{1}{2}x(x - 4) + 2\ln|x - 3| - \frac{1}{2}\tan^{-1}\left(\frac{x}{2}\right) + C$

   (b) $\frac{1}{6}\ln|x + 1| - \frac{1}{2}\ln|x + 2| + \frac{1}{2}\ln|x + 3| - \frac{1}{6}\ln|x + 4| + C$

   (c) $-\frac{1}{120}\ln|1 - x| + \frac{1}{60}\ln|1 - 2x| - \frac{1}{60}\ln|1 - 3x| + \frac{1}{24}\ln|1 - 4x|$
   $-\frac{1}{120}\ln|1 - 5x| + C$

5. $k = \frac{225}{196}$

6. $\ln\left|\dfrac{\sqrt{1 + e^x} - 1}{\sqrt{1 + e^x} + 1}\right| + C$

7. (a) $(x + 1)\ln|1 + \sqrt[3]{x}| - \frac{1}{3}x + \frac{1}{2}(\sqrt[3]{x})^2 - \sqrt[3]{x} + C$

   (b) $x\ln|x + \sqrt[3]{x}| - x + 2\sqrt[3]{x} - 2\tan^{-1}(\sqrt[3]{x}) + C$

11. $\frac{1}{2}x^2 - x + 2\ln|x + 1| + C$

12. $\tan^{-1} x + \frac{1}{3}\tan^{-1}(x^3) + C$

13. $\dfrac{x(x^3 + 4x^2 - 18x + 12)}{2(x - 1)^2} - 6\ln|1 - x| + C$

14. (a) $\dfrac{1}{x^2 + 2x + 2} + \tan^{-1}(x+1) + C$

(b) $\dfrac{15x^5 + 40x^3 + 33x}{48(x^2 + 1)^3} + C$

(c) $-\dfrac{x}{3(x^3 - 1)} + \frac{1}{9}\ln\left|\dfrac{x^2 + x + 1}{(x-1)^2}\right| + \dfrac{2}{3\sqrt{3}}\tan^{-1}\left(\dfrac{2x+1}{\sqrt{3}}\right) + C$

(d) $-\dfrac{x}{(x^3 + 3x + 1)^2} + C$

15. $\frac{2}{3}x - \frac{8}{3}\ln|x - 1| + C$

16. $\frac{1}{2}\ln\left|\dfrac{\cosh x}{\sinh x + 1}\right| + \frac{1}{2}\tan^{-1}(\sinh x) + C$

17. (b) $\frac{1}{4}\ln\left|\dfrac{x+1}{x-1}\right| - \dfrac{x}{2(x^2 - 1)} + C$

18. $\frac{1}{2}\sqrt{x^2 + 2x + 2} - \frac{x}{2} + \ln\left|\sqrt{x^2 + 2x + 2} - x - 2\right|$
$-\frac{1}{2}\ln\left|\sqrt{x^2 + 2x + 2} - x - 1\right| + C$

19. (a) $\mathrm{Li}_{-1}(x) = \dfrac{x}{(1-x)^2}$, $\mathrm{Li}_0 = \dfrac{x}{1-x}$, $\mathrm{Li}_1(x) = -\ln(1-x)$

20. (b) $-\dfrac{5}{4(x+4)^4} + \dfrac{7}{5(x+2)^5} + C$

(d) $\ln|x+2| + \frac{8}{x+2} - \frac{12}{(x+2)^2} + \frac{24}{(x+2)^3} - \frac{32}{(x+2)^4} + \frac{19}{(x+2)^5} - \frac{5}{(x+2)^6} + C$

Chapter 12

2. (a) $\frac{1}{30}\ln\left|\dfrac{3 + 5x}{3 - 5x}\right| + C$

(b) $\frac{1}{24}\ln\left|\dfrac{3 - x}{3 + x}\right| + C$

(c) $\ln|x + \sqrt{x^2 - 7}| + C$

(d) $\dfrac{1}{\sqrt{5}}\sin^{-1}\left(\dfrac{x\sqrt{5}}{\sqrt{3}}\right) + C$

(e) $\frac{1}{2}\ln|2x + \sqrt{4x^2 + 1}| + C$

(f) $\sqrt{2}\sin^{-1}\left(\dfrac{\sqrt{x}}{\sqrt{2}}\right) + C$

(g) $2\ln|\sqrt{x} + \sqrt{8 + x}| + C$

(h) $\sin^{-1}\left(\dfrac{2x + 3}{\sqrt{13}}\right) + C$

(i) $\dfrac{1}{\sqrt{13}}\ln\left|\dfrac{2x+\sqrt{13}+3}{-2x+\sqrt{13}-3}\right|+C$

(j) $\dfrac{1}{\sqrt{3}}\ln|6x-5+2\sqrt{3}\sqrt{3x^2-5x+4}|+C$

(k) $\dfrac{2}{\sqrt{3}}\ln|\sqrt{3x}+\sqrt{5+3x}|+C$

(l) $\dfrac{1}{2}\tan^{-1}\left(\dfrac{x+1}{2}\right)+C$

(m) $\dfrac{1}{12}\ln|2x+1|-\dfrac{1}{12}\ln|5-2x|+C$

(n) $2\ln|x+3+\sqrt{x^2+6x+10}|+C$

(o) $2\ln|\sqrt{x-4}+\sqrt{x+3}|+C$

(p) $\sin^{-1}\left(\dfrac{2x-1}{7}\right)+C$

(q) $\dfrac{1}{2\sqrt{2}}\ln\left|\dfrac{\sqrt{7}-3-x}{\sqrt{7}+3+x}\right|+C$

(r) $\ln|x-3+\sqrt{x^2-6x+13}|+C$

3. (a) $2\ln\left(\dfrac{\sqrt{2}+2}{\sqrt{3}+1}\right)$ (b) $\dfrac{\pi}{4}$ (c) $\dfrac{\pi}{\sqrt{7}}$ (d) $\dfrac{1}{\sqrt{21}}\tan^{-1}\left(\dfrac{3}{\sqrt{21}}\right)$

4. $\alpha=3$

5. (a) $\ln\left(\dfrac{2+\sqrt{5}}{1+\sqrt{2}}\right)$ (b) $\ln(2+\sqrt{3})$ (c) $\dfrac{1}{4a}(\pi-4\tan^{-1}(2))$

6. (a) $\dfrac{1}{4\sqrt{3}}\tan^{-1}\left(\dfrac{x^2+1}{2\sqrt{3}}\right)+C$ (b) $\dfrac{1}{\sqrt{6}}\cos^{-1}\left(\dfrac{12-x}{5x}\right)+C$

7. $\sin^{-1}x-\sqrt{1-x^2}+C$

8. (a) $\dfrac{1}{\sqrt{c}}\ln\left|2cx+b+2\sqrt{c}\sqrt{a+bx+cx^2}\right|+C$

(b) $\dfrac{1}{\sqrt{c}}\sin^{-1}\left(\dfrac{2cx-b}{\sqrt{4ac+b^2}}\right)$

(c) $\dfrac{1}{\sqrt{b^2-4ac}}\ln\left|\dfrac{2cx+b-\sqrt{b^2-4ac}}{2cx+b+\sqrt{b^2-4ac}}\right|+C$, if $4ac-b^2<0$

$\dfrac{2}{\sqrt{4ac-b^2}}\tan^{-1}\left(\dfrac{2cx+b}{\sqrt{4ac-b^2}}\right)+C$, if $4ac-b^2>0$

(d) $\dfrac{1}{\sqrt{b^2+4ac}}\ln\left|\dfrac{\sqrt{b^2+4ac}-b+2cx}{\sqrt{b^2+4ac}+b-2cx}\right|+C$

9. (a) $\dfrac{1}{3}\ln|3x+2|+C$

(b) $2\ln|x^2+4|-\dfrac{1}{2}\tan^{-1}\dfrac{x}{2}+C$

(c) $\frac{2}{9}\sqrt{9x^2 + 16} + \frac{1}{3}\ln|3x + \sqrt{16 + 9x^2}| + C$

(d) $\frac{1}{2}\ln|1 + x^2| + \tan^{-1}x + C$

(e) $\frac{\sqrt{5}-10}{10}\ln|\sqrt{5} + 2 - x| - \frac{10+\sqrt{5}}{10}\ln|\sqrt{5} - 2 + x| + C$

(f) $\ln|x^2 + 2x + 2| - \tan^{-1}(x + 1) + C$

(g) $\sqrt{x^2 + 2x - 1} - 2\ln|x + 1 + \sqrt{x^2 + 2x - 1}| + C$

(h) $2\sqrt{x^2 + 2x + 3} + \ln|x + 1 + \sqrt{x^2 + 2x + 3}| + C$

(i) $\frac{1}{2}\sqrt{2x^2 + 3x + 3} + \frac{1}{4\sqrt{2}}\ln|4x + 3 + 2\sqrt{2}\sqrt{2x^2 + 3x + 4}| + C$

(j) $\frac{1}{2}\ln(x^2 + 6x + 15) - \frac{\sqrt{6}}{2}\tan^{-1}\left(\dfrac{x + 3}{\sqrt{6}}\right) + C$

10. $\alpha = 2$

11. $\dfrac{\pi\ln 3}{2\sqrt{3}}$

12. $\dfrac{\pi}{2n}\ln\left|\dfrac{n + 1}{n - 1}\right|$

13. $\dfrac{1}{\sqrt{5}}\tan^{-1}\left(\dfrac{\tanh(x - 2)}{\sqrt{5}}\right) + C$

14. (a) $(x - 3)\sqrt{\dfrac{x + 2}{3 - x}} + 5\sin^{-1}\left(\sqrt{\dfrac{x + 2}{5}}\right) + C$

(b) $5\cosec^{-1}(5)$

16. (b) $\dfrac{1}{\sqrt{2}}\tan^{-1}(\sqrt{2}) - \sqrt{2}\tan^{-1}\left(\dfrac{1}{\sqrt{2}}\right) + \dfrac{\pi}{4}$

17. (a) $\sin\alpha\tan^{-1}\left(\dfrac{x - \cos\alpha}{\sin\alpha}\right) - \dfrac{1}{2}\cos\alpha\ln(1 - 2x\cos\alpha + x^2) + C$

(b) $\dfrac{\pi}{2\sin\alpha}$

Chapter 13

1. (a) $\cosh^{-1}\left(\dfrac{x}{\sqrt{3}}\right) + C$

(b) $\dfrac{1}{\sqrt{3}}\sinh^{-1}\left(\sqrt{\dfrac{3}{2}}x\right) + C$

(c) $\begin{cases} -\frac{1}{2}\tanh^{-1}\left(\dfrac{x + 3}{2}\right) + C, & |x + 3| < 2 \\[2mm] -\frac{1}{2}\coth^{-1}\left(\dfrac{x + 3}{2}\right) + C, & |x + 3| > 2 \end{cases}$

(d) $\sinh^{-1}\left(\dfrac{x - 1}{2}\right) + C$

(e) $\cosh^{-1}\left(\dfrac{x-3}{3}\right)+C$

(f) $\cosh^{-1}\left(\dfrac{x+3}{3}\right)+C$

2. (a) $\begin{cases} -\frac{1}{\sqrt{2}}\tanh^{-1}(\sqrt{2}x)+C, & |x|<\frac{1}{\sqrt{2}} \\ -\frac{1}{\sqrt{2}}\coth^{-1}(\sqrt{2}x)+C, & |x|>\frac{1}{\sqrt{2}} \end{cases}$ (b) $-\frac{1}{\sqrt{2}}\tanh^{-1}\left(\frac{1}{\sqrt{2}}\right)$

4. (b) $2\tanh^{-1}(\ln x)+C$

5. (a) $\frac{1}{2}x\sqrt{1+x^2}+\frac{1}{2}\sinh^{-1}x+C$
   (b) $\frac{1}{2}x\sqrt{x^2-1}+\frac{1}{2}\cosh^{-1}x+C$
   (c) $\frac{1}{2}(x+3)\sqrt{(x+2)(x+4)}-\cosh^{-1}(x+3)+C$
   (d) $\operatorname{sech}^{-1}x-\sqrt{1-x^2}+C$
   (e) $\frac{1}{32}x(8x^2+1)\sqrt{1+4x^2}-\frac{1}{64}\sinh^{-1}(2x)+C$
   (f) $\frac{3}{8}\sinh^{-1}x+\frac{1}{8}x\sqrt{1+x^2}(2x^2-3)+C$

7. (a) $x\tanh^{-1}x+\frac{1}{2}\ln|1-x^2|+C$
   (b) $x\operatorname{sech}^{-1}x+\sin^{-1}x+C$
   (c) $\frac{x^2}{2}\cosh^{-1}x-\frac{x}{4}\sqrt{x^2-1}-\frac{1}{4}\cosh^{-1}x+C$
   (d) $\frac{1}{2}(x^2-1)\tanh^{-1}x+\frac{x}{2}+C$

8. $\left(x+\frac{1}{2}\right)\sinh^{-1}(\sqrt{x})-\frac{1}{2}\sqrt{x}\sqrt{x+1}+C$

12. (a) $-\dfrac{1}{(x+a)\sqrt{2x+a+b}}$ (b) $-\sqrt{2}\cosh^{-1}\left(\dfrac{x+5}{x+1}\right)+C$

13. $\frac{1}{2}x\sin(\cosh^{-1}x+1)-\frac{1}{2}\sqrt{x^2-1}\cos(\cosh^{-1}x+1)+C$

Chapter 14

1. $\tan x=\dfrac{2t}{1-t^2}, \cot x=\dfrac{1-t^2}{2t}, \sec x=\dfrac{1+t^2}{1-t^2}, \operatorname{cosec} x=\dfrac{1+t^2}{2t}$

3. (a) $\ln|\tan(x/2)|+C$
   (b) $\ln|\tan(x/2)|-\frac{1}{2}\tan^2(x/2)+C$
   (c) $\ln|\sin(x/2)+\cos(x/2)|-\ln|\cos(x/2)|+C$
   (d) $\ln|\sec x+\tan x|+C$

4. (b) $\frac{5}{6}\tan^{-1}(3)-\frac{\pi}{8}$

5. (a) $\dfrac{1}{\sqrt{7}}\ln\left|\dfrac{\sqrt{7}+\tan\frac{x}{2}}{\sqrt{7}-\tan\frac{x}{2}}\right|+C$

   (b) $\dfrac{2}{\sqrt{7}}\tan^{-1}\left(\dfrac{1}{\sqrt{7}}\tan\dfrac{x}{2}\right)+C$

   (c) $\dfrac{x}{4}-\dfrac{3}{4\sqrt{7}}\ln\left|\dfrac{\sqrt{7}+\tan\frac{x}{2}}{\sqrt{7}-\tan\frac{x}{2}}\right|+C$

   (d) $\dfrac{1}{5}\ln\left(\dfrac{3\tan\frac{x}{2}+1}{3-\tan\frac{x}{2}}\right)+C$

   (e) $\ln|1+\sin x|-\dfrac{2}{1+\tan\frac{x}{2}}+C$

   (f) $\dfrac{x}{3}-\dfrac{1}{3\sqrt{5}}\ln\left|\dfrac{2\tan\frac{x}{2}+3+\sqrt{5}}{\sqrt{5}-3-\tan\frac{x}{2}}\right|+C$

6. (a) $\dfrac{\pi}{4}\tan\left(\dfrac{\pi}{8}\right)$ (b) $\dfrac{1}{4}\pi\sqrt{2}$

9. (a) $-x-2\ln\left|\tan\frac{x}{2}-1\right|+C$

   (b) $x\tan\frac{x}{2}+C$

   (c) $\dfrac{x}{2}+\ln\left|\cos\frac{x}{2}\right|+C$

   (d) $\ln|1+\sin x|-\dfrac{2x}{\tan\frac{x}{2}+1}+2\ln\left|\sin\frac{x}{2}+\cos\frac{x}{2}\right|+x+C$

10. (c) $\dfrac{2}{\sqrt{5}}\left[\tan^{-1}\left(\dfrac{1}{\sqrt{5}}\right)-\tan^{-1}\left(\dfrac{1}{\sqrt{15}}\right)\right]$

12. (d) $-\dfrac{3}{\sin x-2}-\dfrac{4}{9}\ln|\sin x-2|+\dfrac{1}{2}\ln|\sin x-1|-\dfrac{1}{18}\ln|\sin x+1|+C$

14. (b) $\dfrac{x}{2}-\dfrac{1}{2}\ln|\sin x+\cos x|+\sqrt{2}\ln\left|\dfrac{\tan\frac{x}{2}-1+\sqrt{2}}{\tan\frac{x}{2}-1-\sqrt{2}}\right|+C$

15. (g) 10π

Chapter 15

1. $\dfrac{\pi}{2}$

2. $-\dfrac{1}{\sin x+\sin 2x}+C$

3. (a) $-\cot x+\operatorname{cosec}x+C$

   (b) $\tan x+C$

   (c) $-\operatorname{cosec}x-\sin x+C$

   (d) $\dfrac{1}{4}\cos 2x-\ln|\cos x|+C$

   (e) $2x-\tan x+C$

(f) $x + \cos x + C$

(g) $x - \sin x \cos x + C$

(h) $4x - 3 \tan x + C$

(i) $-\cot x + \csc x + C$

(j) $-\dfrac{1}{2(\cos x + 1)} - \frac{1}{2} \ln |\csc x + \cot x| + C$

(k) $\dfrac{1}{\sqrt{2}} \ln \left| \dfrac{1 + \sqrt{2} \sin x}{1 - \sqrt{2} \sin x} \right| - \dfrac{1}{2} \ln \left| \dfrac{1 + \sin x}{1 - \sin x} \right| + C$

(l) $-\csc x \sec^2 x + \tan x \sec x + \ln |\sec x + \tan x| + C$

(m) $2 \left(\sin x + \frac{1}{3} \sin 3x + \frac{1}{5} \sin 5x + \frac{1}{7} \sin 7x \right) + C$

(n) $-\frac{1}{5} \cos x - \frac{2}{5} \sin x - \dfrac{4}{5\sqrt{5}} \ln \left| \csc(x + \tan^{-1}(2)) + \cot(x + \tan^{-1}(2)) \right| + C$

(o) $\sqrt{2} \tan^{-1}(\sqrt{2} \tan x) - x + C$

(p) $-\dfrac{1}{1 + \sin x} + C$

(q) $-\frac{1}{2}x - \frac{1}{4} \ln \left| \dfrac{1 - \tan x}{1 + \tan x} \right| + C$

(r) $-\dfrac{1}{2(\sin x + \cos x)} - \dfrac{1}{2\sqrt{2}} \ln \left| \dfrac{\sqrt{2} + \cos x - \sin x}{\sin x + \cos x} \right| + C$

(s) $-\sin x - \sin x \cos x + C$

(t) $\frac{1}{4} \sec 2x + \frac{1}{4} \ln |\tan x| + C$

(u) $\frac{1}{n} \tan^n x + C$

(v) $3 \cos x - \sin x + 2\sqrt{2} \ln \left| \dfrac{\sqrt{2} + \sin x - \cos x}{\sin x + \cos x} \right| + C$

4. (a) $\dfrac{1}{2\sqrt{2}} \ln \left| \dfrac{\sqrt{2} \sin x - 1}{\sqrt{2} \sin x + 1} \right| + C$

6. (a) $A = \dfrac{\sin a}{\sin(a - b)}, \ B = \dfrac{\sin b}{\sin(b - a)}$

   (b) $-\dfrac{\sin a}{\sin(a - b)} \ln |\csc(x - a) + \cot(x - a)| - \dfrac{\sin b}{\sin(b - a)} \ln |\csc(x - b)$
   $+ \cot(x - b)| + C$

7. (a) $\frac{1}{2} \tan^{-1} \left(\dfrac{\sin x + 1}{2} \right) + C$

   (b) $-\frac{1}{7}(\sec x + \tan x)^{-\frac{7}{2}} - \frac{2}{3}(\sec x + \tan x)^{-\frac{3}{2}} + C$

   (c) $\frac{1}{3} \ln |1 + \tan^3 x| + C$

   (d) $\frac{1}{3} \ln |\sin x - 1| - \frac{1}{6} \ln |\sin^2 x + \sin x + 1| + \dfrac{1}{\sqrt{3}} \tan^{-1} \left(\dfrac{2 \sin x + 1}{\sqrt{3}} \right) + C$

(e) $-\frac{1}{17}x + \frac{1}{4}\ln|\sin x| - \frac{1}{68}\ln|\sin x + 4\cos x| + C$

(f) $\cos x \sqrt{\sec 2x} + C$

8. $200\sqrt{2}$

11. (a) $-x + \cot(b-a)\ln\left|\dfrac{\cos(x-b)}{\cos(x-a)}\right| + C$

(b) $x - 2\cot(b-a)\operatorname{cosec}^2(b-a)\ln\left|\dfrac{\cos(x-b)}{\cos(x-a)}\right|$
$+ \cot^2(b-a)\left[\tan(x-a) + \tan(x-b)\right] + C$

13. (a) $\cos 2a(x+a) - \sin 2a \ln|\sin(x+a)| + C$

(b) $2\sin x + 2x\cos a + C$

(c) $4\sin a \ln\left|\cos\dfrac{x+a}{2}\right| + 2\sin x + C$

(d) $\sin 2a \ln|\sin(x+a)| - x\cos 2a + C$

(e) $x\cos 2a + \sin 2a \ln|\cos(x-a)| + C$

(f) $-\cos a \sin^{-1}\left(\dfrac{\cos x}{\cos a}\right) - \sin a \ln\left|\sin x + \sqrt{\sin^2 x - \sin^2 a}\right| + C$

14. $\dfrac{1}{2} - \dfrac{1}{\sqrt{2}}\tan^{-1}\left(\dfrac{1}{\sqrt{2}}\right) + \sqrt{\dfrac{2}{3}}\tan^{-1}\left(\sqrt{\dfrac{2}{3}}\right)$

18. (c) $-\displaystyle\sum_{k=1}^{n}\dfrac{1}{2k-1}\cos(2k-1)x + C$

19. (a) $2\tan^{-1}\left(\dfrac{1+\alpha}{1-\alpha}\tan\dfrac{x}{2}\right) + C$ (b) $\dfrac{x}{2\alpha} - \dfrac{1}{\alpha}\tan^{-1}\left(\dfrac{1+\alpha}{1-\alpha}\tan\dfrac{x}{2}\right) + C$

20. (a) 1 (b) $\dfrac{\alpha}{\sin\alpha}$

21. $J_1 = -\dfrac{1}{\sqrt{2}}\ln\left|\operatorname{cosec}\left(x+\dfrac{\pi}{4}\right) + \cot\left(x+\dfrac{\pi}{4}\right)\right| + C$

$J_2 = x + C$

$J_3 = \dfrac{2}{3}\tan^{-1}(\sin x - \cos x) + \dfrac{2}{3\sqrt{2}}\ln\left|\dfrac{\sqrt{2} - \cos x + \sin x}{\sin x + \cos x}\right| + C$

$J_4 = \dfrac{1}{\sqrt{2}}\tan^{-1}\left(\dfrac{\sqrt{2}\tan x}{1 - \tan^2 x}\right) + C$

$J_6 = -\tan^{-1}(2\cot 2x) + C$

22. $\pi^2/4 - 2\gamma$

23. (c) $\frac{1}{2}\pi\ln 2$

24. (d) $x \cot^{-1}(x^2 + x + 1) + \cot^{-1}(x + 1) + \frac{1}{2}\ln(x^2 + 2x + 2) - \frac{1}{2}\ln(x^2 + 1) + C$

25. $-\dfrac{2\cos x}{\sin^2 x} + C$

26. $\dfrac{1}{1 - a^2}\left(x - a\tan^{-1}(a\tan x)\right) + C$ for $a \neq \pm 1$, and

$\dfrac{x}{2} + \dfrac{1}{4}\sin 2x + C$ for $a = \pm 1$

27. (a) $2\sin x - \sin 2x - \sqrt{3}\ln\left|\dfrac{\sqrt{3} + \tan\frac{x}{2}}{\sqrt{3} - \tan\frac{x}{3}}\right| + C$

   (b) $4\cos x + \cos 2x - \frac{2}{3}\cos 3x - 2\ln|\cos x| + \frac{1}{2}\sec x + C$

28. $\dfrac{2^6}{6}\left[\left(1 + \dfrac{\sqrt{3}}{2}\right)^6 - \left(\dfrac{1}{2} + \sqrt{2}\right)^6\right]$

30. (d) $\frac{2}{5}\ln(\sqrt{2})$

31. (a) $\dfrac{2}{1 + 2^n}\cos\left[(1 + 2^n)\cos^{-1}\left(\dfrac{x}{2}\right)\right] + \dfrac{2}{1 - 2^n}\cos\left[(1 - 2^n)\cos^{-1}\left(\dfrac{x}{2}\right)\right] + C$

33. (a) $\dfrac{2^{n+1}}{2^n + 1}\cos\left[\dfrac{2^n + 1}{2^n}\cos^{-1}\left(\dfrac{x}{2}\right)\right] + \dfrac{2^{n+1}}{2^n - 1}\cos\left[\dfrac{2^n - 1}{2^n}\cos^{-1}\left(\dfrac{x}{2}\right)\right] + C$

   (b) $\dfrac{2^{n-1}}{2^n + 1}\cos\left[\dfrac{2^n + 1}{2^n}\cos^{-1} x\right] + \dfrac{2^{n-1}}{2^n - 1}\cos\left[\dfrac{2^n - 1}{2^n}\cos^{-1} x\right] + C$

34. (e) $\ln\left|\dfrac{\sin x}{\sin 16x}\right| + C$

Chapter 16

1. (a) False   (b) True   (c) True   (d) False

3. All are equal to zero.

6. (a) $\dfrac{1}{2\sqrt{2}}\ln(3 + 2\sqrt{2})$ (b) $\pi/4$

   (c) $\dfrac{\pi}{\sqrt{2}}\ln(\sqrt{2} + 1)$ (d) $\frac{1}{2}\pi(\pi + 3)$

   (e) $\pi/4$ (f) $\frac{1}{2}$

   (g) $\dfrac{\pi}{4\sqrt{2}}\ln\left(\dfrac{\sqrt{2}+1}{\sqrt{2}-1}\right)$ (h) $\pi^2/4$

   (i) 0 (j) 1

   (k) $\pi^2/8$ (l) $\dfrac{3\pi\ln 2}{32}$

7. (a) 1008 (b) $\dfrac{3\,462\,376\,907\,846}{5313}$

(c) 2 (d) $\pi/12$

(e) $\frac{1}{3}\pi \ln 2$ (f) 0

(g) $\pi/12$ (h) $2\sin(2)$

8. (a) $\frac{16}{35}$ (b) 0 (c) $\frac{\pi}{2}$ (d) $\frac{\pi^2}{2}$ (e) π (f) $\frac{\pi^3}{3}$

9. (a) 0 (b) 0

10. $\dfrac{\pi}{2}$

11. $\dfrac{\pi(a+b)}{4}$

13. π if n is odd; 0 if n is even.

14. (b)(i) 2 (ii) $\frac{\pi}{4}$ (iii) $\frac{1}{2}$

15. (b) $\dfrac{\pi^2}{4}$

16. (b) $\dfrac{\pi}{4}$

17. (a) -2 (b) 6 (c) -22

18. $\dfrac{\pi}{8}\ln 2$

19. 0

20. (a) 0

21. $\dfrac{\pi}{4}$

22. (b) 1

24. (c) $\dfrac{\pi}{8a}\ln(2a^2)$

26. (c) $\dfrac{a+b}{a^2+b^2}\dfrac{\pi}{4}$

27. (b) $\pi/4$

28. (b)(i) π (ii) π (iii) $\frac{1}{2}$

29. $\dfrac{u}{v} = \dfrac{1}{4}$

30. (b)(i) $\frac{8}{3}$ (ii) 0 (d)(i) $\frac{\pi}{2}$ (ii) $\frac{\pi}{3}$ (iii) $\frac{\pi}{4}$ (iv) $\frac{3}{4}$

31. (b)(i) π (ii) $3\pi/2$

32. (a) $\dfrac{\pi}{3\sqrt{3}}$ (b) $\frac{1}{4}(\ln 3 - \ln 2)$

33. $\dfrac{\pi}{3\sqrt{3}}$

34. (a) 14   (b) 25   (c) $-\frac{3}{32}$ (d) $\frac{2\sqrt{2}}{3}$ (e) 2

    (f) $\frac{3\sqrt{3}}{2} - 1$ (g) $\frac{23}{3}$ (h) $\frac{5}{2}\ln\left(\frac{8}{3}\right)$

    (i) $2 - \ln 2$ (j) 27

35. $n^2\pi$

37. $k = \pi$

40. $\pi/4$

41. a

43. 1

44. 100

45. (a) 35   (b) $\frac{5}{3}$ (c) 225   (d) $\frac{77}{60}$ (e) $2^a - 1$

46. (a) $\frac{2\pi}{\sqrt{3}}\tan^{-1}\left(\frac{4}{3}\right)$ (b) π (c) $\frac{\pi^3}{8}$ (d) e

47. $\dfrac{a}{\sin a}\ln(\sec a)$

49. (b) π

Chapter 17

2. (a) 0   (b) 1

3. $\frac{2}{3}$

4. 43

10. (a) $\ln 16 - \frac{15}{16}$ (b) $2\pi(\pi + 1)$

11. 0

12. (b)(ii) $\frac{1}{2}\pi^2 - 2$

Chapter 18

1. $\int \operatorname{cosec}^3 x \, dx = -\dfrac{1}{2} \cot x \operatorname{cosec} x - \dfrac{1}{2} \ln|\operatorname{cosec} x + \cot x| + C$, and

   $\int \operatorname{cosec}^4 x \, dx = -\dfrac{1}{3} \cot x \operatorname{cosec}^2 x - \dfrac{2}{3} \cot x + C$

2. (b) $\frac{1}{4} \sin x \cos^3 x + \frac{3}{8} \sin x \cos x + \frac{3}{8} x + C$

   (d) $\int \cos^{-3} x \, dx = \dfrac{1}{2} \sin x \cos^{-2} x + \dfrac{1}{2} \ln|\sec x + \tan x| + C$

   $\int \cos^{-4} x \, dx = \dfrac{1}{3} \sin x \cos^{-3} x + \dfrac{2}{3} \tan x + C$

3. (b) $6 - \frac{16}{e}$

4. (b) $\frac{1}{3} \cosh(1) \sinh^2(1) - \frac{2}{3} \cosh(1) + \frac{2}{3}$

5. (b) $\frac{1}{3} \tan x \sec^2 x + \frac{2}{9} \tan x + C$

6. (b) $\dfrac{128}{195}$

7. (b) $\left(\dfrac{\pi}{2}\right)^6 - 30 \left(\dfrac{\pi}{2}\right)^4 + 360 \left(\dfrac{\pi}{2}\right)^2 - 720$

8. (b) $\frac{11}{2} e - \frac{15}{2}$

9. (a) $\tan^{-1}(\sinh x)$

   (c) $J_3 = \dfrac{1}{2} \tanh x \operatorname{sech} x + \dfrac{1}{2} \tan^{-1}(\sinh x)$

   $J_4 = \dfrac{1}{3} \tanh x \operatorname{sech}^2 x + \dfrac{2}{3} \tanh x$

10. (a) $I_0 = \dfrac{\pi}{3}, I_1 = \dfrac{\pi}{6}$ (c) $I_2 = \dfrac{\pi}{12}, I_3 = \dfrac{\pi}{24}$ (d) $I_n = \dfrac{\pi}{3 \cdot 2^n}$

11. (a) $I_0 = 0$

12. (a) $\dfrac{1}{n+1}$ (d) $\dfrac{1}{213}$

16. (b) $\dfrac{168}{5} - \dfrac{40\sqrt{5}}{3}$

17. (a) $I_0 = 0, I_1 = \pi$

18. (a) $I_0 = a$ (d) $\dfrac{(2^n n!)^2}{(2n+1)!}$

19. (a)(i) $u_0 = 0, u_1 = \pi$

20. (b) 3π

22. (a) 0 when $n = 0$ and π when $n = 1$.

25. (b) $\pi/2$

30. (f) It is irrational

Chapter 19

1. $-\dfrac{1}{x\cos x} + C$

2. (a) $\dfrac{e^x}{1+x}$

 (b) $\dfrac{e^x}{1+x^2} + C$

 (c) $e^x\left(2\ln x - \dfrac{1}{x}\right) + C$

 (d) $e^x - \dfrac{2e^x}{x+1} + C$

 (e) $\dfrac{1+x}{\sqrt{1-x^2}}e^x + C$

 (f) $xe^{x+\frac{1}{x}}$

 (g) $-e^{\cos x}\left(x + \dfrac{1}{\sin x}\right) + C$

3. (b) $\dfrac{e^x\sin x}{\cos x - 1} + C$

4. (a) $x\tan x + C$

 (b) $x\ln^2 x + C$

 (c) $x^x + C$

 (d) $(x^2+1)\tan^{-1} x + C$

 (e) $x\ln(\ln x) + C$

 (f) $e^{\sin x}\cos(\cos x) + C$

5. (b) $\frac{1}{100}\sin(100x)\sin^{100} x + C$

6. $-\dfrac{1}{4}\ln\left(\dfrac{16-\pi^2}{16+\pi^2}\right)\exp\left(-\dfrac{16-\pi^2}{16+\pi^2}\right)$

7. $g(e^e + 1) - g(e)$

8. (a) $\dfrac{x}{x^5+x+1} + C$

 (b) $\dfrac{1-2x}{x^3+1} + C$

(c) $\dfrac{\cos x}{x \cos x - \sin x} + C$

(d) $\dfrac{1}{x(1 - \ln x)} + C$

(e) $\dfrac{1 + x \tan^{-1} x}{\tan^{-1} x - x} + C$

(f) $\dfrac{-x^2 + 3x - \frac{26}{9}}{x^3 - x + 1} + C$

9. $\dfrac{\sin x - x \cos x}{x \sin x + \cos x} + C$

10. (a) $\dfrac{1}{2} \ln \left| \dfrac{x^2 - x + 1}{x^2 + x + 1} \right| + C$

(b) $-\sec^{-1} \left(\dfrac{1 - x^2}{x\sqrt{2}} \right) + C$

(c) $\dfrac{1}{2\sqrt{2018}} \tan^{-1} \left(\dfrac{x^2 - 1}{x\sqrt{2018}} \right) - \dfrac{1}{2\sqrt{2014}} \tan^{-1} \left(\dfrac{x^2 + 1}{x\sqrt{2014}} \right) + C$

(d) $\ln \left| \dfrac{x^2 + 1 + \sqrt{1 + x^4}}{x} \right| + C$

(e) $\dfrac{\sqrt{x^4 - 1}}{x} + C$

(f) $\dfrac{1}{\sqrt{2}} \cos^{-1} \left(\dfrac{x\sqrt{2}}{x^2 + 1} \right) + C$

(g) $\dfrac{x^2 + x + 1}{\sqrt{x^4 + x^2 + 1}} + C$

(h) $\dfrac{2}{\sqrt{11}} \tan^{-1} \left(\dfrac{2x^2 + 3x - 2}{x\sqrt{11}} \right) + C$

11. (a) $\dfrac{1}{\sqrt{2}} \tan^{-1} \left(\dfrac{x^2 - 1}{x\sqrt{2}} \right) + C$

(b) $\dfrac{1}{2\sqrt{2}} \tan^{-1} \left(\dfrac{x^2 - 1}{x\sqrt{2}} \right) + \dfrac{1}{4\sqrt{2}} \ln \left| \dfrac{x^2 - x\sqrt{x} + 1}{x^2 + x\sqrt{2} + 1} \right| + C$

(c) $\dfrac{1}{2\sqrt{2}} \tan^{-1} \left(\dfrac{x^2 - 1}{x\sqrt{2}} \right) - \dfrac{1}{4\sqrt{2}} \ln \left| \dfrac{x_x^2 \sqrt{2} + 1}{x^2 + x\sqrt{2} + 1} \right| + C$

13. (b) $\dfrac{1}{24\sqrt{2}} \left[2\tan^{-1} \left(\dfrac{x^{12} - 1}{x^6\sqrt{2}} \right) + \ln \left| \dfrac{x^{12} - x^6\sqrt{2} + 1}{x^{12} + x^6\sqrt{2} + 1} \right| \right] + C$

14. (a) $I(0) = 2\pi$ (d) $I(x) = 2\pi, I(1) = 2\pi$

15. (a) $\sqrt{\dfrac{1+x}{1-x}} + C$ (b) $\frac{1}{4}\ln^2\left(\dfrac{1+x}{1-x}\right) + C$

   (c) $\dfrac{\pi}{8}\ln 2$ (d) $\dfrac{\pi}{4\sqrt{7}}\tan^{-1}\left(\dfrac{7\sqrt{7}}{13}\right)$ (e) $\dfrac{3}{2}\ln 2$

16. $-\dfrac{3}{16}\ln\left|\dfrac{x-1}{x+1}\right| + \dfrac{3}{8}\tan^{-1}x - \dfrac{x}{4(x^4-1)} + C$

17. (a) $-2\tan^{-1}\left(\sqrt{x + \dfrac{1}{x} + 1}\right) + C$

   (b) $\cos^{-1}\left(\dfrac{x}{x^2+1}\right) + \dfrac{2\sqrt{x^4+x^2+1}}{x^2+1} + C$

18. $\dfrac{5\sin x - x\cos x}{x\sin x + 5\cos x} + C$

Chapter 20

1. (a) Interval is unbounded.
   (b) Integrand is unbounded at $x = \pi/2$.
   (c) Integrand is unbounded at $x = 2$.
   (d) Interval is unbounded.
   (e) Interval is unbounded and the integrand is unbounded at $x = 1$.
   (f) The integrand is unbounded at $x = \pm 1$.

2. No

3. (a) converges, $\frac{1}{12}$
   (b) converges, $\frac{3}{\ln 3}$
   (c) converges, $\frac{1}{25}$
   (d) diverges
   (e) converges, $2\sqrt{3}$
   (f) diverges
   (g) diverges
   (h) converges, $-2/e$
   (i) diverges
   (j) converges, 1
   (k) converges, $\pi^2/8$
   (l) diverges
   (m) converges, 4
   (n) converges, $\frac{2}{\sqrt{5}}\coth^{-1}\left(\frac{3}{\sqrt{5}}\right)$

(o) converges, 2

(p) converges, π

(q) converges, 1

(r) converges, $\frac{\pi}{2} + 1$

(s) converges, $\frac{\pi}{3\sqrt{3}}$

(t) diverges

(u) converges, $(\pi + 2\ln 2)/4$

(v) converges, $\ln 2$

4. (a) $\frac{1}{2} \ln \left| \dfrac{\sqrt{1 + x^2}}{1 + x} \right| + \frac{1}{2} \tan^{-1} x + C$

5. π

6. (a) converges   (b) converges   (c) diverges   (d) converges

   (e) converges   (f) diverges   (g) converges   (h) converges

   (i) converges   (j) converges

7. (a) converges   (b) converges   (c) converges   (d) converges

8. $a = 2(e - 1)$

10. (a) The integrand is unbounded at $x = 0$.

11. (a) $p > 1$ (b) $\frac{1}{p-1}$

13. $n = 4$

16. (b) $\dfrac{\pi \ln a}{2a}$

20. (b) $\dfrac{\pi}{4 \cosh \frac{a}{2}}$

21. (b) $\pi/4$

23. (b) $J_4 = \frac{2}{3}$

24. $\dfrac{2}{\sqrt{1 - a^2}} \tan^{-1} \sqrt{\dfrac{1 - a}{1 + a}}$, when $-1 < a < 1$.

   1, when $a = 1$.

   $\dfrac{2}{\sqrt{a^2 - 1}} \tanh^{-1} \sqrt{\dfrac{a - 1}{a + 1}}$, when $a > 1$.

30. (b) $\frac{2}{3}$ (c) $\frac{7}{15}$

33. (d) $\dfrac{(2n)!\pi}{2^{2n+1}(n!)^2}$

38. (d) $\dfrac{\pi}{\sqrt{2}}$

39. $p < 1$

42. (e) $\dfrac{s}{s^2 + 1}$ (f) $\dfrac{s + 2}{(s + 2)^2 + 1}$

43. $\pi^2/8$

Chapter 21

1. (a) $\sqrt{\pi}$ (b) $\sqrt{2\pi}$ (c) $\sqrt{\pi}$ (d) $\frac{1}{2}\sqrt{\pi}$

   (e) $e\sqrt{\pi}$ (f) $\sqrt{\pi}$ (g) $\sqrt{\pi}$ (h) $\frac{1}{2}\sqrt{\dfrac{\pi}{\ln 2}}$

   (i) $\frac{1}{4}\sqrt{\pi}$ (j) $\frac{3}{8}\sqrt{\pi}$ (k) $\sqrt{\pi}$ (l) $\frac{1}{2}\sqrt{\pi}$

   (m) $\frac{1}{2}\sqrt{\pi} - \frac{1}{2}$ (n) $\frac{3}{4}\sqrt{\pi}$

2. (a) π (b) $\frac{\pi}{2}$ (c) $\frac{1}{2}$ (d) $-\frac{\pi}{2}$ (e) $\frac{\pi}{4}$ (f) $\frac{\pi}{2}$

3. (a) $\frac{\pi}{2n}$ (b) $\frac{\pi}{2}$ when $a > 0$, $-\frac{\pi}{2}$ when $a < 0$.

7. $\mathrm{Si}\,(0) = 0$, $\mathrm{Si}\,(\infty) = \frac{\pi}{2}$, $\mathrm{Si}\,(-\infty) = -\frac{\pi}{2}$, $\mathrm{si}\,(0) = -\frac{\pi}{2}$,
   $\mathrm{si}\,(\infty) = 0$, $\mathrm{si}\,(-\infty) = -\pi$

9. (c) $\mathrm{erf}(0) = 0$, $\mathrm{erf}(\infty) = 1$, $\mathrm{erf}(-\infty) = -1$, $\Phi(0) = \frac{1}{2}$, $\Phi(\infty) = 1$
   (d) $\mathrm{erfc}(0) = 1$, $\mathrm{erfc}(\infty) = 0$, $\mathrm{erfc}(-\infty) = 2$

11. (b)(i) $\sqrt{\pi}$ (b)(ii) $\sqrt{p\pi}$

Appendix A

2. (a) 2   (b) 4   (c) 5   (d) 6

3. (a) $\dfrac{A}{x + 3} + \dfrac{B}{3x + 1}$

   (b) $\dfrac{A}{x} + \dfrac{B}{x + 1} + \dfrac{C}{(x + 1)^2}$

   (c) $\dfrac{A}{x + 4} + \dfrac{B}{x - 1}$

   (d) $\dfrac{A}{x - 1} + \dfrac{Bx + C}{x^2 + x + 1}$

   (e) $\dfrac{A}{x - 1} + \dfrac{B}{x + 1} + \dfrac{Cx + D}{x^2 + 1}$

   (f) $\dfrac{Ax + B}{x^2 + 1} + \dfrac{Cx + D}{x^2 + 4} + \dfrac{Ex + F}{(x^2 + 4)^2}$

4. (a) $\dfrac{1}{x-2} - \dfrac{1}{x-1}$

   (b) $\dfrac{2}{x+2} + \dfrac{3}{x-1}$

   (c) $\dfrac{3}{2x+1} + \dfrac{2}{x-1}$

   (d) $\dfrac{3}{3x+2} - \dfrac{12}{6x-1} + \dfrac{1}{x-1}$

   (e) $\dfrac{2}{x+1} - \dfrac{3}{x+2} + \dfrac{2}{x-8}$

   (f) $\dfrac{3}{x+1} - \dfrac{3}{x-1} + \dfrac{1}{x-3} - \dfrac{1}{x+3}$

   (g) $\dfrac{1}{25x} - \dfrac{1}{25(x+5)} - \dfrac{1}{5(x+5)^2}$

   (h) $\dfrac{1}{(x+4)^2} - \dfrac{4}{(x+4)^3}$

   (i) $\dfrac{1}{x+2} - \dfrac{3}{x+3} + \dfrac{2}{x-2}$

   (j) $\dfrac{x+8}{x^2+4} - \dfrac{2}{2x+1}$

   (k) $\dfrac{1}{5(x+1)} - \dfrac{x-1}{5(x^2+4)} - \dfrac{x-1}{(x^2+4)^2}$

   (l) $\dfrac{x-1}{x^2+1} - \dfrac{2}{x^2+4}$

   (m) $\dfrac{1}{3(x-1)} - \dfrac{x+2}{3(x^2+x+1)}$

   (n) $\dfrac{3(x-2)}{16(x^2-x+2)} + \dfrac{3x+2}{4(x^2-x+2)^2}$

   (o) $\dfrac{2(7x-5)}{(x^2-x+2)^3} + \dfrac{x-2}{x^2-x+2}$

   (p) $\dfrac{x+1}{x^2+x+1} - \dfrac{x-1}{x^2-x+1}$

5. (a) $10x - 5 + \dfrac{4}{x-1} + \dfrac{27}{2x+3}$

   (b) $2x - 3 + \dfrac{1}{x-9} - \dfrac{2}{x-3}$

   (c) $1 - \dfrac{6}{x^2+4}$

   (d) $3x^2 - 2x + \dfrac{x-2}{2x^2+1}$

6. $A = \dfrac{f(a)}{(a-b)(a-c)(a-d)}, B = \dfrac{f(b)}{(b-a)(b-c)(b-d)},$

   $C = \dfrac{f(c)}{(c-a)(c-b)(c-d)}, D = \dfrac{f(d)}{(d-a)(d-b)(d-c)}$

7. (a) $\dfrac{1}{\sqrt{2}} + \dfrac{i}{\sqrt{2}}, -\dfrac{1}{\sqrt{2}} + \dfrac{i}{\sqrt{2}}, \dfrac{1}{\sqrt{2}} - \dfrac{i}{\sqrt{2}}, -\dfrac{1}{\sqrt{2}} - \dfrac{i}{\sqrt{2}}$

8. $f^{(100)}(x) = -\dfrac{100!}{x^{101}} + \dfrac{100!}{(x+1)^{101}} + \dfrac{100!}{(x-1)^{101}}$

9. (a) $\dfrac{1}{x-1} + \dfrac{4}{(x-1)^2}$

  (b) $\dfrac{1}{(x+1)^2} - \dfrac{2}{(x+1)^3} + \dfrac{8}{(x+1)^4}$

  (c) $\dfrac{1}{x-3} + \dfrac{10}{(x-3)^2} + \dfrac{31}{(x-3)^3} + \dfrac{27}{(x-3)^4}$

  (d) $\dfrac{1}{x+2} - \dfrac{10}{(x+2)^2} + \dfrac{21}{(x+2)^3}$

10. (a) $\dfrac{n}{n+1}$　　(b) $\dfrac{n}{2n+1}$　　(c) $\dfrac{n^2-n}{2(n^2+n+1)}$

11. (a) $\dfrac{2}{x+1} + \dfrac{4}{(x+1)^2} - \dfrac{2x+3}{x^2+x+1}$　　(b) $\dfrac{2x-5}{x^2+2x+5} - \dfrac{x+3}{x^2+x+3}$

Index

Printed in the United States
By Bookmasters